FPGAs for Software Programmers

Dirk Koch • Frank Hannig • Daniel Ziener
Editors

FPGAs for Software Programmers

Springer

Editors
Dirk Koch
The University of Manchester
Manchester, United Kingdom

Frank Hannig
Friedrich-Alexander-Universität
Erlangen-Nürnberg (FAU)
Erlangen, Germany

Daniel Ziener
Friedrich-Alexander-Universität
Erlangen-Nürnberg (FAU)
Erlangen, Germany

ISBN 978-3-319-26406-6 ISBN 978-3-319-26408-0 (eBook)
DOI 10.1007/978-3-319-26408-0

Library of Congress Control Number: 2015957420

Springer Cham Heidelberg New York Dordrecht London

Printed on acid-free paper

Springer International Publishing AG Switzerland is part of Springer Science+Business Media (www.springer.com)

Preface

"FPGAs." — "What?"

"Field programmable gate arrays
that are cool programmable chips." — "I see! Programmable. Great.
I have programming skills.
So, this should be an easy task for me."

"Well, not really,
but I can help you."

Dear valued reader, when software engineers and hardware engineers talk about FPGAs, often two worlds are clashing together, each with different mindsets, skills, and expectations. Until recently, FPGA programming was mostly a hardware engineering discipline carried out by specially trained people. The design productivity of FPGA design was rather low, and for the time a hardware engineer was fiddling out a few flipping bits, a software engineer might have implemented his new large-scale cloud service application running on a big data center. Please note that we wrote "*was* rather low!" Like the widespread success of the processor took off with the development of *expressive languages* and corresponding *compilers*, this is currently happening for FPGAs. For decades, FPGA programming was carried out at very low abstraction levels (comparable to assembly language), but during recent years several compilers have emerged that can bring an algorithm to an FPGA at a quality far beyond what human handcrafting would be able to do. Currently, the software and hardware worlds are strongly converging. *Software is going parallel*, which has a long tradition in hardware design and vice versa, hardware gets more and more designed from *expressive languages*. This book aims at closing the gap between these two worlds.

Instead of following a programming model that is tailored to CPUs (following the so-called *von Neumann* computer architecture model), FPGA programming can be considered rather as *molding* the *liquid silicon* [IBE$^+$10] to create a certain desired functionality, and FPGAs allow *computing without processors* [Sin11].

This gives an engineer full control over the machine itself and not only over the program a machine runs. For example, if we want to perform DNA sequencing. Then, let us build a reconfigurable processor which is tailored to small alphabet string matching and achieve 395 million base pairs per second throughput [ALJ15]. Or we need fast AES encryption? Then build a custom compute accelerator on an FPGA, and get 260 Gbps throughput [SS15]. As a reference, the fastest Intel Core i7 Extreme Edition i7-980X CPU does not even yield that memory throughput, and this processor provides three times less AES throughput when using all 12 cores and despite having special hardware support for AES. And if we take an i7 Extreme Edition 975x without AES hardware support, the FPGA would be 52 times faster. Please note that the FPGA delivers its AES throughput on one problem, while the software solutions provide only high *aggregated performance* for AES on different problems. On the same problem, the FPGA will be significantly faster. And not only that the FPGA is much faster, it will do the job at a fraction of the power consumption.

Does implementing AES for an FPGA sound too complicated? Well, tuning the performance for AES on, for example, a GPU (that is optimized for floating-point number-crunching instead of bit-level operations) is likely to be harder, will still be slower than an FPGA implementation, and would burn at least an order of magnitude more power. So why not getting it right in the first place? This is what this book wants to help with.

For making this happen, this book brings together leading experts from the industry and academia across the globe to cover:

Theoretical foundations and background on computer architectures, programming models, and hardware compilation techniques (throughout the first three chapters).

Programming languages and compilers for FPGAs from a rather practical point of view with various case studies. This includes six state-of-the-art and emerging commercial approaches (located in Part I, beginning at page 61) as well as three established academic compilers (located in Part II, beginning at page 173). This shall demonstrate that there is a large ecosystem for FPGA programming that software engineers can choose from to find the best language and compiler environment for virtually any given problem.

FPGA runtime environments are needed for building complex systems and for running and managing them. This provides a higher abstraction in the system allowing to focus more on applications while abstracting from many low-level details of the underlying FPGA hardware (as covered in Part III, beginning at page 225).

Ready-to-go approaches allow building a complex system on a chip (SoC) and using pre-implemented programmable architectures on an FPGA without any need for hardware programming (as revealed in Part IV, starting from page 259).

With all this, we intend making reconfigurable FPGA technology more accessible to software programmers and to moderate between hardware and software engineers. FPGA technology provides maybe the most promising answer for improving performance and energy efficiency of computers at the end of CMOS scaling. To make this happen, it will need hardware and software engineers working harmonically together. Let us start with this today!

This book is written for software engineers who want to know more about FPGAs and how to use them. Consequently, we do not assume any hardware engineering background. In a few cases, where further background information is provided, this book is using terms that are common in computer engineering (e.g., "flip-flop") and compiler construction (e.g., "scheduling") that might not be thoroughly introduced but that are assumed to be clear from the context for most readers. This was done by intention as we do not want to fill the book with all possible background information while eventually hiding the big picture, which is *what are FPGAs good for* and *how can they be programmed and used* by software engineers. Such sections should be seen as optional and should not be seen as essential when reading this book. Those terms are also nicely presented on Wikipedia, if further explanation might be needed. In addition to software engineers, this book can also serve as a textbook in university courses on alternative programming approaches and acceleration techniques. Finally, this book is also relevant for hardware engineers who want to get an overview on how to enhance design productivity with latest languages and tools.

We thank all the chapter authors for their outstanding contributions, their feedback, and their will to compile a comprehensive textbook targeting software engineers. We also thank Springer, especially Charles Glaser for his encouragement to develop this book and Hemalatha Gunasekaran for her support. Furthermore, we thank Mikel Luján for his feedback and suggestions to improve this book. Finally, we thank Philippa Ordnung for providing the Index. All editor's and author's profit of the book sales will be donated to the United Nations Children's Fund (UNICEF).

Manchester, UK — Dirk Koch
Erlangen, Germany — Frank Hannig
Erlangen, Germany — Daniel Ziener

Contributors

Michael Adler C 14
Intel Corporation, VSSAD Group, Hudson, MA, USA

Andreas Agne C 13
University of Paderborn, Paderborn, Germany

Stephan Ahrends C 4
National Instruments, Munich, Germany

Jason Anderson C 10
Department of Electrical and Computer Engineering, University of Toronto, Toronto, ON, Canada

Hugo A. Andrade C 4
National Instruments, Berkeley, CA, USA

Oriol Arcas-Abella C 9
Barcelona Supercomputing Center, Barcelona, Spain;
Universitat Politecnica de Catalunya, Barcelona, Spain

Tobias Becker C 5
Maxeler Technologies, London, UK

Stephen Brown C 10
Department of Electrical and Computer Engineering, University of Toronto, Toronto, ON, Canada

Andrew Canis C 10
Department of Electrical and Computer Engineering, University of Toronto, Toronto, ON, Canada

João M. P. Cardoso C 2
University of Porto, Faculty of Engineering, Porto, Portugal
INESC-TEC, Porto, Portugal

Yu Ting Chen C 10
Department of Electrical and Computer Engineering, University of Toronto, Toronto, ON, Canada

Jongsok Choi C 10
Department of Electrical and Computer Engineering, University of Toronto, Toronto, ON, Canada

Jason Cong C 8
Falcon Computing Solutions, Inc. Los Angeles, CA, USA;
Computer Science Department, University of California, Los Angeles, CA, USA

Adrian Cristal C 7
Barcelona Supercomputing Center, Barcelona, Spain;
Centro Superior de Investigaciones Cientificas (IIIA-CSIC), Barcelona, Spain;
Universitat Politecnica de Catalunya, Barcelona, Spain

Kermin Fleming C 14
Intel Corporation, VSSAD Group, Hudson, MA, USA

Blair Fort C 10
Department of Electrical and Computer Engineering, University of Toronto, Toronto, ON, Canada

Georgi Gaydadjiev C 5
Maxeler Technologies, London, UK

Jeffrey Goeders C 10 C 15
Department of Electrical and Computer Engineering, University of British Columbia, Vancouver, BC, Canada

Robert J. Halstead C 11
University of California, Riverside, CA, USA

Frank Hannig C 1 C 3 C 7 C 12
Friedrich-Alexander-Universität Erlangen-Nürnberg (FAU), Erlangen, Germany

Markus Happe C 13
ETH Zurich, Zurich, Switzerland

Simon Hogg C 4
National Instruments, Austin, TX, USA

Graham M. Holland C 15
School of Engineering Science, Simon Fraser University, Burnaby, BC, Canada

Hsuan Hsiao C 10
Department of Electrical and Computer Engineering, University of Toronto, Toronto, ON, Canada

Muhuan Huang C 8
Falcon Computing Solutions, Inc. Los Angeles, CA, USA

Dirk Koch C 1
The University of Manchester, Manchester, UK

Ruo Long Lian C 10
Department of Electrical and Computer Engineering, University of Toronto, Toronto, ON, Canada

Cheng Liu C 16
Department of Electrical and Electronic Engineering, University of Hong Kong, Pokfulam, Hong Kong

Enno Lübbers C 13
Intel Labs Europe, Munich, Germany

Gorker Alp Malazgirt C 7
Bogazici University, Istanbul, Turkey

Oskar Mencer C 5
Maxeler Technologies, London, UK

Walid A. Najjar C 11
University of California, Riverside, CA, USA

Peichen Pan C 8
Falcon Computing Solutions, Inc. Los Angeles, CA, USA

Marco Platzner C 13
University of Paderborn, Paderborn, Germany

Christian Plessl C 13
University of Paderborn, Paderborn, Germany

Oliver Reiche C 12
Friedrich-Alexander-Universität Erlangen-Nürnberg (FAU), Erlangen, Germany

Moritz Schmid C 7 C 12
Friedrich-Alexander-Universität Erlangen-Nürnberg (FAU), Erlangen, Germany

Christian Schmitt C 7
Friedrich-Alexander-Universität Erlangen-Nürnberg (FAU), Erlangen, Germany

Lesley Shannon C 15
School of Engineering Science, Simon Fraser University, Burnaby, BC, Canada

Deshanand Singh C 6
Altera Corporation, Toronto, ON, Canada

Hayden Kwok-Hay So C 16
Department of Electrical and Electronic Engineering, University of Hong Kong, Pokfulam, Hong Kong

Nehir Sonmez C 7 C 9
Barcelona Supercomputing Center, Barcelona, Spain;
Centro Superior de Investigaciones Cientificas (IIIA-CSIC), Barcelona, Spain;
Universitat Politecnica de Catalunya, Barcelona, Spain

Bain Syrowik C 10
Department of Electrical and Computer Engineering, University of Toronto, Toronto, ON, Canada

Jürgen Teich C 12
Friedrich-Alexander-Universität Erlangen-Nürnberg (FAU), Erlangen, Germany

Jason Villarreal C 11
University of California, Riverside, CA, USA

Yuxin Wang C 8
Falcon Computing Solutions, Inc. Los Angeles, CA, USA

Markus Weinhardt C 2
Osnabrück University of Applied Sciences, Osnabrück, Germany

Steven J.E. Wilton C 15
Department of Electrical and Computer Engineering, University of British Columbia, Vancouver, BC, Canada

Peter Yiannacouras C 6
Altera Corporation, Toronto, ON, Canada

Arda Yurdakul C 7
Bogazici University, Istanbul, Turkey

Peng Zhang C 8
Falcon Computing Solutions, Inc. Los Angeles, CA, USA

Daniel Ziener C 1
Friedrich-Alexander-Universität Erlangen-Nürnberg (FAU), Erlangen, Germany

Contents

List of Acronyms

ALU	Arithmetic Logic Unit
API	Application programming interface
ASIC	Application-specific integrated circuit
ASP	Application-specific processor
AST	Abstract syntax tree
BRAM	Block RAM
BSV	Bluespec SystemVerilog
CAD	Computer-aided design
CDFG	Control data-flow graph
CFG	Control flow graph
CGRA	Coarse-grained reconfigurable array
CHAT	Compiled hardware accelerated threads
CLIP	Component-level IP
CPU	Central processing unit
CUDA	Compute unified device architecture
DAC	Digital to analog converter
DFE	Data-flow engines
DCM	Digital clock manager
DFG	Data flow graph
DFS	Depth-first search
DFT	Discrete Fourier transform
DLL	Dynamic link library
DMA	Direct memory access
DPR	Dynamic partial reconfiguration
DPRAM	Dual-port RAM
DSL	Domain-specific language
DSP	Digital signal processor
ECAD	Electronic computer-aided design
ECC	Error-correcting code
EDA	Electronic design automation
eGPU	Embedded GPU

ESL	Electronic system level
FF	Flip-flop
FIFO	First in, first out
FIR	Finite impulse response
FMA	Fused multiply-add
FPGA	Field programmable gate array
GCD	Greatest common divisor
GPGPU	General purpose GPU
GPU	Graphics processing unit
HAL	Hardware abstraction layer
HDL	Hardware description language
HDR	High dynamic range
HIPA^{cc}	Heterogeneous image processing acceleration
HLL	High-level programming language
HLS	High-level synthesis
HTML	Hypertext markup language
HPC	High-performance computing
HPS	Hard processor system
IDCT	Inverse discrete cosine transform
II	Initiation interval
ILP	Instruction-level parallelism
IP	Intellectual property
IPIN	IP integration nodes
ISP	Image signal processor
LoC	Lines of code
LUT	Lookup table
MAC	Multiply-accumulate
MAD	Multiply-add
MMU	Memory management unit
MPSoC	Multi-processor system-on-chip
MRA	Multi-resolution analysis
NPP	NVIDIA performance primitives
NRE	Non-recurring engineering
OEM	Original equipment manufacturer
OOC	Out-of-context
OpenCL	Open computing language
OpenCV	Open source computer vision
OpenSPL	Open spatial programming language
OSFSM	Operating system synchronization state machine
OSIF	Operating system interface
PCB	Printed circuit board
PDE	Partial differential equation
PE	Processing element
PLL	Phase-locked loop
POSIX	Portable operating system interface

PPnR	Post-place and route
QoR	Quality of results
RGBA	Red green blue alpha
RIO	Reconfigurable I/O
RSoC	Reconfigurable system-on-chip
RTL	Register-transfer level
RTM	Reverse time migration
ROCCC	Riverside optimizing compiler for configurable computing
ROI	Region of interest
S2S	Source-to-source
SCS	Spatial computing substrate
SCTL	Single-cycle timed loops
SDK	Software development kit
SDR	Software-defined radio
SDLC	Software development life cycle
SFU	Special function unit
SIMD	Single instruction multiple data
SMT	Simultaneously multithreading
SoC	System-on-chip
SPL	Software product line
SSA	Static single assignment
TLB	Translation lookaside buffer
TLM	Transaction-level modeling
TRS	Term-rewriting systems
VI	Virtual instrument
VHDL	Very high speed integrated circuit hardware description language
VLIW	Very long instruction word
VLSI	Very-large-scale integration
WHT	Walsh-Hadamard transform

Chapter 1
FPGA Versus Software Programming: Why, When, and How?

Dirk Koch, Daniel Ziener, and Frank Hannig

This chapter provides background information for readers who are interested in the philosophy and technology behind FPGAs. We present this from a software engineer's viewpoint without hiding the hardware specific characteristics of FPGAs. Like it is very often not necessary to understand the internals of a CPU when developing software or applications (as in some cases a developer does not even know on which kind of CPU the code will be executed), this chapter should not be seen as compulsory. However, for performance tuning (an obvious reason for using accelerator hardware), some deeper background behind the used target technology is typically needed.

This chapter compromises on the abstraction used for the presentation and it provides a broader picture on FPGAs, including their compute paradigm, their underlying technology, and how they can be used and programmed. At many places, we have added more detailed information on FPGA technology and references to further background information. The presentation will follow a top-down approach and for the beginning, we can see an FPGA as a programmable chip that, depending on the device capacity, provides thousands to a few million bit-level processing elements (also called *logic cells*) that can implement any elementary logic function (e.g., any logic gate with a hand full inputs). These tiny processing elements are laid out regularly on the chip and a large flexible interconnection network (also called *routing fabric* with in some cases millions of programmable switches and wires

D. Koch (✉)
The University of Manchester, Manchester, UK
e-mail: dirk.koch@manchester.ac.uk

D. Ziener • F. Hannig
Friedrich-Alexander-Universität Erlangen-Nürnberg (FAU), Erlangen, Germany
e-mail: daniel.ziener@fau.de; frank.hannig@fau.de

D. Koch et al. (eds.), *FPGAs for Software Programmers*,
DOI 10.1007/978-3-319-26408-0_1

between these processing elements) allows to build more complex digital circuits from those logic cells. This allows implementing *any* digital circuit, as long as it fits the FPGA capacity.

1.1 Software and Hardware Programmability

Programmability allows it to adapt a device to perform different tasks using the same physical real estate (regardless of whether it is a CPU or FPGA). This is useful for *sharing resources* among different tasks and therefore for *reusing resources* over time, for *using the same device for various applications* (and hence allowing for lower device cost due to higher volume mass production), and for allowing *adaptations and customizations* in the field (e.g., fixing bugs or installing new applications).

1.1.1 Compute Paradigms

We all have heard about CPUs, GPUs, FPGAs and probably a couple of further ways to perform computations. The following paragraphs will discuss the distinct models, also known as *compute paradigms*, for the most popular processing options in more detail.

Traditional von Neumann Model

The most popular compute paradigm is the *von Neumann model* where a processor fetches instructions (stated by a program counter), decodes them, fetches operands, processes them by an Arithmetic Logic Unit (ALU) and writes them to a register file or memory. This takes (1) considerable resources (i.e. real estate) on the chip for the instruction fetch and decode units, (2) power these units take, and (3) I/O throughput for getting the instruction stream into the processor. This is known as the *von Neuman bottleneck* that in particular arises if instructions and data transfers to memory share the same I/O bus. This might need throttling down the machine because instructions cannot be fetched fast enough. All modern processors follow a *Harvard architecture* with separate caches for instructions and data which, however, still has many of the shortcomings of the von Neumann model. An illustration of the von Neumann model is given in Fig. 1.1a.

Vector Processing and SIMD

One major approach applied to many popular processors for improving the work done per instruction was adding vector units. As shown in Fig. 1.1b, a vector unit

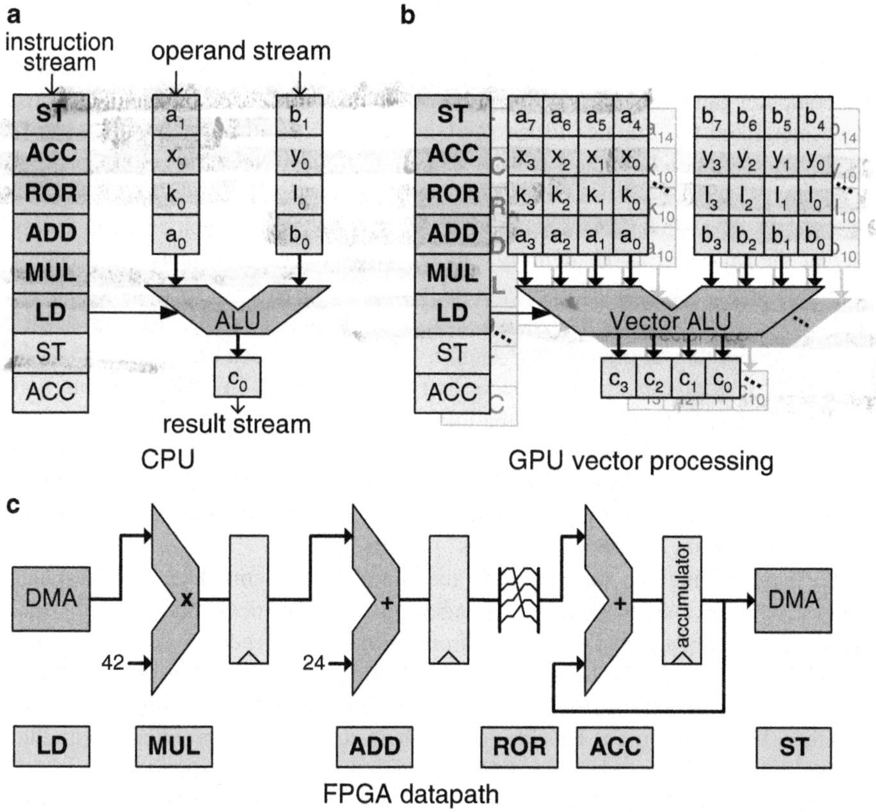

Fig. 1.1 Compute paradigms. (**a**) Traditional von Neumann model with a CPU decoding an instruction stream, fetching operands, processing them and writing results. (**b**) Vector processing achieves higher throughput by processing multiple values in parallel by sharing the same instruction stream (i.e. the same control flow). Multiple vector units can work in parallel on different data items, which is the model deployed in GPUs. (**c**) Reconfigurable datapath on an FPGA. Instead of performing different instructions on a single ALU, a chain of dedicated processing stages is used in a pipelined fashion. If resources and I/O throughput permit, the pipeline might also use vector processing or multiple pipelines may process in parallel

performs multiple identical operations on a set of input operands simultaneously, so the effort for instruction fetching and decoding is amortized on more actual processing work. Famous examples for vector extensions are the MMX technology introduced by Intel in their Pentium processors in 1997, followed by AMD with 3DNow! and ARM with the NEON vector extension in the ARM Cortex-A series. Vector units are also referred to as Single Instruction Multiple Data (SIMD) units. While vector processing is promising from an architectural point of view, it is challenging for compilers to harness this kind of computing and much of the code using vector units is written by hand using assembly language.

GPU Processing

For further improving processing performance (mostly driven by video processing for computer games), large amounts of lightweight vector processing elements have been integrated in Graphics Processor Units (GPUs). Here, lightweight means a simple instruction fetch and decode unit for controlling the vector units for improved performance-per-energy levels. However, communication between cores (or with the host machine) is done through global shared memory (i.e., each core can access any of the on-board GPU memory). As a consequence, with many hundreds of vector units on a single GPU, the required memory bandwidth of GPUs is enormous and despite a significant amount of on-chip caches. High-end GPUs factor this in by providing hundreds of gigabytes of memory throughput which takes a considerable amount of energy. For example, according to Vermij et al. [VFHB14], about half the energy is used by the shared memory, the L2 Cache, on-chip communication, and by the memory controllers over a series of benchmarks running on a GPU. As a reference, the corresponding power breakdown for a Xeon E5-2630 showed that "around 50% of the energy goes to the three levels of cache, and another 20% to 30% is spent on the memory controllers" [VFHB14].

GPU processing can work well on data parallel problems, like most graphics processing tasks that work on independent pixels or vertices, but problems that require computing a global result, synchronization will be needed as described in the following Sect. 1.1.2. For example in Fig. 1.1, the executed program is an unrolled loop (meaning we skipped any loop control instructions) that accumulates a result together. While we might be able to distribute the compute problem to parallel vector units, as illustrated in Fig. 1.1b (meaning that different processing elements compute a different part of the problem), we still have to accumulate a final result over all used vector units (see the ACC accumulate instruction in the examples that computes $R_O = R_O + R_I$). Depending on the problem, the synchronization effort can actually exceed the processing time and the GPU cores will spend more time on synchronization spinlocking than actually on providing useful work.

FPGA Datapath

As FPGAs can implement any digital circuit, they can implement architectures that follow the von Neuman, the vector, or the GPU model. This is in many cases a reasonable way to use FPGAs technology to implement another programmable architecture on top of the FPGA fabric. The FPGA will then mimic another programmable core (e.g., a well known CPU or a specialized programmable device) that is then programmed with rather conventional software developing techniques. This concept is known as *hardware overlays* and is covered in more detail in Chap. 15 for a systolic processor array. This is a computer architecture using small CPU-like processing elements that typically have little local memory and a next neighbor communication infrastructure. Note that the capacity of today's high density FPGAs allows easily to host several hundred 32-bit processors including a suitable infrastructure (with memories, on chip networks, and I/O peripherals) on a single device. A vector overlay is available with the MXP Matrix Processor from VectorBlox Computing Inc. That vector processor can deliver for some applications a 100× speedup over a baseline FPGA softcore processor and the design process is purely carried out using software programming.[1]

A way we can look at a processor is similar to having a universal worker who works off a job list (its program). However, if we take this analogy and compare it with how we perform industrial mass production, then we see that a different approach has been proven to perform much better: the *assembly belt*. The way an assembly belt production is orchestrated is to locate highly specialized and optimized workers (or machines) along a network of conveyor belts that move the product to produce (e.g., a car) and materials across the factory. This is exactly the idea behind the *FPGA datapath paradigm* where specialized processing elements form a processing pipeline and instead of material, data is streamed between and through these processing elements. As shown in Fig. 1.1c, there is no instruction stream and the original instruction sequence is decomposed in a chain of simple arithmetic operators. Consequently, there is no need for instruction fetch or decode, this is encoded directly into the structure of the datapath.

When arranging a factory for efficient production, this can include a very complex arrangement of workers, machines, and conveyor belts. Similarly, mapping certain computations onto an FPGA might need a complex arrangement of processing elements (or accelerator modules) with multiple processing pipelines that might split or join. Due to the enormous flexibility of the reconfigurable interconnection network provided by the FPGA, we can map virtually any communication dependency between processing elements (e.g., streaming, broadcast, multicast, bus-based communication, or even networks on chip). To summarize this: with FPGAs, we can tailor the machine to the problem, while with CPUs/GPUs we can only tailor the implementation to the machine.

While the clock frequency of such a processing pipeline on an FPGA is rather low with commonly only a few 100 MHz, we can start a new iteration in each operational cycle. A CPU, in contrast, has to rush through all instructions of a loop body before the next iteration can be started. So with longer pipelines, the speedup of a datapath implemented on an FPGA can be enormous. For example, the Maxeler OpenSPL compiler (which is introduced in Chap. 5) uses several of the High-Level Synthesis (HLS) techniques that are described in more detail in Chap. 2. Starting from a Java-like description, this compiler can automatically generate processing pipelines with eventually over a thousand pipeline stages on a single FPGA. And the Maxeler compiler is further able to exploit multiple chained FPGAs for implementing even larger pipelines.

As the clock frequency of processors is not likely to increase significantly and because the improvement in the micro architecture will probably not result in major performance increases, we cannot expect to see substantial single core

[1] http://vectorblox.com/about-technology/applications/.

(or thread) performance boosts in the near future. However, to fulfill demands for more performance, parallel execution has to be applied. In the last paragraph, it was described that parallel processing might often work well on GPUs for computing data parallel problems by distributing the problem to many GPU cores that then perform the same program but on different input data. This can be seen as *horizontal scaling* (or *horizontal parallelism*), because instead of running sequentially through a piece of code on one core, multiple cores are working side-by-side in this model. This can, of course, be applied to FPGAs. In addition, FPGAs allow for further *vertical scaling* (or *vertical parallelism*) where different processing elements perform different operations (typically in a pipelined fashion). If we assume, for example, a sequence of instructions in a loop body, it is not easy to parallelize the loop body by executing the first half of the instructions on one CPU and the other half of the instructions on another CPU. On an FPGA, however, this can be applied to any extend and all the way to individual instructions, as it is the case in an FPGA datapath. In other words, parallelizing an algorithm to FPGAs can in some cases be much easier performed than it would be possible for a multi core processors or GPUs. And with a rising demand for parallelizing algorithms it is more and more worth looking into FPGA programming.

1.1.2 Flexibility and Customization

Programming is used to solve specific tasks and the target platform (e.g., a processor or an FPGA) has to support running all tasks. We could say that virtually all available programmable platforms are Turing complete and therefore that we are (at least in theory) able to solve any task on virtually any programmable platform (assuming that we are not bound by memory). However, depending on the problem or algorithm that we want to solve, different programmable platforms are better suited than others. This is of course the reason for the large variety of programmable platforms. The following list states the benefits of some programmable architectures:

CPUs are specifically good when dealing with control-dominant problems. This is for example the case when some code contains many `if-then-else` or `case` statements. A lot of operating systems, GUIs, and all sorts of complex state machines fall into this category. Furthermore, CPUs are very adaptive and good at executing a large variety of different tasks reasonably well. Moreover, CPUs can change the executed program in just a couple of clock cycles (which is typically much less than a microsecond).

GPUs perform well on data-parallel problems that can be vectorized (for using the single instruction multiple data (SIMD) units available on the GPU thread processing elements) and for problems that need only little control flow and little synchronization with other threads or tasks. This holds in particular for floating-point computations, as heavily used in graphics processing. However for synchronization, typically spinlocking techniques are used that can take considerable performance and programming effort [RNPL15]

FPGAs achieve outstanding performance on stream processing problems and whenever pipelining can be applied on large data sets. FPGAs allow for a tight synchronized operation of data movement (including chip I/O) and processing of various different algorithms in a daisy-chained fashion. This often greatly reduces the need for data movement. For example, in a video processing system, we might decompress a video stream using only on-FPGA resources while streaming the result over to various further video acceleration modules without ever needing off-chip communication. This not only removes possible performance drops if we are I/O bound (e.g., if we have to occupy the memory subsystem for storing temporary data in a CPU-based system) but also helps in saving power. Please note that very often it takes much more power to move operands (or any kind of data) into a processing chip than performing computations on those operands.

Another important property of FPGAs is that they can implement any kind of register manipulation instruction directly. For example, most cryptographic algorithms require simple bit fiddling operations that can take a CPU tens to hundreds of cycles for performing all the needed bit-mask and shift operations (for, lets say, a 32-bit register value). On FPGAs however, each bit is directly accessible without any instruction and constant shift operations are implemented by the interconnection network and shift operations do not even take any FPGA logic resources (e.g., logic cells). In other words, FPGAs do not follow the restrictions of a pre-defined instruction set and computer organization. FPGAs allow us thinking in mathematical expressions and all sorts of operations rather than in a fixed register file and a collection of predefined instructions (as in all CPUs). For example, if we want to implement a modulo 10 Gray-code counter, we can build this directly in logic and use a small 4-bit register for storing the count state.

As can be seen, there is not a clear winning compute platform and depending on the problem and requirements, one or the other is better. In practice, there is also a human factor involved as it needs skilled people to program the different compute platforms. While there are many software developers, there are far less GPU or FPGA programmers, as can be observed on basically any job portal website. The intention of this book is to tackle exactly this issue and helping in making FPGAs accessible to software engineers.

A big difference in hardware and software programming platforms exists in the ability to control the structure and architecture of the programming platform. If we take a CPU or GPU, then a vendor (e.g., Intel, AMD, NVIDIA, or ARM) took decisions on the instruction set architectures, the organization and sizes of caches, and the capabilities of the I/O subsystem. In contrast when using FPGAs, there is much more freedom. If we need, for example, an operation for computing the correlation of a motion estimate vector, we can simply generate the corresponding logic and the correlation of one or many pixels can be computed in a single cycle, if needed. If an algorithm requires a certain amount of on-chip memory, we can allocate the exact corresponding number of memory blocks. For FPGAs, we are basically only limited by the available resources, but much less by architectural decisions of the FPGA vendor. And with devices providing millions of logic cells, thousands of arithmetic blocks, and tens of megabytes of on-chip memory, the limit is very low today.[2]

The need for flexibility can be discussed from a more system level perspective: with the exponential rise of capacity available on a single chip, we typically want to build more complex systems which implies the need for more heterogeneous compute platforms. This can be observed for the various compute resources including multicore CPUs, GPUs, and accelerators that are available in recent SoCs (System on Chips) targeting mobile phones or tablet PCs. However, by predefining an SoC architecture, we define its major characteristics and limit the level at which an equipment manufacturer can differentiate the final product. FPGAs, however, provide much more flexibility and they allow tailoring a large and complex SoC exactly to the needs of the application. This is relevant when considering very large chips in the future that should serve various different application domains. In this situation, FPGAs might utilize much better the available real estate on the chip due to their great customization abilities.

1.1.3 The Cost of Programmability

Unfortunately, programmability does not come for free. In a CPU, for example, we might have 100 possible instructions, but we will only start one machine instruction in each operational cycle, hence leaving probably most parts of the ALU underutilized.

Similarly on an FPGA, we typically cannot use all available functional blocks of the FPGA fabric in real-world applications and the routing between these blocks is carried out by a switchable interconnection network consisting of eventually many millions of programmable multiplexers that would not be needed when implementing the interconnection of a non-programmable digital circuit (i.e. an ASIC). In addition to real estate on the chip, all these switches will slow down the circuit and draw some extra power. All this would not be needed when connecting gates and functional blocks directly by wires, as it is the case for dedicated integrated circuits (i.e. ASICs).

This technology gap between FPGAs and ASICs was quantified by Kuon and Rose in [KR06]. Considering the same 90 nm fabrication process, an FPGA

[2]The limit is probably less what is available, but what is affordable. The high capacity flagship devices of the two major FPGA vendors Xilinx and Altera typically cost over 10 K US$. So the limit is often not a technological, but an economical one. However when Xilinx for example introduced its new Series-7 FPGAs, the smallest Kintex-7 device (XC7K70T) provided at a price tag below 100 US$ significantly more resources than its Virtex-2 flagship (XC2V8000) a decade ago. So in other words, what is far too expensive today might be well affordable very soon.

consisting of logic cells (based on lookup tables and flip-flops) would be on average 35× bigger (in terms of area) as the ASIC counterpart over a large set of examined benchmark circuits. However, by introducing additional dedicated functional blocks, such as multipliers and memory blocks, the gap in area can be reduced to ∼18×. For the performance it was found that FPGAs are 3–4× slower than their ASIC counterparts and they also draw about 10× more power.

However, due to high volume mass production, more advanced processes are commonly used that result in a lower monetary cost, better energy efficiency, and higher performance which all can offset much of the cost related to FPGA programmability. Furthermore, programmability removes underutilization if some tasks are used only very seldom. Due to re-programmability, the same physical resources can be shared and consequently resources can be better utilized as it would be possible with dedicated circuits for each task.[3] On a more macroscopic level, this shares some ideas behind cloud computing. The virtualization of servers in a cloud data center comes at a price that pays off because it commonly allows running a lower number of physical machines for a given load scenario.

The Impact of Technology Scaling

The probably most important driving factors in semiconductor technology today is *energy efficiency* and *power density*. We all know about the astonishing progress in silicon process technology which was firstly described by Moore's law [Moo65] that predicts an exponential growth of the transistor count on a chip (i.e. a doubling approximately every 18 months). Even given that chips for computing cannot be directly compared with memory chips, we can get a glimpse of what is possible today when calculating the number of transistors in a memory card that we plug into our phones, which can easily exceed a 100 billion transistors. However, while the density was (and still is) growing exponentially, power consumption per area stayed constant, which is known as Dennard scaling [DGY$^+$74]. In other words, the energy efficiency per logic cell was also exponentially improving. Unfortunately,

[3] Most available FPGAs in these days can be partially reconfigured. This allows some parts of the FPGA to be operational (e.g., a CPU controlling and managing the system and a memory controller) while changing some accelerator modules. *Partial reconfiguration* can be used for a time-multiplexing of FPGA resources (e.g., if a problem does not fit the device). This helps for better utilizing an FPGA (and hence allowing for using a smaller and therefore cheaper and less power hungry device). This technique can also be used to speed-up computations. Then, instead of providing a chain of smaller modules concurrently on the FPGA, we might use reconfiguration to load only the module currently needed, such that we have more resources available for each module. This can then be used for exploiting a higher level of parallelization, which in turn can eventually speedup processing significantly. Finally, partial reconfiguration can be used for implementing dynamically instantiating hardware modules in a system (e.g., for creating a hardware thread at runtime). More comprehensive information on partial runtime reconfiguration is provided in [Koc13].

this has slowed down in recent years and the power density (i.e. power per surface) of many chips today is way beyond the power density of the hops that we use for cooking in our kitchens.

As the power density cannot exceed the thermal design of the chip and the surrounding system (the power used by a chip ends up in heat), the important question today is not only how many transistors can be squeezed on a chip, but how many of them are allowed to be powered and operational (or at what speed) without exceeding a given thermal budget. This phenomena is known as dark silicon [GHSV+11]. However, if we are restricted by power rather than by transistor count on a chip, the cost of reconfigurability in terms of area is becoming a minor issue as all the unused resources make a programmable chip kind of dark anyway.

Economical Aspects

There are economical aspects that can pay for the extra cost of programmability. First of all, programmability allows faster time-to-market for products because being able to quickly develop, test, modify, and produce a product allows participating in higher profitable market windows. Programmability also allows for an after-sales business (e.g., by providing updates) and longer operation of equipment. The latter aspect was actually one of the major economical enablers for the success of FPGA technology. The internet hype and with it the fast development of technology and standards made it necessary to update network equipment in the field. However, for this application, CPUs had been far too slow and ASICs took too long to design and were not flexible enough. To overcome this, FPGAs had been (and are still) the ideal platform to power major parts of the internet infrastructure.

1.2 What Are FPGAs?

FPGAs (Field Programmable Gate Arrays) are digital chips that can be programmed for implementing arbitrary digital circuits. This means that FPGAs have first to be programmed with a so called *configuration* (often called a configuration bitstream) to set the desired behavior of the used functional elements of the FPGA. This can be compared to processors which are able to execute arbitrary programs after loading the corresponding machine code into some memory. Without loading any executable code, the processor is unable to perform any useful operation. Translated to FPGAs, an FPGA without a proper configuration does nothing. Only after the configuration step, the FPGA is able to operate with the desired behavior encoded into the configuration. FPGAs are an example for *PLDs* (Programmable Logic Devices). FPGAs have a significant market segment in the microelectronics and, particularly in the embedded system area. For example, FPGAs are commonly used in network equipment, avionics,[4] automotive, automation, various kinds of test equipment, medical devices, just to name some application domains. While currently only a tiny share of FPGAs are used for data processing in data centers, this is likely going to change in the near future as FPGAs not only provide very high performance, but they are also extremely energy efficient computing devices. This holds in particular when considering big data processing. For example, Microsoft recently demonstrated a doubling of the ranking throughput of their Bing search engine by equipping 1632 servers with FPGA accelerators which only added an extra 10 % in power consumption [PCC+14]. In other words, Microsoft was able to improve the energy efficiency by 77 % while providing faster response times due to introducing FPGAs in their data centers.

Sometimes, FPGAs are referred to as *FPGA technology*. This expresses what target technology for implementing a digital circuit is used. In other words, it allows distinguishing if the target is an ASIC with logic gates as the building blocks or an FPGA that provides programmable logic cells. It is quite common to implement a digital circuit firstly on an FPGA for testing purposes before changing to ASIC technology when targeting high volume markets.

1.3 How FPGAs Work

So far we have talked mainly about programming an FPGA and what an FPGA is. This section will reveal its operation. Oversimplified, an FPGA consists of multiplexers and configuration memory cells that control those multiplexers and some wiring between them. This is illustrated in Fig. 1.2. It shows a multiplexer, the most important building block of an FPGA[5] and the operation of a multiplexer.

1.3.1 Lookup Table-Based Logic Cells

By connecting configuration memory cells to the select inputs of the multiplexer, a reconfigurable switch as part of a switch matrix is built, as shown in Fig. 1.2c. For implementing the actual reconfigurable logic cells, we connect configuration memory cells to the data inputs. For example, if we want to implement an AND gate, we will set the top configuration memory cell (connected to the input `11`) to `1`

---

[4]In an article from EETimes it is stated that "Microsemi already has over 1000 FPGAs in every Airbus A380" (http://www.electronics-eetimes.com/?cmp_id=7&news_id=222914228).

[5]Despite that FPGAs are basically made from multiplexers, it is a bit ironic that they are not very good at implementing multiplexers using their reconfigurable logic cells. In the case we want to implement multiplexers with many inputs, this would cost a considerable number of logic cells.

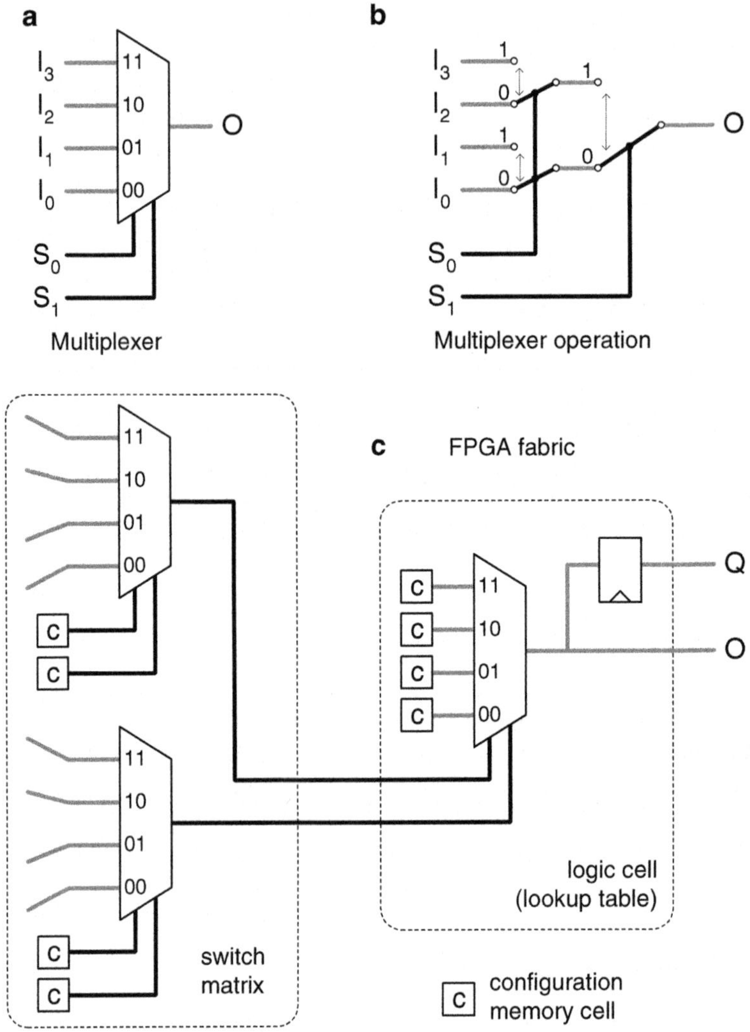

Fig. 1.2 FPGA fabric illustration. (**a**) The basic building block of an FPGA is a multiplexer. (**b**) Its internal operation can be compared with switches of a rail network and exactly one of the inputs (`I`) is connected with the output (`O`) depending on the select input (`S`). (**c**) (*left*) By connecting the select inputs to configuration memory cells, a switchable routing network is implemented for the interconnection between the logic cells. (**c**) (*right*) By connecting configuration memory cells to the multiplexer inputs, a lookup table is built that can implement any Boolean logic function for the lookup table inputs that are connected to the select inputs. For implementing larger combinatorial circuits, the `O` output can be connected to further logic cells and for storing states, a register is provided on the output `Q`. The values of the configuration memory cells are written during the configuration of the FPGA. By fitting large quantities of these basic blocks consisting of switch matrices and logic cells on a chip and connecting them together, we can build a reconfigurable FPGA fabric that can host any digital circuit

and the other three cells to `0`. Consequently, each logic cell provides a lookup table that stores the result for the desired Boolean function.

A lookup table (often abbreviated as *LUT*) can perform any Boolean function. The only limit is the size of the table, that is given by the number of inputs (with $k = 2$ in the figure). If larger functions have to be implemented than what fits in a single lookup table, the functions are implemented by cascading multiple logic cells (and therefore multiple lookup tables). For example, if we want to implement an AND gate with three instead of two inputs, then we implement the AND gate for two inputs, as described in the previous paragraph. The output is then connected to one input of another logic cell that again implements an AND gate. The other input of the second AND gate is then the third input of our 3-input AND gate.

The size of lookup tables found in commercial FPGAs varies from 3 to 6 inputs depending on the vendor and device family. Figure 1.3 gives more examples for truth tables of Boolean functions. For a table with k inputs, it takes 2^k configuration memory cells to build the logic cell. The examples in Fig. 1.3 show that the same logic function can be implemented by permuting inputs. This is used for simplifying the routing step (see Sect. 1.4.2). Instead of routing to a specific LUT input, it is sufficient to route to any input while eventually adjusting the table entries.

Usually, the lookup tables in the logic cells are combined with state flip-flops. These flip-flops are used for storing states in FPGA-based circuits (e.g., the state of an n-bit counter can be stored in the corresponding flip-flops of n logic cells).

function	$A_0 \cdot A_1 \cdot A_2 \cdot A_3$	$(A_0 \cdot A_1) + A_2$	$(A_2 \cdot A_3) + A_1$	A_3
A3 A2 A1 A0	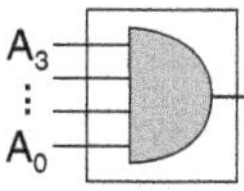	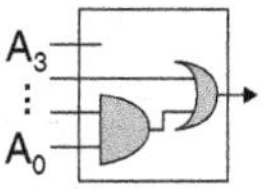	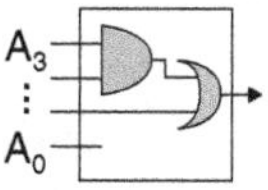	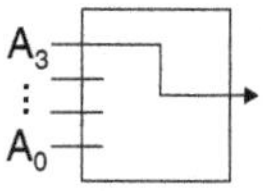
0: 0 0 0 0	0	0	0	0
1: 0 0 0 1	0	0	0	0
2: 0 0 1 0	0	0	1	0
3: 0 0 1 1	0	1	1	0
4: 0 1 0 0	0	1	0	0
5: 0 1 0 1	0	1	0	0
6: 0 1 1 0	0	1	1	0
7: 0 1 1 1	0	1	1	0
8: 1 0 0 0	0	0	0	1
9: 1 0 0 1	0	0	0	1
A: 1 0 1 0	0	0	1	1
B: 1 0 1 1	0	1	1	1
C: 1 1 0 0	0	1	1	1
D: 1 1 0 1	0	1	1	1
E: 1 1 1 0	0	1	1	1
F: 1 1 1 1	1	1	1	1

Fig. 1.3 Lookup table configuration examples (partly taken from [Koc13])

1.3.2 Configuration Memory Cells

Depending on the FPGA vendor, the configuration memory cells can be provided as SRAM cells, which allows fast and unlimited configuration data write processes. This is the technology used by the FPGA vendors Altera, Lattice, and Xilinx. SRAM-based FPGAs provide the highest logic densities and best performance. As SRAM cells are a volatile memory technology, it requires always a device configuration when powering the device up (which can be seen as a boot phase). The size of the configuration binary (i.e. the bitstream) can range from a few tens of kilobytes to a few tens of megabytes. As a consequence, the configuration of a large FPGA device can take over a second in some systems.

The FPGA vendor Microsemi provides FPGAs that are based on non-volatile configuration memory cells (Flash or antifuse). These devices are suited for extreme environments including space or avionics where electronic components are exposed to higher levels of ionizing radiation (which might impact the state of configuration memory cells of SRAM-based devices). In addition, these FPGAs are very power efficient, start immediately, and have better options for implementing cryptographic systems, because secrets can be stored persistent and securely on the FPGA device itself. However, the capacity and performance is lower than that of their SRAM counterparts.

1.3.3 Interconnection Network (Routing Fabric)

If we look at the technical parameters of an FPGA, we will find much information about the logic cells, including capacity, features, and performance. However, with respect to the area occupied on the FPGA die, the interconnection fabric with all its multiplexers is actually taking more resources than the logic cells themselves and this information is mostly hidden by the FPGA vendors. Also the performance of a circuit mapped to an FPGA is much related to the interconnection network. As a rule of thumb for present FPGAs, we can say that about 2/3 of the latency of a critical path is spent on reconfigurable interconnections. Note that the allowed clock frequency is the reciprocal of the *critical path delay*. This delay is the time needed for a signal to propagate from a flip-flop output through eventually multiple levels of logic cells and routing to a flip-flop input (this process has to be completed before the next clock edge arrives at the flip-flops). The critical path delay is then the slowest of all paths in a circuit mapped to an FPGA.

In practice, the multiplexers used for carrying out the switching between the logic cells will be much bigger than the ones shown in Fig. 1.2, with multiplexers typically providing between 10 and 30 inputs. Furthermore, the interconnection network provides additional switches for intermediate routing (and not only for connecting just the logic cell inputs). The FPGA vendors typically provide enough routing resources to carry out the routing of virtually any design. However, in some cases, it might be needed to work with lower utilization levels such that the ratio of routing resources per logic cells is higher which then permits routing the design.

1.3.4 Further Blocks

Today, most FPGAs provide not only logic cells, but a variety of building blocks (also called *primitives*). Please note that modules that are implemented in the FPGA fabric are often called *soft-logic* or *soft-IPs* (i.e. intellectual property cores implemented in the FPGA user logic). Respectively, cores that are provided with the FPGA fabric are called *hardened-logic* or *hard-IPs*.

Commonly provided hard-IPs on FPGAs include I/O blocks (e.g., for connections to high-speed transceivers which in some cases operate beyond 10 Gbit), dedicated memory blocks, and primitives providing multipliers (or even small ALUs). Furthermore, there could be special hardened functional blocks, like for example a PCIe connection core, DDR memory controllers, or even complete CPUs. Using hardened CPUs on an FPGA is covered in Sect. 15.2.3 (on page 268). The idea behind providing hardened IP blocks is their better area, performance and power ratio, when these blocks are used (because they do not need a costly interconnection network). However, if a certain design leaves these blocks unused, they are basically wasted real estate (see also Sect. 1.1.3). FPGA vendors provide a selection of such blocks in order to meet the requirements of most customers.

1.3.5 How Much Logic is Needed to Implement Certain Logic Functions

When we develop an algorithm in software for a CPU, we are interested in its execution time and the memory requirements. In the case of FPGAs, we are not bound to a temporal domain only, but we have also a spatial domain, which means that we are also interested in how much real estate a certain logic function or algorithm will take. The implementation cost is often not easy to answer and depends on many design factors. For example, a fully featured 32-bit processor requires about 2000 6-input lookup tables on a modern SRAM-based FPGA. However, there are area optimized 32-bit processors that use bit-serial processing and such a processor can be as small as 200 lookup tables [RVS$^+$10]. Such a processor would easily fit a few thousand times on a recent FPGA. In bit-serial operation, we are only computing one bit per clock cycle. Consequently, a 32-bit bit-serial processor is at least 32× slower than its parallel counterpart when operating at the same clock speed. This example shows that in hardware design it is often possible to trade processing speed for real estate. Some behavioral compilers incorporate this automatically for the developer. These tools allow specifying a certain compute throughput and the tool will minimize resource usage, or vice versa, it is possible to define the available resources and the tool maximizes performance.

For integer arithmetic, the cost for addition and subtraction scales linear with the size of the operands (in terms of bits) and is about one LUT per bit. Please note that it is common in hardware design to use arbitrarily sized bit vectors (e.g., there is no need to provide a 32 bit datapath when implementing only a modulo-10 counter). Comparators scale also linear with about 0.5 LUTs per bit to compare. Multipliers scale linear to quadratic, depending if we use sequential or full parallel operation. However, because multipliers can be expensive, there are multiplier blocks available on most FPGAs in these days. Division and square root is particularly expensive and should be avoided if possible. Floating point arithmetic is possible on FPGAs, but here addition and subtraction is more expensive due to the need for normalization (which is a comma shift operation that is needed if the exponents of two floating point numbers differ). However, many high-level synthesis tools support floating point data types and can exploit various optimizations for keeping implementation cost low. Many high level synthesis tools provide resource (and performance) estimators to give instantaneous feedback when programming for an FPGA.

This section is not meant to be complete, but is intended to give an introduction to how hardware can be made reprogrammable. Like with driving a car, where we do not have to understand thermodynamics, it is not necessary to understand these low-level aspects when programming an FPGA. The interested reader might find useful further information in the following books: [HD07, Bob07, Koc13].

1.4 FPGA Design Flow

The *FPGA design flow* comprises different *abstraction levels* which represent the description of an algorithm or an accelerator core over multiple transformation steps all the way to the final configuration bitstream to be sent to the FPGA. The design flow for FPGAs (and ASICs) can be divided into (1) a *fronted phase* and (2) a *backend phase*. The latter phase is typically carried out automatically by CAD tools (similar to the compilation process for a CPU). A short description about the backend flow is given in Sect. 2.5 on page 46 and more detailed further down in this section.

1.4.1 FPGA Front End Design Phase

Similar to the software world, there exists a wide variety of possibilities for generating a hardware description for FPGAs. All these approaches belong to the front end design phase and there exists a large ecosystem with languages, libraries, and tools targeting the front end design. One of the main aims of this book is to give a broad overview so that a software engineer can quickly narrow a search in this ecosystem.

Model-Driven and Domain-Specific Design

FPGAs can follow a model driven design approach which is commonly supported with comfortable and productive GUIs. For example, MATLAB Simulink from The MathWorks, Inc. and LabVIEW from National Instruments (see also Chap. 4) allow the generation of FPGA designs and running those designs pretty much entirely using a computer mouse only. These are out-of-the-box solutions targeting measurement instruments and control systems of virtually any complexity.

FPGA vendors and third-party suppliers provide large IP core libraries for various application domains. This will be covered in Chap. 15 for rapid SoC design using comfortable wizards.

Traditional Programming Languages

FPGAs can be programmed in traditional programming languages or in dialects of those. An overview of this is given in Chap. 3 along with more detailed examples in consecutive chapters. An example of a full environment with hardware platforms (from desktop systems to large data centers) and compilers for Java-like FPGA programming is provided in Chap. 5. The vast majority of high language design tools are for C/C++ or its dialects, which is covered in several chapters. This includes compilers from FPGA vendors in Chap. 6 (OpenCL from Altera Inc.) and Chap. 7 (Vivado HLS from Xilinx Inc.), as well as chapters on academic open source tools. The LegUp tool (developed at the University of Toronto) is presented in Chap. 10 and the ROCCC toolset (developed at The University of California, Riverside) is covered in Chap. 11, respectively.

Source-to-Source Compilation

For improving design productivity in specific domains (e.g., linear algebra), there exist source-to-source compilers that generate from an even higher abstraction level (commonly) C/C++ code that will then be further compiled by tools as described in the previous paragraph. Examples of this kind of solution are provided in Chap. 8 with the Merlin Compiler from Falcon Computing Solutions and in Chap. 12 with the HIPA^{cc} tool (developed at the Friedrich-Alexander University Erlangen-Nürnberg).

One big advantage of approaches operating at higher abstraction levels is not only that specifying a system is easier and can be done faster; there is also much less test effort needed. This is because many tools provide correct-by-construction transformations for specifications provided by the designer.

Low-Level Design

At the other end of the spectrum, there are the lower level Hardware Description Languages (HDLs) with Verilog and VHDL being the most popular ones. These languages give full control over the generated hardware. However, these languages are typically used by trained hardware design engineers and it needs a strong hardware background to harness the full potential of this approach. A solution that provides a higher level of abstraction (while still maintaining strong control over the generated hardware in an easier way than traditional HDLs) comes with the Bluespec SystemVerilog language and compiler, as revealed in Chap. 9.

The result of the front end design phase is a Register-transfer level (RTL) description that is passed to the backend flow. All this is described in more detail in the next section.

1.4.2 FPGA Backend Design Flow

The backend design flow for FPGAs and ASICs looks similar from a distance, as illustrated in Fig. 1.4. The figure shows a general design flow for electronic designs with FPGA and ASIC target technologies. The different abstraction levels were introduced by Gaijski et al. in [GDWL92] and include:

RTL level *The Register-Transfer Level* is basically an abstraction where all state information is stored in registers (or whatever kind of memory) and where logic (which includes arithmetic) between the registers is used to generate or compute new states. Consequently, the RTL level describes all memory elements (e.g., flip-flops, registers, or memories) and the used logic as well as the connections between the different memory and logic elements and therefore the flow of data through a circuit.

RTL specifications are commonly done using hardware description languages (HDLs). Those languages provide data types, logical and arithmetic operations,

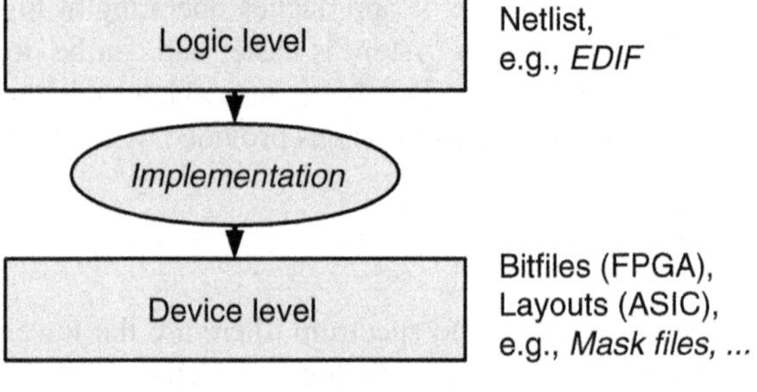

Fig. 1.4 A general design flow for FPGA and ASIC designs with the synthesis and implementation steps and the different abstraction levels

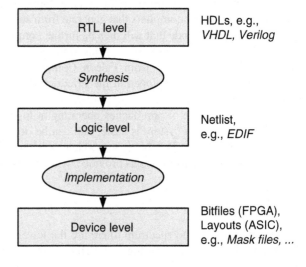

hierarchies, and many programming constructs (e.g., loops) that are known from software programming languages such as C. However, hardware is inherent parallel and there is not a single instruction executed at a point in time, but many modules that will work concurrently together in a system, which is well supported by HDLs.

We could see the RTL level in analogy to a CPU, where all state information is stored in registers (or some memory hierarchy) and where the ALU is in charge of changing the value of registers (or memories). The abstraction level of RTL is rather low and in an analogy to software, we could say it is comparable to an abstraction in the range somewhere from assembly to C. However, RTL code is still well readable and maintainable. As with assembly code, RTL code might be generated by other compilation processes as an intermediate representation and might not even be seen by an FPGA developer.

Logic level A synthesis tool will take the RTL description and translate it into a netlist which is the *logic abstraction level.* Here, any behavioral description from the RTL level is translated into registers and logic gates for implementing elementary Boolean functions. The netlist is a graph where the nodes denote registers and gates and the edges denote the connecting signals. For example, a compare between two registers A and B (let's say from an expression `if` $A = B$ `then ...`) will result in an XNOR gate for each bit of the two values A and B (an XNOR is exactly `1` if both inputs are identical) and an AND gate (or a tree of AND gates) to check if all the XNOR gates deliver `1` at the output. This process is entirely carried out by logic compilers but can eventually be guided (hardware engineers prefer the term "to constrain a design or implementation") by the user (e.g., for either generating a small cost-efficient implementation or a high-performance implementation).

In the case of FPGAs as the synthesis target, there are specific synthesis options to control which kind of primitive on the FPGA fabric will later be used to implement a specific memory or piece of logic. For example, a shift register (which is basically a chain of flip-flops) can be implemented with the flip-flops after the lookup table function generators, with lookup table primitives that provide memory functionality, or with a dual-ported memory surrounded with some address counter logic. Similarly, multiplications can be done by dedicated multiplier blocks or by generating an array of bit-level multiplier cells (for example if we run short on multiplier blocks). In most cases, the default settings for the synthesis tool will work fine and many higher level tools are able to adjust the logic synthesis process automatically without bothering the designer.

Device Level The netlist from the previous level undergoes several optimization processes and is now to be implemented by the primitives (lookup tables, memory blocks, I/O cells, etc.) and the interconnection network of the FPGA.

In a first step, called *technology mapping*, the logic gates are fitted into the lookup tables. For this process, there exist algorithms for minimizing the number of lookup tables needed (e.g., Chortle [FRC90]). Alternatively there are algorithms for minimizing the depth of the tree when larger Boolean circuits have to be fitted into the given LUTs (e.g., FlowMap [CD94]). The last option results in fewer lookup tables to pass and consequently in lower propagation delay and therefore in a faster possible clock frequency.

After technology mapping, the primitives will get placed on the FPGA fabric. This is done with clustering algorithms and simulated annealing in order to minimize congestion and the distance between connected primitives.

The placed primitives get now routed and the difficulty here is that a physical wire of the FPGA can only be used exclusively for carrying out the linking of one signal path. A well-known algorithm for this is called Pathfinder [ME95]. In a nutshell, this algorithm routes each path individually and increases the cost for each wire segment according to how often it was used. Then by iteratively restarting the routing process, popular wires get more expensive which eventually resolves the conflict that multiple paths want to use the same wire.

The size of the place and route problem is quite large with more than one million primitives (i.e., logic cells, multipliers, memory blocks, etc.) and much more than ten million wires, each with about 10–30 possible programmable connections on recent devices. The tool time for computing all these steps can in some cases exceed a full day for such large devices and requires a few tens of gigabytes memory. Luckily, the process runs fully automated and there are techniques like incremental compilation and design preservation that allow keeping parts of the physical implementation untouched if only changes were done in a design. This will then significantly speed up design respins. In addition, there are tools that support a component-based design flow for plugging fully pre-implemented modules together to a working system. One approach for this method is known as hardware linking that works (in analogy to software linking) by stitching together configuration bitstreams of individual modules [KBT08]. This allows rapidly building systems without any time consuming logic synthesis, placement, and routing steps, but requires building a corresponding module library (e.g., [YKL15]).

The final mapped, placed, and routed netlist is then translated into a configuration bitstream that contains the settings for each primitive and switch matrix multiplexer (i.e. the values of the configuration memory cells shown in Fig. 1.2). More details on the low-level implementation can be found in [BRM99, Bob07, HD07, Koc13].

1.5 Overview

Most chapters in this book have already been introduced throughout this introduction. This introduction provides a discussion about compute paradigms and a brief introduction to FPGA technology. This is followed in Chap. 2 by a presentation of the theoretical background behind high-level synthesis which allows the generation of digital circuits directly from languages such as Java or C/C++. Chapter 3 gives a classification of various HLS approaches and Chaps. 4–12 are devoted to specific languages and tools including several case studies and small code examples.

Using FPGAs in a programming environment requires operating system services, some kind of infrastructure, and methods to couple this with a software subsystem. ReconOS (developed at the University of Paderborn) provides this functionality well for embedded systems, as presented in Chap. 13, while the LEAP FPGA operating system is targeting compute accelerators for x86 machines (Chap. 14).

The last two chapters are for software engineers who want to use FPGAs without designing any hardware by themselves. Chapter 15 provides examples on how complex systems can be built and programmed on FPGAs following a library approach. After this, Chap. 16 shows how FPGA technology can be used to host another programmable architecture that is directly software programmable. The overall goal of this book is to give software engineers a better understanding on accelerator technologies in general and FPGAs in particular. We also want to postulate the message that FPGAs are not only a vehicle for highly skilled hardware design engineers, but that there are alternative paths for software engineers to harness the performance and energy efficiency of FPGAs.

Chapter 2
High-Level Synthesis

João M.P. Cardoso and Markus Weinhardt

2.1 Introduction

The compilation of high-level languages, such as software programming languages, to FPGAs (Field-Programmable Logic Arrays) [BR96] is of paramount importance to the mainstream adoption of FPGAs. An efficient compilation process will improve designer productivity and will make the use of FPGA technology viable for software programmers.

This chapter focuses on the compilation of computations (algorithms) targeting FPGAs in a process known as High-Level Synthesis (HLS) for FPGAs. HLS is the translation of *algorithmic descriptions* to application-specific architectures, usually implemented as digital systems consisting of a datapath and a control unit or finite state machine (FSM) [GDWL92, Mic94].

In this chapter, we use ANSI-C [KR88] as a widely used language for specifying the input behavioral description (e.g., a function). Other input languages such as Java, C++, MATLAB or even specific hardware description languages (HDLs) as SystemC or algorithmic-level behavioral VHDL could be used in a similar manner. As opposed to these input descriptions, the generated digital systems are described at the register-transfer level (RTL). These RTL descriptions are then mapped to physical hardware (ASICs or FPGAs) using backend tools, cf. Sect. 2.5.

The main difference between algorithmic and RTL specifications is the level of detail: algorithmic descriptions only define the order in which operations are to be performed, whereas an RTL design specifies their exact timing. Furthermore, data

J.M.P. Cardoso (✉)
University of Porto, Faculty of Engineering, Porto, Portugal
INESC-TEC, Porto, Portugal
e-mail: jmpc@acm.org

M. Weinhardt
Osnabrück University of Applied Sciences, Osnabrück, Germany

D. Koch et al. (eds.), *FPGAs for Software Programmers*,
DOI 10.1007/978-3-319-26408-0_2

Fig. 2.1 C function of the greatest common divisor (GCD) algorithm

```
int gcd(int a, int b) { //  1
  while (a!=b) {        //  2
    if (a>b)            //  3
      a = a-b;          //  4
    else                //  5
      b = b-a;          //  6
    /* end if */        //  7
  }                     //  8
  return a;             //  9
}                       // 10
```

and operations are mapped to concrete hardware units (register, adders, multipliers, etc.). The resulting design at RTL describes for each control step, which values are transferred to which storage locations (registers). In other words, the specification now contains details of a digital system implementation.

In summary, HLS needs to solve the following tasks:

- **Scheduling:** Assigning the operations to *control steps*, i.e., clock cycles of synchronous hardware designs.
- **Allocation:** Selecting the number and type of hardware units used for implementing the functionality, i.e., arithmetical operators, storage (registers and memories) and interconnect resources (buses, multiplexers, etc.).
- **Binding:** Mapping of all operations to the allocated hardware units.

These tasks will be illustrated in detail in the following sections, but first we illustrate the use of HLS with a simple example. The well-known Euclidean algorithm computes the greatest common divisor (GCD) of two natural numbers. Figure 2.1 shows an implementation in C.

Figure 2.2 shows a digital circuit which could be generated by HLS from the C function in Fig. 2.1. The two D-registers at the top (with clock signal *clk*) store the variables a and b. For simplicity, it is assumed that the input values for a and b are stored in the registers by a special mechanism not shown in Fig. 2.2 before the control signal *run* is set to 1 (i.e., before the circuit starts executing). Then, the entire loop body (including the evaluation of the loop control) is performed in one clock cycle, i.e. all operations are scheduled to the same control step. In this implementation, the function is synthesized for maximum parallelism. To achieve this, four operators (equality tester, greater-than comparator, and two subtracters) have to be allocated so that each operation can be bound to its own hardware unit. The two subtracters compute the then- and else-part of the if-statement in parallel, and the multiplexers at the bottom select the correct (i.e., new or unchanged) values for a and b, controlled by the comparator. Finally, the equality tester disables the registers ($EN = 0$) and sets the *done* signal to 1 as soon as a and b are equal. Then the hardware circuit has finished its execution and the result can be read at output *a_out*.

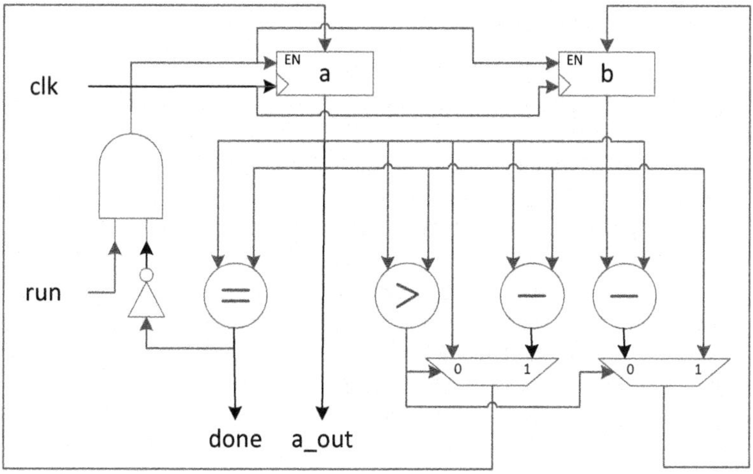

Fig. 2.2 Digital system computing the greatest common divisor (GCD) between two numbers

Note that in a real HLS system, the generated circuit, represented graphically in Fig. 2.2, is specified by an RTL HDL description, e.g., in VHDL. In a next stage, these RTL descriptions are input to specific tools able to generate the control and datapath units (in a process known as RTL and logic synthesis [Mic94]). Finally, mapping, placement and routing is applied and the binaries to program the FPGA resources are generated.

2.2 From Behavior to Graph Representation

This section outlines how a behavioral description, e.g., a C function, is translated to a graph representation called *control/data flow graph (CDFG)* serving as intermediate representation of a HLS tool. The following Sect. 2.3 explains how an RTL design is synthesized from a CDFG.

2.2.1 Compiler Frontend

The frontend (i.e., the analysis phase) of a HLS system is similar to the one existent in a traditional software compiler [ALSU06]. In the case of C as input language, it is nearly identical. It involves common lexical and syntactic analysis, and generates a high-level intermediate representation (IR) such as an abstract syntax tree (AST). This data structure is then semantically analyzed and attributed with type and other information. Before being converted to the CDFG representation, several machine-independent optimizations (e.g., dead-code elimination, strength reduction, and constant propagation and folding) are applied to the AST or to a lower-level intermediate representation. For optimal FPGA mapping, several hardware-specific optimizations and loop transformations are applied, cf. Sect. 2.4.

2.2.2 *Control Flow Graphs*

The first step in generating a CDFG is the generation of a *control flow graph (CFG)* of the function [ALSU06, Sect. 8.4]. The nodes of this graph are the function's *basic blocks*, and the arcs indicate the flow of control between the basic blocks, i.e. the execution of the program. A basic block (BB) is defined as a sequence of instructions which have a single entry point (beginning) and which are executed to the end of the BB without any exit or branch instructions in the middle (i.e., with a single exit point). Branch nodes which evaluate a condition can form a separate BB. They have two or more output arcs representing alternative execution paths of the program, depending on the condition. The CFG's arcs represent *control dependences* which *must* be obeyed for correct program execution.

A CFG can be easily constructed from the AST as shown in Fig. 2.3. For loops are usually converted to while loops. Similarly, case statements are just an extension of if statements (conditions) with several output arcs. Note that node BB5 (a join node) in Fig. 2.3b can be combined with subsequent nodes since it does not contain any instructions.

The CFG for the GCD example (Fig. 2.1) is shown in Fig. 2.4. Note that all basic blocks consist of a single instruction in this simple example. For clarity, the empty basic block BB7/8 was maintained in Fig. 2.4 though it could be further merged

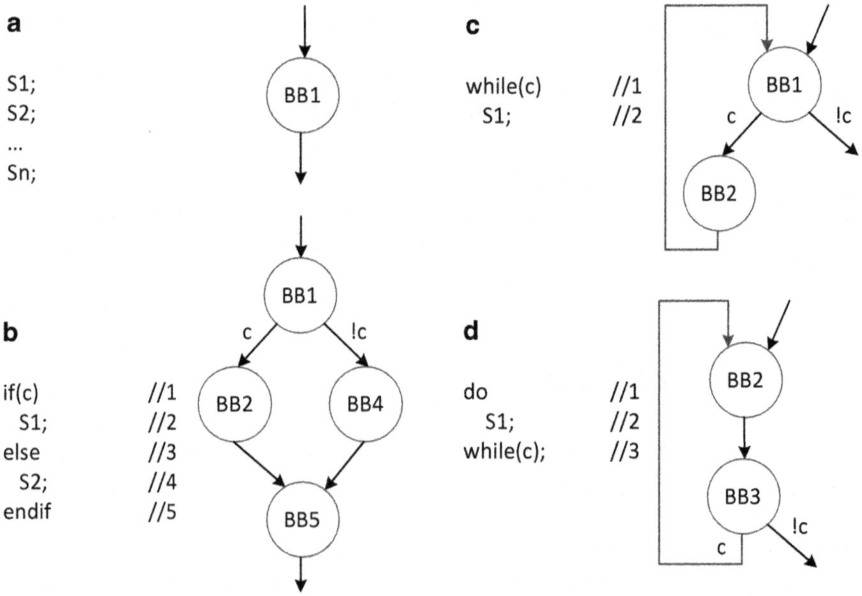

Fig. 2.3 Construction of control flow graphs. (**a**) Sequence, (**b**) condition, (**c**) while loop, (**d**) do-while loop

with BB2. In BB1 (function entry) the function parameters are input, and in BB9 (function return) the return value is output and the function completion is indicated.

To represent function calls, the CFGs can be extended with *call nodes* which transfer control to another function's CFG. There can be also CFG extensions to represent parallel regions of code. It is also possible and suitable to use a top-level graph representation showing tasks, their dependences and parallel characteristics. Note that the compilation to FPGAs may also take advantage of the concurrent nature of hardware and the large number of hardware resources to implement designs taking advantage of coarse-grained parallelism (e.g., executing concurrent tasks in parallel).

2.2.3 Data Flow Graphs

As mentioned above, the execution order of the basic blocks is determined by the CFG. However, there is more freedom within a BB: the instructions can be swapped or even executed in parallel as long as no *data dependences* prevent this. A data dependence exists if the output of one instruction is used as input of another instruction. This information is explicitly represented in the *data flow graph (DFG)* of a BB, comparable to the DAG representation in software compilers [ALSU06, Sect. 8.5]. The BB's operations are represented by the DFG's nodes. An arc is drawn from node i to node j if j directly uses i's output as its input. Arcs are usually drawn in top-down direction. If assignments use complicated terms, the DFG shows the terms' operator trees. The input nodes are variable reads, and the output nodes are variable writes. Only variables live at the end of a BB, i.e., values which may be used in subsequent BBs, need to be written. Note that aliases (arising from array or pointer accesses or from procedure calls) hinder the construction of DFGs.

Since the GCD example only contains minimal BBs, we introduce a second example to illustrate DFGs: the *diffeq* function in Fig. 2.5 [PK89] solves a specific differential equation. Neglecting the function entry and exit, this example contains only two BBs marked by comments in Fig. 2.5. BB1 contains the loop condition and BB2 the loop body. Hence its CFG is equivalent to the CFG in Fig. 2.3c.

The DFGs for BB1 and BB2 are displayed in Fig. 2.6.[1] Obviously there are no data dependences between the multiplications 3*x, u*dx, 3*y and the addition x+dx. Hence they can be executed in arbitrary order or in parallel. This is not true for the remaining operations which use the results of the mentioned operations. Generally speaking, any topological sorting of the nodes is a valid schedule for computing the results of a DFG.

[1] Since addition and multiplication are commutative operations, the operand order has been changed for some nodes to improve the graph layout.

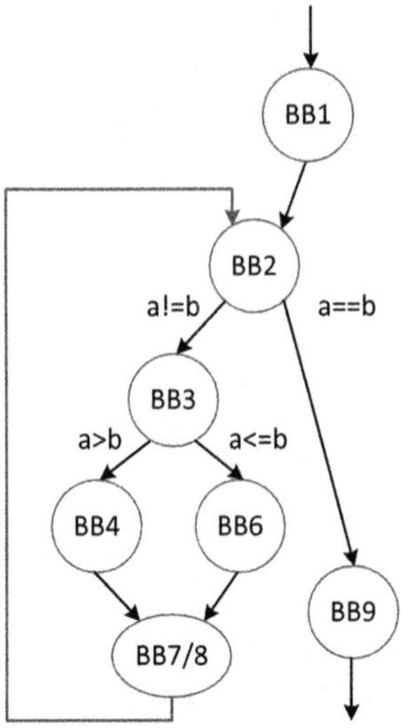

Fig. 2.4 CFG of the GCD function

```
int diffeq(int x, int u, int y, int a, int dx) {
  int x1, u1, y1;
  while (x < a) {     // BB1
    x1 = x + dx;      // BB2 beginning
    u1 = u - (3 * x) * (u * dx) - (3 * y) * dx;
    y1 = y + u * dx;
    x = x1; u = u1; y = y1; // BB2 end
  }
  return y;
}
```

Fig. 2.5 C function for solving the differential equation $y'' + 3xy' + 3y = 0$

2.2.4 Control/Data Flow Graphs

Finally, the CFG of a function is combined with the DFGs of its basic blocks to construct its CDFG. In contrast to the CFGs, branch nodes are represented by triangles in CDFGs. Figure 2.7 shows the CDFG for function *diffeq*. For simplicity, the DFGs of Fig. 2.6 are not repeated.

The CDFG contains all the input program's information in a suitable form for the subsequent synthesis steps.

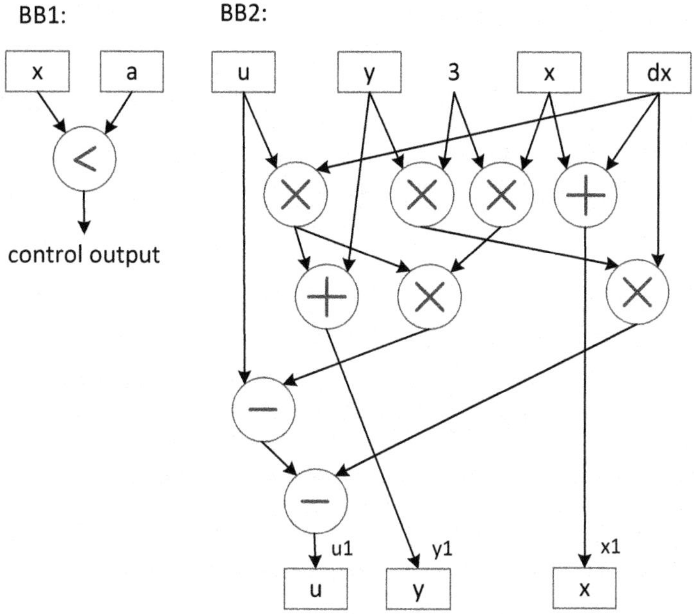

Fig. 2.6 Data flow graphs (DFGs) for the BBs of the *diffeq* function

Fig. 2.7 Control/data flow graph (CDFG) for the diffeq function

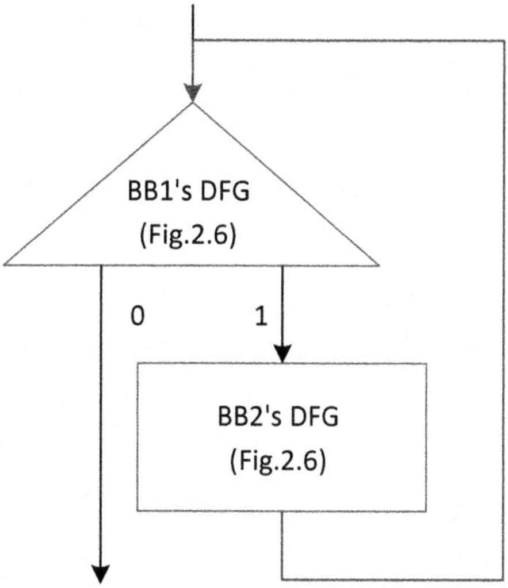

2.3 Basic Synthesis Tasks

As outlined in Sect. 2.1, the basic synthesis tasks are scheduling, allocation and binding. Since these tasks are all interdependent, they should be solved together for optimal results. However, since this is not feasible in practice, they are executed one after another which may lead to suboptimal results.

2.3.1 *Scheduling*

During scheduling, all operations of the input CDFG are mapped to control steps. Each control step corresponds to a state of the circuit's controller or FSM which activates the states in accordance with the CFG's control dependences and the basic blocks' data dependences. Scheduling can be performed individually for each DFG or combined for the entire CDFG. For simplicity, we illustrate the effect of several scheduling algorithms on BB2 of Fig. 2.6. Figure 2.8 shows one legal schedule requiring four control steps CS0–CS3. It is assumed that every operator executes in one control step.

All operations mapped to the same control step are executed in parallel, i.e., they exploit instruction-level parallelism (ILP). Therefore every operation requires its own operator (functional unit). For example, CS0 requires three multipliers and

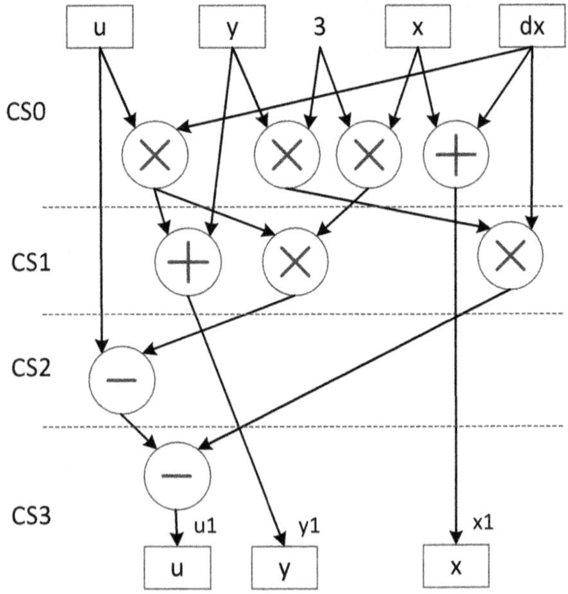

Fig. 2.8 Scheduled loop body (BB2) of the *diffeq* function (ASAP schedule)

one adder.[2] On the other hand, operations in different control steps can share the same operator. Hence the chosen schedule has a strong impact on the unit allocation. There is a tradeoff between the design goals of high performance (few control steps) and small circuit area (few allocated units). A point in the design space cannot optimize all goals together.

Figure 2.8 shows a schedule with the smallest possible number of control steps: every operation is scheduled as soon as possible (ASAP). The algorithm is therefore called ASAP scheduling. The same performance is achieved by an ALAP (as late as possible) schedule as shown in Fig. 2.9. However, the number and type of the required hardware operators may be different. The ASAP schedule in Fig. 2.8 requires at least three multipliers, an adder and a subtracter. In our example, the ALAP schedule requires at least two multipliers, two adders and one subtracter (with operator sharing).

By comparing the ASAP and ALAP schedules, it is obvious that the DFG leaves some flexibility for some nodes. The *mobility* M of an operator is computed as the difference between the numbers of the ALAP and ASAP control steps. They are marked in Fig. 2.9 unless the value is zero.

When variables (e.g., arrays) are mapped to memory components (e.g., Block RAMs) the load/stores need to be scheduled and the datapath also needs to include the hardware required to interface the memory ports to the other datapath units.

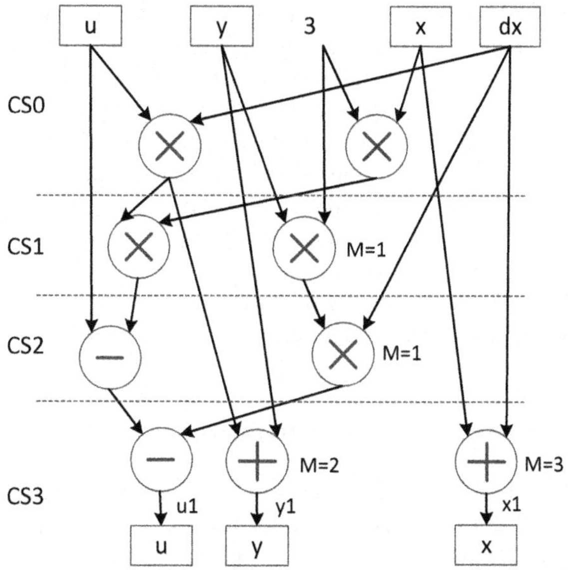

Fig. 2.9 Scheduled loop body (BB2) of the *diffeq* function (ALAP schedule)

[2]The adder could also be executed by a combined adder/subtracter or by an integer ALU.

Time-Constrained Scheduling

For many applications, e.g. real-time systems, high performance is more important than low resource usage. Therefore, *time-constrained scheduling algorithms* are used. They generate schedules with a restricted number of control steps (the minimum being the lengths of the unrestricted ASAP and ALAP schedules) using the smallest number of hardware units. Unfortunately, exact optimization algorithms may lead to exponential runtimes which is not feasible for large DFGs. Therefore, most approaches use *force-directed scheduling (FDS)* [Mic94], a constructive heuristic algorithm.

For FDS, the ASAP and ALAP schedules for the given number of control steps and the operators' mobility are computed first. Then, for each operator type, the estimated operator cost (EOC) is computed for each control step using probabilities. For operators with zero mobility, the probability is one in their fixed control step and zero in all others. For mobile operators, their probability is distributed over their mobility range. For example, the adder computing $x1$ in Fig. 2.9 with M = 3 could occur in all four control steps. Therefore its probability is 0.25 in all steps. This number is also the EOC for adders in CS0 since there are no other adders which could be scheduled in this control step. However, in CS1, there could be another adder (computing $y1$) with M = 2 and probability 0.33. Hence the overall EOC for adders in CS1 is 0.58, (i.e., the sum of the individual probabilities of the additions that can be scheduled in this control step). The maximum EOC over all control steps is used as the estimated overall cost for this operator type. Finally, the sum over all operator types (weighted with the operator cost, i.e., area) is the total cost of a (partial) schedule. FDS then evaluates the impact on the total cost for all possible assignments of all unassigned (mobile) operations. The assignment with the highest impact is chosen and fixed. For the ASAP and ALAP schedules in Figs. 2.8 and 2.9, this is achieved by assigning the multiplier computing 3 * y to CS1. This also assigns the multiplier depending on its output to CS2. After that, only additions and subtractions need to be assigned. For simplicity, we use combined adder/subtracters for both these operations. The resulting FDS schedule is shown in Fig. 2.10. It only uses two multipliers and one adder/subtracter, less than both the ASAP and ALAP schedules.

Resource-Constrained Scheduling

Resource-constrained scheduling algorithms optimize the number of control steps required for a given set of operators. A frequently used algorithm is *list scheduling* [Mic94]. In list scheduling, all operations which are ready for scheduling (i.e., those with all predecessors already scheduled) are entered into a priority list and scheduled considering the priority order. For each operation scheduled, the ready list is updated with the possible new ready nodes. When it is not possible to schedule any of the nodes in the list in the current control step (i.e., there is no free resource for these operations), the scheduling proceeds with a new control step. Different priority

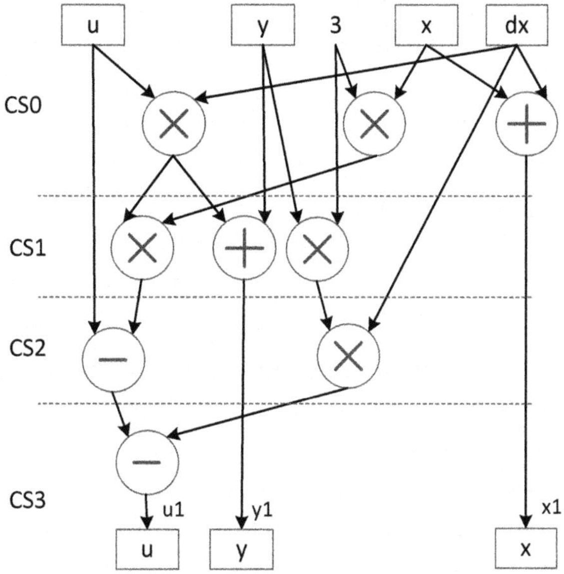

Fig. 2.10 Force-directed scheduling (FDS) for BB2 of the *diffeq* function

criteria can be used, e.g. the operations with the lowest mobility are scheduled first. Let us apply list scheduling for BB2 of the *diffeq* example. If only one multiplier and one adder/subtracter are available, the result is the schedule shown in Fig. 2.11 which requires six control steps.

Scheduling with Varying Delays

So far we assumed that all operators have the same speed, i.e., have the same combinatorial delay. But this is not realistic. For example, a 32-bit multiplier has a longer delay than a 32-bit adder. Since the duration of a control step (or clock cycle) depends on the *slowest operator* in all steps, the adders waste much of the cycle time in our previous examples. To improve this situation, there are two possibilities: (a) slow operators are scheduled over several cycles (multi-cycle operators), or (b) several fast operators are executed within one cycle (operator chaining). Figure 2.12 shows examples for these situations where an adder is about twice as fast as a multiplier. Both methods can significantly improve the performance of a digital design since the total execution time is the number of cycles times the duration of a cycle.

Overall, the maximum clock frequency at which the design can operate will be dependent of the critical path delay, i.e., the longest delay of the existing combinatorial paths (consisting of a sequence of one or more functional units without registers).

Obviously the scheduling algorithms introduced above need to be adapted to exploit these methods.

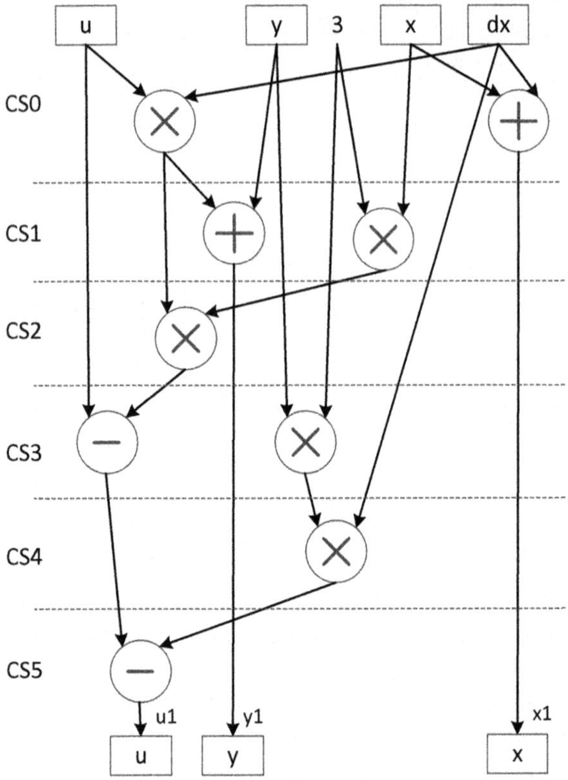

Fig. 2.11 List-based schedule for BB2 of the *diffeq* function

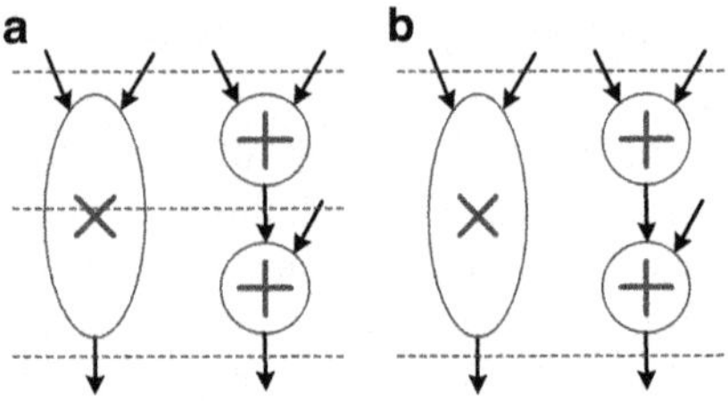

Fig. 2.12 Multi-cycle operators (**a**) and operator chaining (**b**)

2.3.2 *Allocation and Binding*

After scheduling, hardware operators are selected for the operations, and registers for the intermediate values *(allocation)*. The minimum number of operators is determined by the number of operations scheduled in one control step, and the minimum number of registers is determined by the number of values which need to be stored across control steps. Then, the individual operations and values are mapped to the operators and registers, respectively *(binding)*. If an operator or register is shared among several operations or values, respectively, the correct input values have to be selected by multiplexers in every control step.

Let us take the FDS schedule introduced in Fig. 2.10 as an example. The schedule implies the allocation of two multipliers (MUL1 and MUL2) and one simple ALU for additions and subtractions. In Fig. 2.13a, all operations bound to the same hardware operator are drawn in one column. Additionally, five registers storing the newly computed values for each control step are allocated. Registers can only be reused if their values are not live anymore, i.e. not used subsequently. Note that four registers would be sufficient, but a fifth register was chosen because combining MUL and ALU outputs in one register would require another multiplexer at the register input and would add delay to the circuit. On FPGAs with plenty of flip-flops, it makes more sense to use an extra register. Figure 2.13b shows the resulting datapath (without the controller). The hardware operators select their respective inputs according to the current control step CS. The control step also defines the functionality of the ALU (add or sub) and whether a new register value is stored or not. In the figure, decoded CS signals (CS0–CS3) are used for simplicity.

Note however, that register and operator sharing is not always the best solution. Especially on FPGAs, the size of 2:1 multiplexers is comparable to adders or subtracters. Therefore, additional adders may be worthwhile even if they are not used in every cycle.

2.3.3 Controller Synthesis

Finally, the sequencing of control steps must be implemented by a control unit (FSM) where each step corresponds to a state. In our example, CS0 to CS3 are executed sequentially. As shown in Fig. 2.13b, the control inputs of the multiplexers and register must be set accordingly. The controller is implemented by standard logic synthesis methods, cf. Sect. 2.5 [Mic94]. After the last control step, the result values (x, y and u in Fig. 2.13) are available in their respective registers.

2.4 Compiler Optimizations

The computing in space provided by FPGAs leverages the large number of hardware resources able to execute in parallel, hardware structures dedicated to stream processing, and the multiple on-chip memories accessed at the same time. In order to take advantage of these resources, compilation to FPGAs [NI14] needs to expose high-levels of fine-grained as well as coarse-grained parallelism, through the application of architecture-driven optimizations and architecture-neutral code transformations (e.g., loop transformations).

For achieving very efficient FPGA implementations an advanced compiler needs to apply an extensive portfolio of optimizations [NI14]. Those optimizations include

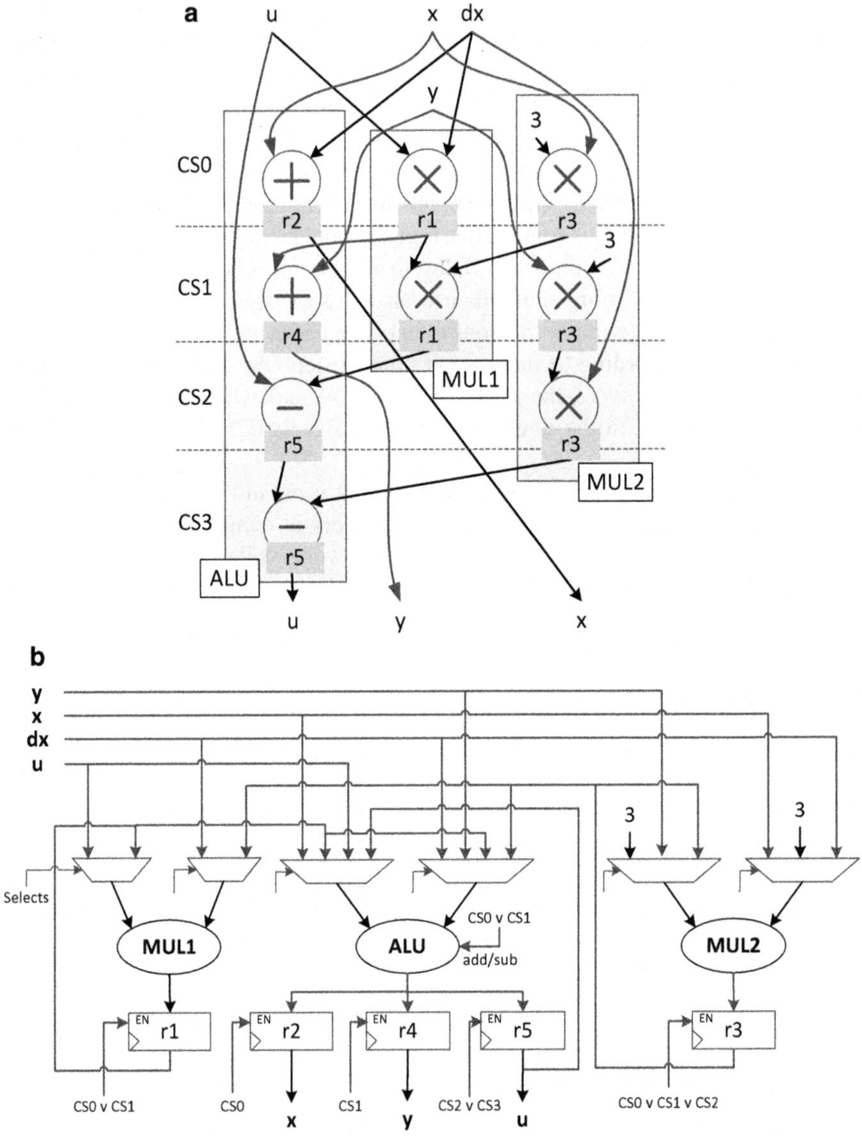

Fig. 2.13 Allocation and binding of BB2 of diffeq, using FDS schedule from Fig. 2.10: (**a**) operator and register binding; (**b**) resulting datapath

target independent (but target aware) optimizations such as constant propagation, constant folding, loop transformations (e.g., unrolling, tiling, fusion, distribution, strip-mining) [ALSU06], common subexpression elimination, scalar replacement, strength reduction, code motion, and elimination of memory accesses using register promotion. We suggest [CDW10] for more details about optimizations for compiling to FPGAs.

The next sections describe the application of some of these compiler optimizations using illustrative examples.

2.4.1 Code Transformations

Code transformations are important compiler optimizations for generating highly optimized FPGA implementations. They focus on the restructuring of code in order to expose a more suitable specification and to increase the potential for further compiler optimizations.

Let us consider how the optimizations mentioned above can be applied to an example program, the small finite-impulse response (FIR) filter in Fig. 2.14a. There are a couple of standard compiler optimizations that can be applied to this example.

a

```
// x is an input array of ints
// y is an output array of ints
#define c0 2
#define c1 4
#define c2 4
#define c3 2
#define M 256 // no. of samples
#define N  4  // no. of coeff.s
int c[N] = {c0, c1, c2, c3};
...
  // Loop 1:
  for(int j=N-1; j<M; j++) {
    output=0;
    // Loop 2:
    for(int i=0; i<N; i++) {
      output+=c[i]*x[j-i];
    }
    y[j] = output;
  }
...
```

b

```
#define c0 2
#define
c1 4
#define c2 4
#define c3 2
#define M 256
#define N  4
int c[N] = {c0, c1, c2, c3};
...
  // Loop 1:
  for(int j=3; j<M; j++) {
    output=0;
    output+=c[0]*x[j-0];
    output+=c[1]*x[j-1];
    output+=c[2]*x[j-2];
    output+=c[3]*x[j-3];
    y[j] = output;
  }
...
```

Fig. 2.14 FIR example: (**a**) original code; (**b**) transformed code after loop unrolling

a
```
...
  // Loop 1:
  for(int j=3; j<M; j++) {
    output =c0*x[j];
    output+=c1*x[j-1];
    output+=c2*x[j-2];
    output+=c3*x[j-3];
    y[j] = output;
  }
...
```

b
```
for(int j=3; j<M; j++) {
  x_3=x[j];
  x_2=x[j-1];
  x_1=x[j-2];
  x_0=x[j-3];
  output=c0*x_3;
  output+=c1*x_2;
  output+=c2*x_1;
  output+=c3*x_0;
  y[j] = output;
}
```

Fig. 2.15 FIR example: (**a**) code after loop unrolling and constant replacement; (**b**) intermediate code using scalars to store loaded values

For example, *Loop 2* is a good candidate for full loop unrolling as the number of iterations is small (4) and statically known. The result of applying full loop unrolling to *Loop 2* is presented in Fig. 2.14b. We can now see that we can apply an algebraic simplification to $j - 0$ (resulting in j), constant propagation and algebraic simplification to the first use of output, and constant propagation of the elements of the c array (an array of constants). Figure 2.15a shows the result of those optimizations. Note also that the c array is no more needed and it was removed and loads to its elements were substituted by the respective constant values. At this point we can also perform operator strength reduction in each of the multiplications by the c's coefficients, as c0*x_3 is in fact 2*x_3 and this can be performed by a shift by a constant instead of using a multiplication: x_3 << 1. This is an optimization which is usually important for FPGA compilation as these shifts do not require hardware components and can be fully implemented by appropriate wiring.

As we can see in Fig. 2.15b, there are three elements of x that can be reused in every iteration of *Loop 1*. Figure 2.16a shows a possibility to transform the code to eliminate redundant memory accesses (loads of x, in this case) by data reuse. Figure 2.16b does the same but moves the code responsible for loading the first three elements to x_2, x_1, and x_0 from the first iteration of *Loop 1* to code before the loop.

Figure 2.16c shows a further transformation that can be helpful. It considers the execution of the FIR example in a streaming model of computation. In this case, we assume that the input samples of x are not stored but arrive in an input port (e.g., a FIFO channel) and the output results are not stored in array variable y but are written to an output port. As soon as data arrives, computations are performed, and results are output. This type of implementation can thus achieve high throughputs (e.g., one new output per clock cycle) which are sometimes required.

Note that transforming an application expressed in a non-streaming language to a version exposing streaming can be very difficult. In real life examples the

a
```
for(int j=3; j<M; j++) {
  x_3=x[j];
  if(j==3) {
    x_2=x[j-1];
    x_1=x[j-2];
    x_0=x[j-3];
  }
  output=c0*x_3;
  output+=c1*x_2;
  output+=c2*x_1;
  output+=c3*x_0;
  x_0=x_1;
  x_1=x_2;
  x_2=x_3;
  y[j] = output;
}
```

b
```
x_0=x[0];
x_1=x[1];
x_2=x[2];

for(int j=3; j<M; j++) {
  x_3=x[j];
  output=c0*x_3;
  output+=c1*x_2;
  output+=c2*x_1;
  output+=c3*x_0;
  x_0=x_1;
  x_1=x_2;
  x_2=x_3;
  y[j] = output;
}
```

c
```
x_0= receive(PORT_A);
x_1= receive(PORT_A);
x_2= receive(PORT_A);

for(int j=3; j<M; j++) {
  x_3=receive(PORT_A);
  output=c0*x_3;
  output+=c1*x_2;
  output+=c2*x_1;
  output+=c3*x_0;
  x_0=x_1;
  x_1=x_2;
  x_2=x_3;
  send(PORT_B, output);
}
```

Fig. 2.16 Data reuse and streaming transformations to FIR example: (**a**) code with data reuse; (**b**) code after code motion; (**c**) code transformed to a streaming model

transformation might be too complex to be performed automatically and might have to be done manually.

2.4.2 Hardware-Specific Optimizations

These are optimizations more specific to hardware compilation which may be of limited interest for a software compiler as they mainly focus on customization optimizations. The customization comes from, e.g., specialized data types, and bit-width narrowing. The ultimate goal is to leverage the hardware customization provided by specific functional units, memory, and pipelining structures. For instance, conversions from floating- to fixed-point data types can be used as fixed-point data types enable cheaper hardware implementations than the ones for dealing with floating-point data types. It can be also an option to use custom floating-point data types.

A suitable customization optimization is bit-width narrowing for each variable used in the code. In a target architecture where variables can be stored in custom registers, e.g., the use of 3 bits instead of 8 saves hardware not only in the register to store the values but also in the hardware components to perform operations as this custom bit-widths can be now propagated to the operations over the data. A very simple example is the use of 8 bits for *j* and 2 bits for *i* in the code in Fig. 2.14a.

Another very important optimization is if-conversion [GDWL92] which converts control flow to data flow. In the case of hardware compilation, if-conversion is not limited to the support of predicates, but may enable the execution of operations even if they are in branches not taken due to the FPGA spatial model of computation. Figure 2.17a presents a simple example with an if-else which can be translated to the predicated code of Fig. 2.17b. The hardware implementation is shown in Fig. 2.17c. We can see that both branches of the if-else and the comparison are all executed in parallel. The correct result is then selected by a multiplexer. An interesting aspect of this optimization is that it does not need to be controlled by the control unit (FSM).

The critical path of the generated datapath may increase the combinatorial delay, i.e., the duration of a clock cycle. In order to reduce the critical path, some optimizations can usually be applied. Reassociation of operations and algebraic

a

```
// int f=abs(int n)
if (n >= 0)
  f = n;
else
  f = -n;
```

b

```
// int f=abs(int n)
p1 = (n >= 0);
f = n;          <p1>
f = -n;         <!p1>
```

c

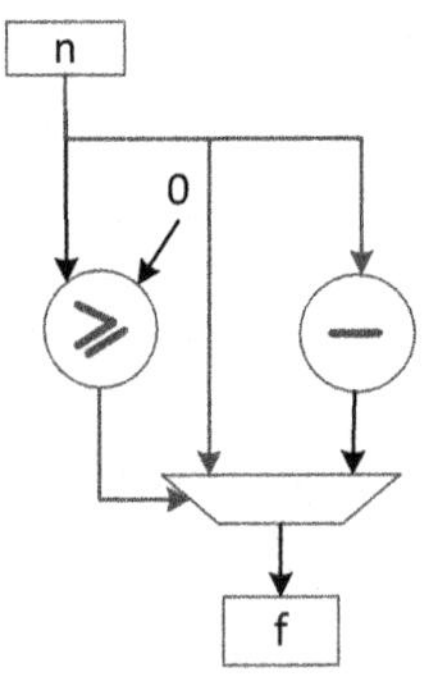

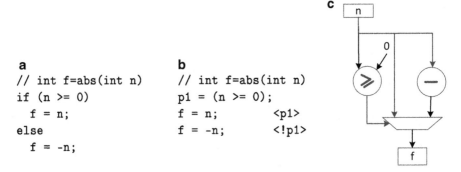

Fig. 2.17 Example to determine the absolute value of an input number: (**a**) C code; (**b**) predicated example after if-conversion; (**c**) dataflow graph

a

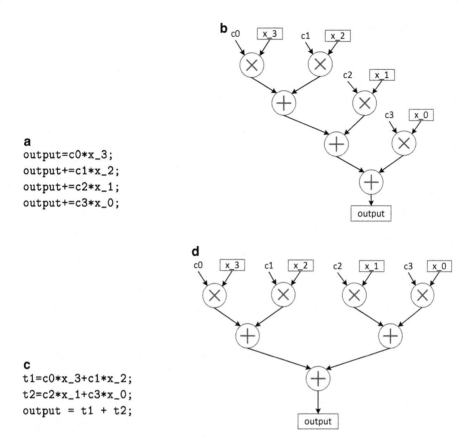

```
output=c0*x_3;
output+=c1*x_2;
output+=c2*x_1;
output+=c3*x_0;
```

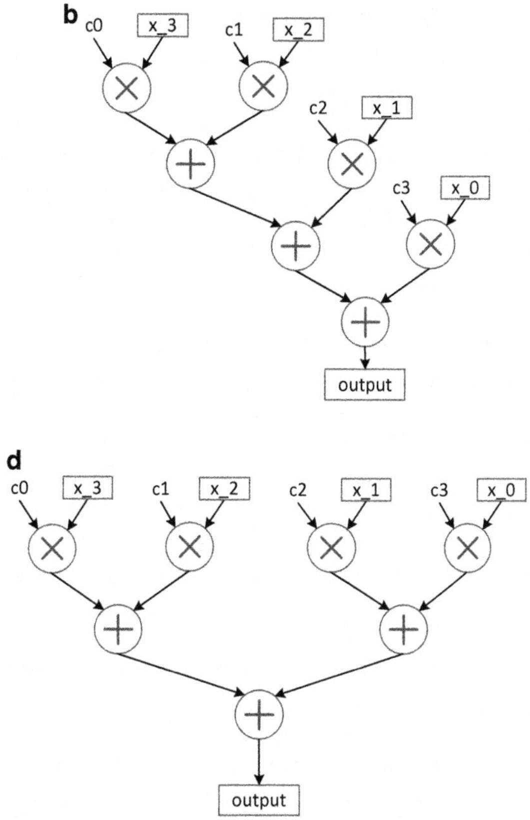

c

```
t1=c0*x_3+c1*x_2;
t2=c2*x_1+c3*x_0;
output = t1 + t2;
```

Fig. 2.18 Statements amenable to tree-height reduction: (**a**) C code; (**b**) dataflow graph; (**c**) C code after reassociation of operations for tree-height reduction; (**d**) resultant dataflow graph

properties can be used to reorganize expressions in order to reduce the number of operations in the critical path or the number of required control steps. One example is the use of tree-height reduction which reassociates the operations in an expression in order to reduce the height of the tree. Figure 2.18 shows the application of tree-height reduction to the code of Fig. 2.15b responsible for the calculation of each output value. The original code in Fig. 2.18a leads to the dataflow graph illustrated in Fig. 2.18b which includes 4 levels of arithmetic operators. The reassociation of operations presented in Fig. 2.18c leads to the tree-height reduction shown in Fig. 2.18d which includes 3 levels of arithmetic operations, a reduction of 1.

2.4.3 Software Pipelining

Software pipelining [GDWL92] is one of the traditional compiler optimizations used to parallelize loop execution by overlapping the execution of subsequent iterations. Software pipelining is related to the hardware techniques loop pipelining and loop folding [Rau94]. The optimization leads to three stages of loop execution: a prologue ("fill the pipe"), a steady state (also known as kernel), and an epilogue ("drain the pipe"). The common goal is to reduce the number of cycles to execute each iteration of the steady state, known as initiation interval (II), which represents the number of cycles between the start of successive iterations. Loops with II = 1 represent cycle execution with each loop iteration performed in a single clock cycle and thus maximized in terms of throughput. When compiling to FPGAs, however, a trade-off between the value of II, the maximum clock frequency, and the required hardware resources is usually explored.

Figure 2.19 presents a simple example, the addition of two vectors, and two variants of code transformed in order to accomplish software pipelining. If we assume that all arithmetic and load/store operations execute in one cycle and all hardware resources required for parallel execution are available (including three memory ports or one distinct memory per array variable), II is equal to 2 for the loop pipelining example described by the code in Fig. 2.19b since the multiplication and the assignment to array *C* require two sequential cycles for each kernel iteration. The code now requires 1+9*2+1 = 20 clock cycles instead of 3*10 = 30 clock cycles for the original code without software pipelining. Figure 2.19c presents the same software pipelining optimization as Fig. 2.19b, but now using predicates to control the prologue, steady state, and epilogue stages. In order to achieve II = 1, we need to move more operations to the prologue and epilogue and to have all the operations in the loop body executing simultaneously per loop iteration. Figure 2.20 illustrates

a

```
for(int i=0; i<10; i++) {
  C[i] = A[i] * B[i];
}
```

b

```
a_tmp = A[0];
b_tmp = B[0];
for(int i=0; i<9; i++) {
  C[i] = a_tmp * b_tmp;
  a_tmp = A[i+1];
  b_tmp = B[i+1];
}
C[9] = a_tmp * b_tmp;
```

c

```
for(int i=0; i<11; i++) {
  C[i-1] = a_tmp * b_tmp; <i!=0>
  a_tmp = A[i];           <i!=10>
  b_tmp = B[i];           <i!=10>
}
```

Fig. 2.19 Simple example to illustrate software pipelining: (**a**) original code; (**b**) code after software pipelining with prologue and epilogue moved to outside the loop; (**c**) code after software pipelining with prologue and epilogue controlled by predicates

Fig. 2.20 Simple example to illustrate software pipelining with II = 1

```
a = A[0]; b = B[0];
ab = a * b; a = A[1]; b = B[1];
for(int i=0; i<8; i++) {
  C[i]=ab; ab=a * b; a=A[i+1]; b=B[i+1];
}
C[8] = ab, ab = a * b;
C[9] = ab;
```

the code that represents the application of software pipelining for II = 1. In this case, the latency achieved equals 2+8*1+2 = 12 clock cycles. Although we illustrate software pipelining using code transformations, an automatic implementation is not performed at this level but works on the compiler intermediate representations (e.g., data dependence graph and reservation table [Muc97]) instead.

There are many algorithms for software pipelining, the most widely used one being iterative modulo scheduling [Rau96, Rau94]. The algorithm starts by considering the minimum value of II (MII) which is the maximum value of the resource constrained (ResMII) and the recurrence constrained (RecMII) initiation interval. ResMII is calculated based on the resource usage required in each iteration and RecMII represents the latencies imposed by cycles in the DDG (data dependence graph). Then, the algorithm tries to schedule the loop (e.g., using list scheduling or other scheduling algorithm) with MII. If it fails, it means that the loop cannot be scheduled using this value of the MII and the algorithm iterates by increasing the MII value by one until a modulo schedule is obtained (i.e., a value of II that allows the scheduling of the kernel).

Independent of the II value, the kernel always needs to perform all the operations of the loop body. In order to avoid limitations imposed by ResMII and to be able to execute all the operations in the same clock cycle, one needs to have as many hardware components as the number of operations in the loop body. If there is only 1 hardware multiplier but 2 multiplications in the body, the II needs to have a minimum value of 2 (ResMII = 2/1). Thus, in order to avoid constraints in terms of ResMII, 2 hardware multipliers are required. A similar thought is used for all operations, including load and store operations. The number of memories and memory ports are the constraints that limit the number of simultaneous load/store operations.

The other constraint for II is the RecMII which is computed by considering the delay of the longest loop-carried data dependence (also known as latency of the recurrences or cycles in the DDG) divided by the dependence distance. This represents the number of cycles which a new value requires to appear at the input of the circuit.

We now show a possible software pipelining for the FIR example with code shown in Fig. 2.15b. The example in Fig. 2.15b imposes 4 simultaneous accesses to the memory storing the x array in order to achieve an II = 1. Using the direct support of two ports provided by on-chip FPGA RAMs, the minimum II equals 2 and the

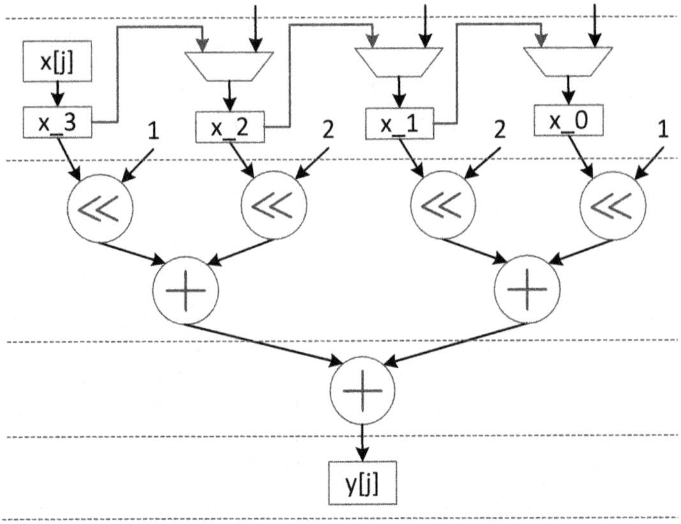

Fig. 2.21 Dataflow graph for the loop body of the FIR example in Fig. 2.16b. The *dotted lines* illustrate possible pipelining stages which would imply additional registers in the outputs of the two adders in the second stage

latency of the example is will be around 512 clock cycles, achieving a throughput of 1 sample per two clock cycles. If we reuse data as illustrated in Fig. 2.16b, the ResMII is no more 2 as there is only one access to array variable x per iteration and the II achieved is 1, resulting in a latency of around 256 clock cycles (a throughput equal to 1 sample per cycle). The dataflow graph of the loop body is shown in Fig. 2.21.

Note that FPGA Block RAMs (on-chip RAMs) have two ports, but the loop body of Fig. 2.16b is reading/writing only one value per iteration and thus uses only one port. In order to use the other port, one can add more accesses to array x and y to the loop body. For that, we can unroll the remaining outer *Loop 1* by a factor of 2 (see Fig. 2.22). In this case we still achieve an II of 1 with a latency of about 128 (a throughput equal to 2 samples per cycle). Given the restriction on the number of ports, the unrolling of the loop by factors higher than 2 would only contribute to higher II values and to the use of more hardware resources. Note, however, that it can be possible to customize on-chip memory resources to provide memories with more ports or to use more memories and replicate the data.

The value of II and the number of pipelining stages included in the datapath are also dependent on the required target clock frequency. Usually, a design-space exploration needs to be performed in order to achieve the best implementation given the input design requirements.

Fig. 2.22 Loop unrolling by two, applied to the FIR example in Fig. 2.16b

```
for(int j=3; j<M; j+=2) {
  x_3=x[j];
  output=c0*x_3;
  output+=c1*x_2;
  output+=c2*x_1;
  output+=c3*x_0;
  x_0=x_1;
  x_1=x_2;
  x_2=x_3;
  y[j] = output;

  x_3=x[j+1];
  output=c0*x_3;
  output+=c1*x_2;
  output+=c2*x_1;
  output+=c3*x_0;
  x_0=x_1;
  x_1=x_2;
  x_2=x_3;
  y[j+1] = output;
}
```

2.4.4 Mapping Data Structures

The mapping of the data-structures onto the memory resources on the FPGA-based computing platform is an important step of an FPGA compiler. That mapping binds the data structures (e.g., array variables) used in the program to the FIFOs and memories in the target architecture. The memory resources include the RAMs internal to the FPGA and the RAMs externally coupled to each FPGA.

In the previous examples we assumed the use of one on-chip memory per array variable. However, advanced transformations can be used. They include the partitioning and distribution of the data structures (e.g., array variables) in order to enable concurrent memory accesses to elements of the same data structure, especially as the target FPGAs include several on-chip memories and possibly more than one external memory. It can also be an option to use data replication in order to access data elements of the same data structure at the same time.

2.5 Backend Stages

This section summarizes how an FPGA bitstream, i.e., binary data which configures a circuit in an FPGA, is generated from a hardware design at RTL. This process requires a number of stages (see, e.g., [CCP06, HB06]). Figure 2.23 shows a general design flow commonly used as the backend of an HLS tool. Mature tools for the backend stages are provided by the FPGA vendors.

The first step performs lower-level synthesis: RTL and logic synthesis generate the circuits for the control unit, add multiplexers, assign registers, minimize boolean functions, and implement arithmetical and logical operators. Registers are implemented with D-type flip-flops which are available in large numbers on modern FPGAs. The result is a netlist, i.e. a graph representation of hardware components such as gates and connecting signals.

The next step maps the netlist to the hardware resources of the FPGA device. A particular feature of modern FPGAs is that they do not implement boolean logic as individual AND, OR and inverter gates, but use LEs (logic elements) or CLBs (configurable logic blocks) which consist of small look-up tables (LUTs) which are capable of implementing all boolean functions with a small number (4–6) of inputs. Therefore, the boolean functions have to be mapped to these LUTs. The mapper should also exploit special features like fast carry-chains for adders. Some complex arithmetic operators (like multipliers) do not need to be assembled from LUTs but are available as fixed circuits (hard macros) which should also be considered by the mapper. Small memories can be mapped to on-chip block RAMs. They allow much faster access than off-chip memory.

Fig. 2.23 FPGA design flow

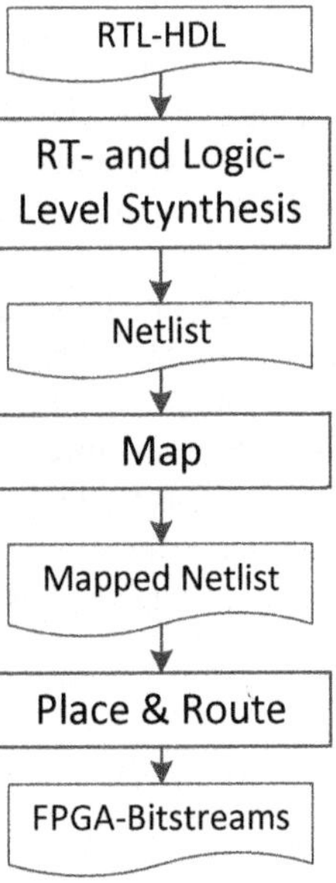

Finally, the last step of the design flow, Placement and Routing (P&R), places the hardware units of the mapped netlist to the FPGA components and connects them using the configurable routing resources. Connected hardware units should be placed close to each other so that the resulting connections are as short as possible. This ensures short signal delays in the connections and enables a short clock cycle, i.e., a high clock frequency for the circuit. The P&R process is very computing intensive and may take a long time depending on the complexities of the design and FPGA. For example, millions of gates and other hardware components have to be placed and routed considering also the millions of FPGA hardware resources.

All the component and connection information is stored in a bitstream file. This file is then used to program or configure an FPGA. After configuration, the FPGA executes the functionality of the original behavioral description in hardware.

2.6 Conclusions

This chapter presented some of the main tasks performed by high-level FPGA compilers. It is intended to help software programmers and novices in high-level synthesis to acquire the main concepts, illustrated by several examples. But it is neither intended to be a literature survey nor to introduce advanced and specific optimization techniques. Instead, it prepares the reader for the following book chapters and for reading further literature about compilation to FPGAs, we recommend the references [CH02, CDW10, NI14, GDWL92, Mic94, Wol96].

Chapter 3
A Quick Tour of High-Level Synthesis Solutions for FPGAs

Frank Hannig

3.1 Introduction

In Electronic design automation (EDA), HLS denotes the systematic transformation of a specification at a high abstraction level, e.g., a behavioral description in form of a mathematical expression or program notation, into a structural representation at RTL that realizes the given behavior. Traditionally, HLS involves the three major steps of *allocation*, *scheduling*, and *binding* as described in Chap. 2. However, depending on the starting point, HLS might involve also analysis, parallelization, and optimization techniques such as used in compiler design (see Sect. 2.4) for uni- and multi-processors. But, compared to parallel processor architectures, FPGA designs offer a much higher degree of freedom to exploit both *temporal parallelism* (e.g., very deep processing pipelines) and *spatial parallelism* (e.g., as many operations and hardware modules as possible or as the I/O bandwidth allows can be instantiated). In addition, thanks to the interconnect flexibility in FPGAs, these two types of parallelism can be arbitrarily combined with each other.

HLS is not a new discipline and has over 30 years of tradition, which is also covered by several articles. For instance, in [GB08], Rajesh Gupta and Forrest Brewer present a retrospective of HLS. Similarly, Grant Martin and Gary Smith provide an overview of the evolution of HLS in their general interest article "High-Level Synthesis: Past, Present, and Future" [MS09] in the IEEE Magazine on Design & Test of Computers. Here, Martin and Smith categorize the evolution of HLS into three generations plus a prehistoric epoch in the 1970s with pioneering works by Barbacci and Siewiorek [BS73, Bar76]. The first generation in the 1980s until mid-1990s was characterized by vivid research with many seminal papers on HLS. Raul Camposano and Wayne Wolf [CW91], Daniel Gajski et al. [GDWL92], and

F. Hannig (✉)
Friedrich-Alexander-Universität Erlangen-Nürnberg (FAU), Erlangen, Germany
e-mail: frank.hannig@fau.de

D. Koch et al. (eds.), *FPGAs for Software Programmers*,
DOI 10.1007/978-3-319-26408-0_3

Giovanni De Micheli [Mic94] to name a few of the most active researchers and list their textbooks that still form the basis for education and ongoing research. A survey of HLS tools of this early decade is presented in [WC91] by Walker and Camposano.

In the second generation (mid-1990s to early 2000s), the first commercial tools, such as Synopsys' Behavioral Compiler came into the market. But, this generation failed because both promises and expectations were too high. One erroneous belief was that HLS could replace RTL synthesis. In addition, the design entry in form of behavioral Hardware description languages (HDLs) was mainly tailored for RTL designers, and thus difficult to access for algorithm developers and software engineers. Finally, the obtained results had often poor quality, the synthesis process was not comprehensible and difficult to control. From the early 2000s, the third generation of HLS tools hit the market and there has been a lot of tool movement between the vendors. Nowadays, each major EDA company has a matured HLS tool in its product portfolio.

3.2 Categories and Properties

In the following, we give an overview of currently available HLS tools with an emphasis on FPGA technology. We make a point of the fact that the discussed tools are either commercially or as open-source software available at the point of time when writing this book—whereby this list of tools does not claim to be exhaustive. Furthermore, we do not aim at going into tool specifics but rather want to classify them with respect to certain properties. As tool properties, we consider the *availability*, the supported *target architectures*, the level of *programming abstraction and parallelization support*, as well as the way of *design entry*.

Availability denotes whether a design framework is commercially or freely accessible. That is, in the first case, the tool has to be acquired, for example in form of a *royalty bearing license*. Sometimes, in case of academic design tools, software is made available only upon request under individually agreed terms and conditions. Whereas, the second case denotes *free software*, i.e., with a *free license*, which is commonly referred to as *open source*.

Target architectures denotes the spectrum of platforms and technologies that a development tool is aiming at. Some HLS tools are technology independent. That is, they solely support the synthesis of a behavioral description into a structural one that is still technology independent. Whereas, other tools target either Application-specific integrated circuits (ASICs), FPGAs, or both of them. Furthermore, it is important to which extent predefined building blocks and IP components are provided and can be reused, respectively. Finally, some design environments are tailored to support only selected FPGA boards. This has the advantage that the entire board support infrastructure, such as communication and memory interfaces, is optimally matched to the design environment, and the software engineer can focus on application development.

Type of computation differentiates whether a tool targets dataflow-dominated or control-dominated applications.

Programming abstraction and parallelization support denotes different scales of computing irrespective of the actual programming language used for design entry (see next item). Although some design tools start from a High-level programming language (HLL) such as C or Scala, they are rather a hardware description language than an HLS approach. For instance, Chisel [BVR+12] is a hardware construction language that is embedded into the modern Scala programming language [OSV11], which is both a functional and object-oriented language. Chisel itself purely describes the structure of low-level hardware blocks that can be hierarchically composed to larger structural components. In addition, control flow in form of conditional execution and finite state machines can be described. Thus, we call such a level of abstraction *low* in Fig. 3.1, even though the approach has a high expressiveness, that means, it is very general and can be used for a wide range of applications. However, the programming effort is rather high and a high Quality of results (QoR) demands also great programming skills.

Since parallel execution is the most natural and important way of execution in hardware, in case of modeling dataflow-dominated applications, the support for efficiently synthesizing a behavioral description into a parallel RTL structure is of utmost importance. One obvious way is to map the internal representation in form of a Data flow graph (DFG) directly to multiple functional units of appropriate type that can work concurrently (see Sect. 2.2.3). The *breadth* and the *depth* of this directed acyclic graph is a measure for the degree of

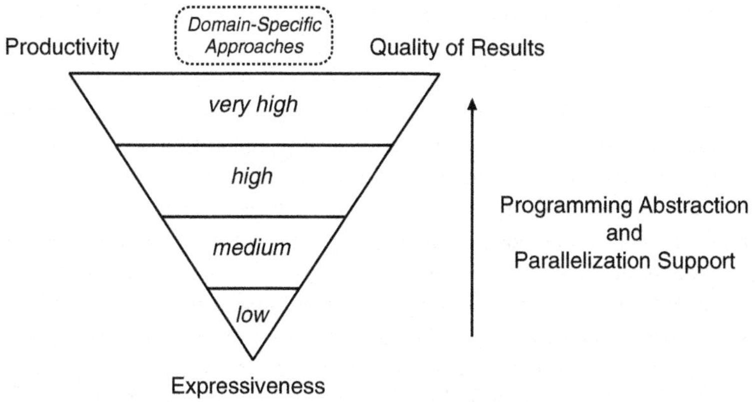

Fig. 3.1 The three measures of effectiveness for HLS approaches: Expressiveness, productivity, and performance. Each approach makes a compromise between these measures. Domain-specific approaches trade off expressiveness for productivity and QoR

Instruction-level parallelism (ILP) and the pipeline depth, respectively. The support of this type of parallelism, we call *medium* (see Fig. 3.1).

The next level of parallelism is *loop-level parallelism*. Loop unrolling (see Sect. 2.4.1) is a prominent technique to increase the ILP. Another technique, which is related to scheduling a loop program, is *software pipelining*. It allows to interleave the execution of multiple loop iterations simultaneously (see Sect. 2.4.3), and hence can further increase the pipeline depth and consequently throughput. If an HLS framework is able to employ these types of parallelism either automatically or user-guided, e.g., in form of program annotations or script-based, we call this level of abstraction and parallelization support *high* in the following.

Building blocks, templates, or advanced code transformations for efficient *data reuse* or *loop tiling* techniques to further increase the degree of parallelism may result in a very high QoR (upper right corner in the triangle of Fig. 3.1). If both QoR and productivity are important, *domain-specific approaches* are a valid option. In order to achieve both goals alike, such approaches have a limited expressiveness and are focused on a particular domain. Here, *productivity* is a measure of the design effort needed to specify and design a certain application. In case of programming languages, this can be, for instance, the number of Lines of code (LoC).

Such programming languages are called Domain-specific languages (DSLs). They allow domain experts to describe a problem they want to solve in a declarative, abstract, and compact way, using concepts and terms that are familiar to them. Because no implementation details need to be specified, parallelization can be supported by HLS, exploiting generic parallel execution patterns in the domain. Since all the details are derived by the HLS approach itself, DSL programs are much more portable and scalable. Support for new hardware features (e.g., FPGA primitives) only takes an extension of the DSL-based synthesis approach, but not of all the DSL programs. Due to the aforementioned advantages of DSLs, we assign them the highest abstraction and parallelization level in Fig. 3.1 (*very high*). Domain-specific languages can be mainly subdivided into two categories: *Internal (embedded)* and *external* DSLs. Internal DSLs use the syntax of a host general-purpose programming language and extend or restrict it by introducing new domain-specific language elements like special data types, routines or macros. External DSLs introduce a completely new syntax and semantics. Thus, in general, they are more flexible and expressive than internal ones at the cost of a higher design effort. Just like general-purpose programming languages, DSLs may be represented in a textual or in a graphical way. Very often, however, graphical programming languages like LabVIEW and MATLAB Simulink are DSLs in reality. Their visual representation often is the most natural way of describing scenarios in certain domains, and thus further eases the development process.

Design entry denotes the way how an application is specified. Common examples include the usage of *general-purpose programming languages* such as C, C++, or SystemC. However, typically not the entire language standard is supported, i.e., only a restricted subset can be used. In the other direction, these languages are often extended by certain data types specific to hardware design, or they are augmented by entire models of computation such as dataflow models. Other options include the usage of the aforementioned *DSLs*, *specification languages* such as UML, or *proprietary programming languages* that often realize a declarative programming paradigm.

3.3 Classification of High-Level Synthesis Tools

In Table 3.1, an overview of selected HLS tools is given, according to the before outlined properties. In addition, each tool is briefly described in the following.

Altera SDK for OpenCL allows software programmers to write FPGA-accelerated kernel functions in OpenCL C, an ANSI C-based language. For more details, we refer to Chap. 6 and Altera's website.[1]

Bluespec is a commercial HLS tool developed and offered by Bluespec, Inc., an EDA company co-founded by Prof. Arvind of MIT in 2003. For design entry Bluespec SystemVerilog (BSV) is used, which includes both (a) behavioral modeling of complex concurrent systems by using the formal specification concept of *atomic rules and interfaces* (also known as *guarded atomic actions*), and (b) powerful abstraction mechanisms for describing and organizing the structure of large and complex systems. The advantage is that parallelism and correctness are inherently provided by such a formal specification language, and, thus, may reduce verification time significantly. The Bluespec HLS tool generates either Verilog RTL or executable SystemC models as output. Since BSV is closely based on SystemVerilog, which is a combined HDL and hardware verification language, it might be rather suited for developers with a background in RTL design than for software engineers. For further details, we refer to [Nik08], the introduction in Chap. 9, and Bluespec's website.[2]

Catapult is an entire product family of technically matured HLS tools from Calypto Design Systems. Initially, it was developed and introduced by Mentor Graphics as Catapult C [Bol08] more than 10 years ago, before it was acquired by Calypto Design Systems in 2011. Catapult supports a subset of ANSI C++ and SystemC for design entry, from which it generates structural RTL descriptions in Very high speed integrated circuit hardware description language (VHDL), Verilog, or SystemC. It provides a rich set of data types in form of C++ classes, including integers of arbitrary length, fixed-point, floating-point, and complex data types. Further, it features testbench generation, complex SoC interconnects, Transaction-level modeling (TLM) and Electronic system level

[1] https://www.altera.com.

[2] http://www.bluespec.com/high-level-synthesis-tools.html.

Table 3.1 Comparison of selected HLS framework for FPGAs

Tool Name	Availability: Commercial	Availability: Open Source	Target Architecture: ASIC	Target Architecture: FPGA	Target Architecture: Specific FPGA Board	Computation Type: Dataflow	Computation Type: Control Flow	Abstraction & Parallelization: low, medium, high, very high	Design Entry
Altera SDK/OpenCL	✓			✓	✓	✓	(✓)	□□□□■□□	OpenCL C
Bluespec	✓		✓	✓		✓	✓	□□□□□■□	BSV
Catapult	✓		✓	✓		✓	✓	□□□□■□□	ANSI C++, SystemC
CHC Compiler	✓			✓		✓	(✓)	□□□■□□□	Standard C
C-to-Silicon	✓		✓	(✓)		✓	✓	□□□■□□□	C, C++, SystemC
CyberWorkBench	✓		✓	✓		✓	✓	□□□□■□□	ANSI C, SystemC
Cynthesizer	✓		✓	✓		✓	✓	□□□□■□□	SystemC
GAUT		(✓)	✓	✓		✓	(✓)	□□■□□□□	C
HDL Coder	✓		(✓)	✓	✓	✓	(✓)	□□□□□■□	MATLAB, Simulink
HIPAcc		✓		✓		✓		□□□□□□■	C++ embedded DSL
Impulse C	✓			✓	✓	✓	✓	□□□■□□□	ANSI C
LabVIEW FPGA	✓				✓	✓	✓	□□□□□■□	G
LegUp		✓		✓		✓	✓	□□■□□□□	ANSI C
MaxCompiler	✓				✓	✓		□□□□□■□	MaxJ, OpenSPL
Merlin Compiler	✓			✓		✓	(✓)	□□□□□■□	C and C++
PARO		(✓)		✓		✓		□□□□□□■	PAULA
ROCCC		✓		✓	(✓)	✓	(✓)	□□□□□■□	C
SDAccel	✓				✓	✓	(✓)	□□□□■□□	OpenCL, C, C++
SPARK		(✓)		✓		(✓)	✓	□□□■□□□	ANSI C
SPIRAL	✓	✓	✓	✓		✓		□□□□□□■	SPL
Trident		✓		✓		✓	(✓)	□□■□□□□	C
Synphony C Comp.	✓		✓	✓		✓	✓	□□□□□■□	C and C++
Vivado HLS	✓			✓	(✓)	✓	✓	□□□□■□□	C, C++, SystemC

(ESL) flows, as well as low-power optimization. Catapult covers a wide range of both dataflow and control-flow applications. Optimization and parallelization has to be done user-guided by providing constraints during the synthesis, and thus, require in-depth hardware knowledge to achieve a decent QoR. For further details on Catapult, we refer to the overview and the step-by-step approach provided in [Bol08] and [Fin10], respectively. Most recent advancements and features might be directly obtained on the corresponding product web page of Calypto Design Systems.[3]

[3] http://www.calypto.com.

C-to-Hardware Compiler, also known as **CHC Compiler**, is an HLS tool for FPGAs offered by Altium. The CHC Compiler uses standard C (ISO C) as input language and produces a synthesizable hardware file in the proprietary hardware assembly language HASM, which can be translated to hardware description languages such as VHDL The CHC Compiler is closely tied to Altium Designer, a development environment for Printed circuit board (PCB), embedded software and FPGA design. The HLS tool can be mainly used in two ways: (1) To compile an embedded program to software and selected functions to hardware, which can still be called from the software. In this way components of an Application-specific processor (ASP) can be created. (2) To synthesize a hardware block that can be used as a component in the FPGA. In contrast to above, it does not require an ASP. The transformation and optimization in the compilation flow can be either controlled by pragmas, compiler flags, or automatically. For further details on the C-to-Hardware Compiler, we refer to the user manual [Alt13b] and the web page of Altium.[4]

C-to-Silicon is an HLS tool from Cadence, which mainly focuses on ASIC design and partly on FPGA targets. As input design language it uses C, C++, SystemC, and OSCI TLM 1.0, it generates Verilog code. C-to-Silicon can tightly interact with Cadence's ASIC design flow, the RTL Compiler. In case of C or C++ input specifications, a SystemC wrapper is generated, which unfortunately can be simulated only with the Cadence tool chain. Furthermore, C-to-Silicon provides a collection of design IP such as math functions, fixed-point and floating-point data types, as well as several IP cores for standardized interconnects, as well as testbench generation. The entire HLS design flows has to be controlled by user-specified constraints in form of Tcl scripts. Further C-to-Silicon details can be accessed on the corresponding product web page from Cadence.[5]

CyberWorkBench, sometimes also referred to as *All-in-C*, is a synthesis and verification tool by NEC that has been developed since more than 25 years. As input ANSI C and SystemC programs are accepted from which optimized synthesizable RTL code in Verilog or VHLD is generated for FPGAs or ASICs. The design flow supports both control-dominated circuits and datapath modules. CyberWorkBench features automatic pipelining, power optimization, and parallelism extraction. These optimization are guided by attributes in the C code and global synthesis options. Attributes denote, for instance, function inlining or loop optimizations such as loop merging or unrolling. Global options include scheduling policies such as speculative scheduling, ASAP, or ALAP. Formal design verification can be done by using assertions and properties. For further information, we refer to [WO00, WS08] and NEC's website.[6]

---

[4]http://www.altium.com.

[5]http://www.cadence.com.

[6]http://www.nec.com/en/global/prod/cwb/.

Cynthesizer is Cadence HLS counterpart of Catapult with a similar function range. Originally, it was developed by Forte Design Systems, which was recently acquired by Cadence. As input language, untimed behavioral descriptions in SystemC are used, from which a subset can be synthesized to RTL. The Cynthesizer is very versatile, it supports testbench generation, TLM simulation, effective IP reuse support, and low-power design. The framework is very powerful, parallelization techniques (e.g., loop unrolling or pipelining) have to be specified manually by directives in the code, and thus, require a deep understanding of the RTL design in order to achieve high-performance results. For further details on Cynthesizer, we refer to the introduction provided in [Mer08] or directly to Cadence's website.[7]

GAUT is an academic open source HLS framework, developed at Université de Bretagne-Sud, with emphasis on digital signal processing applications. The framework is based on the GNU Compiler Collection (GCC) and its GIMPLE intermediate representation. Starting from a restricted C function, GAUT extracts the potential instruction-level parallelism, and performs scheduling and binding of operations, but also memory and communication interface synthesis, and finally generates target-independent VHDL code. For further details on GAUT, we refer to the overview provided in [CCB+08] or directly the project web page.[8]

HDL Coder is a code generator from The MathWorks, Inc., which can generate synthesizable HDL code from MATLAB functions and Simulink models. It offers RTL code generation in the form of VHDL or Verilog in two variants, target-independent and optimized back ends for both Xilinx and Altera FPGAs and SoCs. HDL Coder can generate target-specific HDL code on the basis of Xilinx LogiCORE IP blocks (e.g., Xilinx floating-point arithmetic functions) and Xilinx Zynq-7000 SoC devices. Similarly, it can be seamlessly combined with Altera's DSP Builder and Altera SoC devices. For verification, the original MATLAB or Simulink model can be used as a system-level testbench and compared against an HDL simulation, using third-party RTL simulators. For more details, we refer to MathWorks' website.[9]

HIPA^{cc} that has been mainly developed at the Friedrich-Alexander University Erlangen-Nürnberg, is an open-source DSL and compiler for a wide range of accelerator technologies, including also a design trajectory for FPGAs. For more details, we refer to Chap. 12 and to its website.[10]

Impulse C is an HLS tool offered by Impulse Accelerated Technologies, Incorporated. It targets both software engineers as well as FPGA designers, and uses ANSI C as input language from which VHDL and Verilog code can be generated. As in case of many other C-based HLS tools, Impulse C requires

[7]http://www.cadence.com.

[8]http://hls-labsticc.univ-ubs.fr/.

[9]http://www.mathworks.com/solutions/fpga-design/solutions.html.

[10]http://www.hipacc-lang.org.

prior hardware design experience to achieve a reasonable QoR by the pragma-based parallelization and compilation approach. Advantageous is that the tools offers direct support for selected FPGA platforms, and can automatically generate host/FPGA interfaces for selected FPGA boards. For more details, we refer to Impulse's website.[11]

LabVIEW FPGA Module is a solution offered by National Instruments (NI) that specifically targets Xilinx-based hardware platforms offered by NI. For design entry, the LabVIEW graphical programming language G is used. For more details, we refer to Chap. 4 and to its website.[12]

LegUp is an open-source HLS tool developed at the University of Toronto. For more details, we refer to Chap. 10 and to its website.[13]

MaxCompiler is a compilation environment from Maxeler Technologies for their corresponding FPGA-based acceleration systems. It offers an all-in-one solution for developing compute kernels, manager configurations, and host applications for Maxeler's dataflow computing platforms. Kernels are dataflow-dominated programs that can be described either in MaxJ, an extension of the Java programming language, or by the recently introduced Open Spatial Programming Language (OpenSPL). For further details, we refer to Chap. 5 and to Maxeler's website.[14]

Merlin is a source-to-source compiler developed by Falcon Computing Solutions, a startup from UCLA. For further details, we refer to Chap. 8.

SDAccel is Xilinx HLS counterpart of the OpenCL-based approach offered by Altera. It is a complete software development environment, which allows software engineers with little FPGA knowledge to develop and optimize applications written in OpenCL, C, and C++ on FPGA platforms (e.g., as accelerator in data-centers) by completely abstracting from the underlying hardware. Furthermore, large applications with multiple kernels are split up and can be loaded on-demand to an FPGA by using the concept of partial reconfiguration. For further details on SDAccel, we refer to Xilinx' corresponding web page.[15]

SPARK is a for non-commercial purposes freely available HLS tool developed at the University of California, San Diego. A very restricted subset of C is used as input (no support for pointers, function recursion, multi-dimensional arrays, etc.) from which structural VHDL (RTL code) is generated. In SPARK, many compiler transformations have been re-instrumented for synthesis by incorporating ideas of mutual exclusive operations, resource sharing and hardware cost models. SPARK is particularly targeted to control-intensive microprocessor

[11]http://www.impulseaccelerated.com.

[12]http://www.ni.com/labview/fpga/.

[13]http://www.legup.org.

[14]http://www.maxeler.com.

[15]http://www.xilinx.com/products/design-tools/software-zone/sdaccel.html.

functional blocks, as well as multimedia and image processing applications. A hardware/software partitioning algorithm that distributes applications to a CPU and an FPGA is also contained in the framework. Further information can be obtained from the "SPARK" textbook [GGDN04a] or the website,[16] where also the software can be downloaded. However, further development seems to be stalled since 10 years.

SPIRAL is a domain-specific library generation framework mainly developed at Carnegie Mellon University. More specifically, SPIRAL [FVM+08] considers the domain of linear transformations, including Discrete Fourier transform (DFT), Walsh-Hadamard transform (WHT), FIR and IIR filters, discrete cosine and sine transforms, and discrete wavelet transform. A declarative, mathematical DSL called SPL (Signal Processing Language) is used for problem specification. Subsequently, this input is structurally optimized through rewriting and eventually translated into C and Verilog code in case of targeting parallel processors and FPGAs, respectively. SPIRAL uses evolutionary search methods, dynamic programming, and techniques from artificial intelligence to learn which choice of algorithm is best. Code generators for some of the aforementioned linear transformations are available under GNU GPL license (commercial licenses can be obtained on request) on SPIRAL's project website.[17]

Trident is an FPGA compiler framework for floating-point algorithms based on LLVM. It starts from a restricted algorithm description in C, followed by optimization and scheduling phases, and finally generates VHDL code. We refer to [TGP07] and the corresponding SourceForge web page[18] for further details and download, respectively.

PARO is an HLS tool developed at the University of Erlangen-Nürnberg [HRDT08]. It is capable of automatically generating highly parallel hardware accelerators for a broad variety of compute-intensive applications with multi-dimensional dataflow. An external DSL, called PAULA [HRT08], is used as input language, which is a functional programming language that is very closely related to a mathematical problem description. PAULA allows to describe algorithms as systems of multi-dimensional recurrence equations with affine data dependences defined over polyhedral iteration spaces, but can handle also runtime-dependent conditions to a limited extent. Furthermore, PAULA offers compact language constructs for multi-dimensional reductions and offers domain-specific augmentations for image processing [STH+14]. The compilation process relies heavily on well-known compiler optimizations but also on powerful loop transformations in the polyhedron model (e.g., loop tiling), as well as on mixed integer programming techniques for determining an optimal software-pipelined schedule under resource constraints. As output, PARO generates target-independent synthesizable VHDL code, or programs for

[16] http://mesl.ucsd.edu/spark/download.shtml.

[17] http://spiral.net/hardware.html.

[18] http://trident.sourceforge.net.

programmable processor arrays, such as TCPAs [HLB+14]. Disadvantageously, the PARO framework is only available upon request under individual terms and conditions for selected partners.

ROCCC is an open-source HLS tool developed at the University of California, Riverside. For more details, we refer to Chap. 11 and to its website.[19]

Synphony C Compiler was originally developed by Synfora, a spin-off of HP Labs, under the name PICO Express, before Synfora was acquired by Synopsys in 2010. The Synphony C Compiler takes C and C++ as input and generates RTL code in VHDL or Verilog. It can closely interact with other tools from Synopsys' product family: The Design Compiler for RTL synthesis (ASIC designs), Synplify for FPGA synthesis, as well as simulators and equivalence checkers. The HLS can exploit parallelism at multiple levels including ILP, loop-level parallelism, and task-level parallelism. For further details on PICO and the Synphony C Compiler, we refer to [KAS+02, AK08] and Synopsys' website,[20] respectively.

Vivado High-Level Synthesis or short **Vivado HLS** is an HLS tool offered by Xilinx. For more details, we refer to Chap. 7 and to the Xilinx' website.[21]

As mentioned before, the above list of currently available HLS tools is without any claim to comprehensiveness and can give only glimpses into the different approaches and their features. For further readings, we recommend—of course—the corresponding chapters in this book, but also the following literature (in reverse chronological order):

- The overview article on HLS tools as per 2012, given by Meeus et al. in the Journal of Design Automation for Embedded Systems [MVG+12].
- The survey "Compiling for Reconfigurable Computing" by Cardoso, Diniz, and Weinhardt in the ACM Computing Surveys [CDW10].
- Springer's book on "High-Level Synthesis: From Algorithm to Digital Circuit" [CM08].

---

[19] http://roccc.cs.ucr.edu.

[20] http://www.synopsys.com.

[21] http://www.xilinx.com.

Part I
Commercial HLS Solutions

Chapter 4
Making FPGAs Accessible with LabVIEW

Hugo A. Andrade, Stephan Ahrends, and Simon Hogg

In this chapter we present a graphical programming framework, the NI LabVIEW software, and associated language and libraries, as well as programming techniques and patterns that we have found useful in making FPGAs accessible to scientists and engineers who are typically domain expert software programmers.

4.1 Introduction

Many scientists and engineers do software programming as a (significant) part of their daily job, either setting up experiments, taking measurements, developing prototypes, testing their designs, embedding software as part of larger systems, or modeling and analyzing the world around them. The type of software that they develop varies significantly with their domain or task that they are performing, and many times is categorized as scientific programming, embedded programming, test software development, and more recently cyber-physical systems programming, etc. Many of these scientist and engineers are domain experts in other fields like medicine, biology, chemistry, mechanical engineering, etc. and are not formally trained as computer scientists or electrical engineers, but have learned software programming well enough to be proficient in their own domain and use software

H.A. Andrade (✉)
National Instruments, Berkeley, CA, USA
e-mail: hugo.andrade@ni.com

S. Ahrends
National Instruments, Munich, Germany

S. Hogg
National Instruments, Austin, TX, USA

D. Koch et al. (eds.), *FPGAs for Software Programmers*,
DOI 10.1007/978-3-319-26408-0_4

programming to their advantage. So having to go even further to understand how to program an FPGA is a foreign concept.

For such scientists and engineers having a programming framework that makes their programming job easier is a welcome help. In this chapter we present one such framework, LabVIEW (Laboratory Virtual Instrumentation Engineering Workbench), specifically targeted to scientists and engineers, who need to program as part of their jobs. In addition to highlighting the merits of a visual framework and graphical programming language, we will also show the importance of intuitive and precise access to I/O, and a rich set of library components so that reuse can be exploited. We will also show how such framework has evolved from a programming environment to a system-level development framework for heterogeneous targets. Software programmers will find familiar elements in this flow, and will be eased into more hardware centric elements as needed/beneficial for specific requirements, in particular interfacing to real-world I/O at several levels.

The framework presents a platform-based design methodology, which supports multiple models of computation to capture the application at the right level of abstraction, and then helps map that application into several pre-qualified platform targets. These heterogeneous targets range from personal computers running desktop operating systems, to embedded real-time processors, to micro-controllers and FPGAs.

This chapter is subdivided into the following sections: In Sect. 4.2 we describe the general LabVIEW framework in technical detail, including key language and environment features and the set of libraries supported. In Sect. 4.3 we show how this environment can make development for software programmers trying to use FPGAs much easier. Section 4.4 shows a supporting use case. Section 4.5 summarizes our conclusions.

4.2 The LabVIEW Software Development Framework

LabVIEW is a graphical application programming environment originally developed in the mid 1980s by National Instruments for the Data Acquisition (DAQ), Test and Measurement (T&M) and the Industrial Automation (IA) markets. It is composed of several well integrated sub-tools targeted at making the development and prototyping of science and engineering applications that require interaction with real-world data and signals very simple and efficient. These sub-tools can be categorized as shown in Fig. 4.1 [Hog15], and are described in the following subsections.

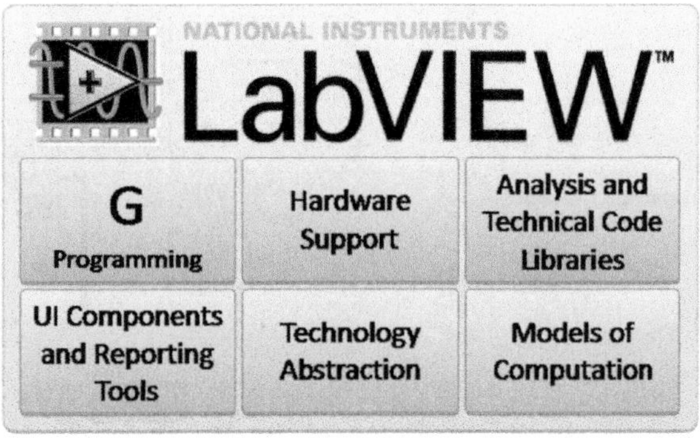

Fig. 4.1 LabVIEW subtools by category

4.2.1 G Programming Language

One of the key subtools is a compiler for the graphical programming language called G. G is a dataflow language that due to its intuitive graphical representation and programmatic syntax has been well accepted in the instrumentation industry, especially by scientists and engineers who are familiar with programming concepts but are not professional software developers and rather domain experts. Using it, they can quickly tie together data acquisition, analysis, and logical operations and understand how data is being modified. Though it is easy to use and flexible, it is built on an elegant and practical model of computation.

The idea behind G was to provide an intuitive block diagram view to the domain expert software programmer, since most scientists and engineers understood that concept; and it became the primary syntactical element in LabVIEW. The semantics follow homogeneous structured dynamic dataflow, which combines constructs from imperative and functional languages [AK98], where actor nodes (operations or functions) called Virtual instruments (VIs) operate on data as soon as it becomes available, rather than in the sequential line-by-line manner that most textual programming languages employ. The VIs are connected via unidirectional edges (called "wires") as shown on Fig. 4.2. VIs are either primitives built into G or sub-VIs written in the G language. Dataflow allows the user to not worry about memory allocation, garbage collection, and concurrency. The focus is on data, who produces and who consumes it, and what operation is performed on the data, rather than on how to pass data from one place to the other, or determine when to send or receive it.

The user interface is presented through a "front panel" that provides "controls" and "indicators" through which the user sends and receives information. The programmatic hierarchical interface is specified through the connector pane.

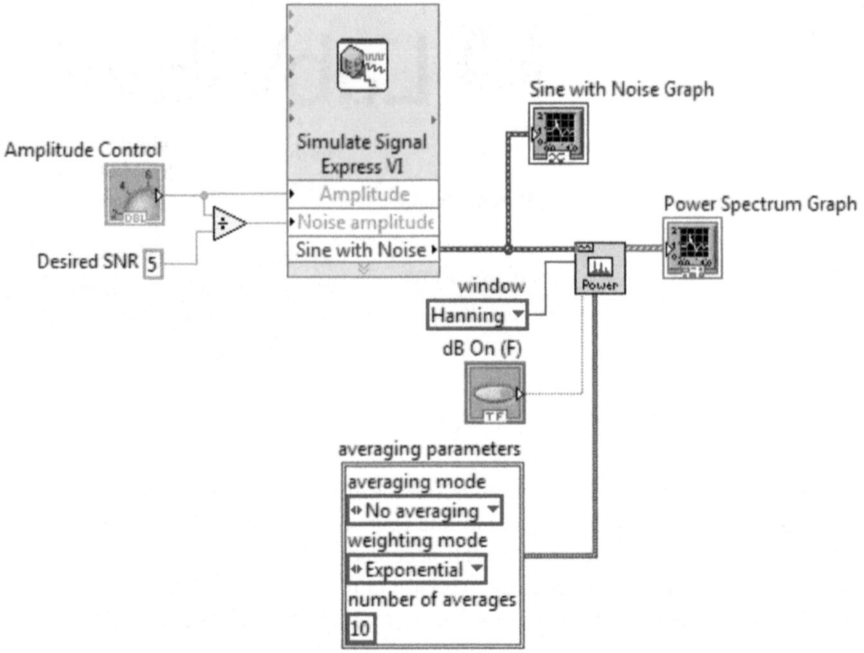

Fig. 4.2 LabVIEW block diagram

4.2.2 Hardware Support

Typically, integrating different hardware devices can be a major pain point when automating any test, measurement, or control system. Worse yet, not integrating the different hardware pieces leads to the hugely inefficient and error-prone process of manually taking individual measurements and then trying to correlate, process, and tabulate data by hand.

LabVIEW makes the process of integrating hardware much easier by using a consistent programming approach no matter what hardware is being used. The same initialize-configure-read/write-close pattern is repeated for a wide variety of hardware devices, data is always returned in a format compatible with the analysis and reporting functions, and the user is not forced to search through instrument programming manuals to find low-level message and register-based communication protocols unless specifically needed.

LabVIEW has freely available drivers for thousands of NI and third-party hardware devices (e.g. scientific instruments, data acquisition devices, sensors, cameras, motors and actuators, etc.) In the rare case that a LabVIEW driver does not already exist, the user has tools to create his/her own, reuse a Dynamic link library (DLL) or another driver not related to LabVIEW, or use low-level communication mechanisms to operate hardware without a driver.

The cross-platform nature of LabVIEW also allows the user to deploy his code to many different computing platforms. In addition to the desktop OSs (Windows, OS X, and Linux), LabVIEW can target embedded real-time controllers, ARM microprocessors, and FPGAs, so the user can quickly prototype and deploy to the most appropriate hardware platform without having to learn new tool chains.

4.2.3 Analysis and Technical Code Libraries

LabVIEW tailors the G programming language to engineering and scientific use by incorporating hundreds of specialized functions and algorithms that are not typically included with general-purpose programming languages.

In addition to the standard programming language constructs, LabVIEW contains functions for:

- String, array, and waveform manipulation
- Signal processing, including filters, windowing, spectral analysis, and transforms
- Mathematical analysis, including curve fitting, statistics, differential equations, linear algebra, and interpolation
- Communication, including high-level protocols, HTTP, SMTP, FTP, TCP, UDP, Serial, and Bluetooth
- Report generation, file I/O, and database connectivity
- Add-on packages for specialized disciplines, such as:
  - Control design and simulation
  - Sound and vibration analysis
  - Machine vision and image processing
  - RF and communication

All of the included functions in LabVIEW work seamlessly with the data acquired from the supported hardware, and no special conversion or data movement setup is needed.

4.2.4 UI Components and Reporting Tools

Every LabVIEW block diagram also has an associated front panel, which is the user interface of the application. On the front panel generic controls and indicators such as strings, numbers, and buttons or technical controls and indicators such as graphs, charts, tables, thermometers, dials, and scales can be placed. These are designed for engineering use, meaning the user can enter SI units such as 4M instead of 4,000,000, change the scale, export data to tools such as NI DIAdem and Microsoft Excel, and they can be completely customized.

In addition to displaying data as an application is running, LabVIEW also contains several options for generating reports from test or acquired data. The user can send simple reports directly to a printer or Hypertext markup language (HTML) file, programmatically generate Microsoft Office documents, or integrate with DIAdem for more advanced reporting. Remote front panels and Web service support can be used to publish data over the Internet with the built-in Web server.

4.2.5 Technology Abstraction

LabVIEW quickly adopts technology advances in personal and embedded computing in such a way that the domain expert user gets the new capabilities without having to learn new paradigms. Examples of this approach include how LabVIEW is able to automatically generate multithreaded code for execution on multicore processors or program FPGAs to gain the speed and reliability of custom hardware chips without the user needing to learn the underlying details of multithreading, or hardware description languages for FPGAs. The same applies to new OSs, networking protocols, etc.

4.2.6 Models of Computation

When LabVIEW was first released, G was the only way to define the user functionality. Much has changed since then. The user can now pick the most efficient approach to solve the problem at hand. Examine the following considerations:

- Graphical data flow is the default model of computation for LabVIEW.
- Statecharts provide a higher level of abstraction for state-based (control) applications.
- Simulation diagrams are a familiar way of modeling and analyzing dynamic systems.
- Formula Nodes put simple mathematical formulas in line with the G code. MathScript is math-oriented, textual programming for LabVIEW that can be used to call .m files without the need for extra software.
- DLL calls, ActiveX/.NET communication, and the inline C node let the user reuse existing ANSI C/C++ code and code from other programming languages.
- Component-level IP (CLIP) and IP integration nodes import FPGA intellectual property based on VHDL or other HDLs.
- Single-cycle timed loops (SCTL) provide a synchronous-reactive model of computation for RTL hardware design.

These flexible models of computation allow the domain expert to pick the right abstraction for the particular problem being solved. In any given application the developer is likely to use more than one approach, and LabVIEW is an effective tool to quickly integrate different models and execute them together.

4.2.7 Levels of Abstraction

LabVIEW can express functionality at several levels that enable the user to trade off productivity or ease of use with performance of the resulting systems. It provides several ways to accomplish similar tasks, so the user can balance simplicity and customization on a task-by-task basis.

Express VIs offer quick and easy configuration of VIs, but are somewhat limited in flexibility. Many of these types of VIs generate lower level G code that can be made accessible to the user as a starting point to develop more flexible code. Third party developers can themselves offer Express VIs to end-users.

Non-express VIs, where the Application programming interface (API) is exposed directly, are categorized as productive (high-level) or low-level. The most common way to program in LabVIEW is using high-level functions that strike a balance between abstracting the unnecessary administrative tasks such as memory management and format conversion, but keep the flexibility of being able to customize almost every aspect of whatever task you need to accomplish. In contrast, when one needs to be able to completely define every detail of a task, LabVIEW offers the same low-level access as available in traditional programming languages.

VIs at all levels of abstraction are provided as part of the main libraries, examples, additional libraries, third party libraries, and, in a growing number, by an open community. To deal with the large number of options a comprehensive VI search facility and hyper-linked documentation are provided in the framework.

4.3 Enabling FPGAs for Domain Expert Software Programmers

In this section we present some framework extensions (LabVIEW FPGA Module [Nat15b]) as well as techniques and patterns that focus on making FPGAs more accessible to LabVIEW software programmers, in particular those that are domain experts, focusing on protocol-aware testing, software-defined instruments, and embedded and cyber-physical systems.

4.3.1 Integrated Synthesis and Implementation Tools

The Xilinx synthesis tools are integrated with LabVIEW FPGA. They are invoked automatically on hidden generated code to create a bitfile to download and run on the FPGA.

4.3.2 Reference Target Platform

In order to make targeting FPGAs more consistent, we defined a canonical FPGA platform that is not limited to just a fabric, but includes an instruction processor and attached I/O. It is centered on the common use case of real-world I/O access, and was specialized for an FPGA-based computing platform. We call it the LabVIEW Reconfigurable I/O (RIO) architecture [Nat15c], as shown in Fig. 4.3.

There are three main computing engines: an externally attached or internal instruction processor, typically running a real-time operating system, an (optional) remote host computer, typically running a desktop operating system, and the FPGA fabric. All of these computing engines are programmable by LabVIEW. Depending on the application and the knowledge or experience of the programmer, parts of the code may run on the host computer, the real-time computer or the FPGA. The framework attempts to show a consistent programming environment, yet has not hidden the boundaries between these components.

I/O attached to the FPGA is viewed as part of a RIO target, and has first class representation within the LabVIEW FPGA environment. The nodes present an interface to the program in terms closest to the domain expert, e.g., analog input in engineering units, as opposed to digital protocols to specific analog to digital converters. Similarly control of the timing of the critical I/O operations is done in high-level terms.

4.3.3 Pre-configured FPGA Bit Personalities with I/O

As we have mentioned, access to real-world I/O is one of most important requirements for the type of domain experts that we are targeting. Yet, many of them do not want to program an FPGA just to have synchronous deterministic access to I/O. The NI Scan Engine [Nat15e] enables efficient access to coherent sets of I/O channels attached to the FPGA, using a scan that stores data in a global memory map and updates all values at a single rate, known as the scan period. To do this a specialized IP is placed on the FPGA that interfaces to I/O modules connected to the FPGA. Yet, at the same time, it allows the rest of the fabric to be available for further FPGA computing or connecting to other types of specialized I/O that may not only interface to the Scan Engine.

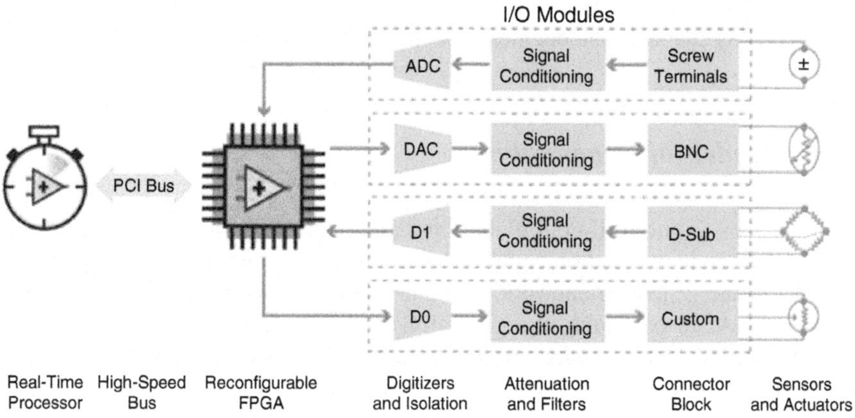

Fig. 4.3 LabVIEW RIO architecture

4.3.4 FPGA Extensions

Whenever possible we try to make I/O functionality directly available to the instruction processor, either by using the Scan Engine or other "relaying" techniques. This gives the programmer the most flexibility. In doing so, we make choices on levels of API supported. For example, in the case of RIO-enabled RF instruments, we have pre-built bit personalities onto the FPGA that allow the board to appear as a more traditional instrument, by supporting standard APIs for that class of instrument. Yet, domain experts want to make changes and specialize such a software-defined instrument. Such a change in the FPGAs makes it difficult to preserve the investment of code that has been made on the host already, which uses the standard API; matching the original validated FPGA design as well. So a set of side-band interfaces were defined that allow most of the matched FPGA-Host code to continue to operate as originally intended, but allowing for extensibility. The side-band interface includes feature or signature export from the FPGA and import on the host, as well as communication through host-FPGA boundary.

4.3.5 Effective Use of Hardware-Software Communication

In order to be able to start with a design that has most of the code on the instruction processor, and migrate only relevant parts to the fabric, we need to minimize the development and run-time cost of doing so. This means that the inter-process communication has to be independent of target and has to be efficient. The LabVIEW RIO framework supports both queue and memory register abstractions to communicate between tasks on the same processor or FPGA, as well as across these entities. If the entities are across a bus, it automatically instantiates Direct memory access (DMA) channels and engines to support the operation.

4.3.6 Models of Computation

Different jobs require different tools. Similarly, depending on the tasks, the user may want to code application components using different models of computation. Just like in the general framework, multiple models of computation are available for the FPGA. In particular, the general G structured homogeneous dataflow is the default model of computation. The single-cycle Timed Loop sub-framework discussed below follows a synchronous reactive model of computation consistent with execution on clock ticks. For control applications we also offer a modal model that follows StateCharts syntax and semantics, including hierarchical and concurrent states. Some models of computation like continuous time (mainly for physical subsystem modeling) are not available for implementation on FPGA yet, but are available for co-simulation.

4.3.7 Consistent IP and Programming Model

As part of making software components portable between targets it is important to not only have a consistent programming language, but also consistent set of libraries across most targets. Each of these libraries is optimized and the resource utilization is made clear, either in documentation or as part of the framework information system.

4.3.8 Optimization

Algorithmic IP and High-Level Synthesis

Optimizing an algorithm for an FPGA can be a complex task since the concurrency and connectivity options are fairly large. As an add-on to the LabVIEW FPGA Module, LabVIEW FPGA IP Builder generates high-performance FPGA IP by leveraging HLS technology. HLS generates efficient hardware designs from three components: (1) an algorithm implemented in a high-level language such as LabVIEW, (2) visual synthesis directives specifying the performance and resource objectives, and (3) an FPGA target with known characteristics. It can be used to automatically optimize FPGA VIs and easily port desktop code to the FPGA. The user is able to quickly explore designs with rapid performance and resource estimates. He/she can reuse IP, unmodified, to adapt to different application requirements.

Low-Level IP and Graphical RTL Abstraction

Sometimes in order to optimize an algorithm or to connect to detailed I/O protocols it is necessary to describe the operations at the cycle or RTL. LabVIEW provides the single-cycle Timed Loop model of computation in which the user, when needed, can explicitly specify the operations that must execute in a single cycle. Within this environment the compiler takes advantage of the fact that data needs to only propagate between adjacent registers and removes part of the logic that would normally ensure that the regular dataflow semantics are followed between VIs.

4.3.9 External IP

Even if LabVIEW FPGA already provides many high level component and IP libraries, sometimes the end-user either has legacy IP or has available to them IP written in different programming or hardware description languages. In such a case, it is important to be able to import that IP, and make it appear in a consistent and native form to the LabVIEW framework. CLIP and IP integration nodes (IPIN) are two ways to import external IP into LabVIEW FPGA. A CLIP Node executes independently and in parallel to the IP developed in LabVIEW FPGA. In addition, CLIP can interface directly with the FPGA clocks and I/O pins. In contrast, the IP Integration Node is inserted into the LabVIEW FPGA block diagram and executes as defined by the dataflow of the LabVIEW VI. As part of the LabVIEW dataflow execution, the IP integration Node provides the ability to verify the overall application behavior and timing using the cycle-accurate simulation tools.

4.3.10 Soft Computing Targets

FPGAs are very flexible, including the development of instruction processors on the fabric. These processors are complete but are not the most powerful, yet the fact that the user can dedicate a processor to a specific I/O component is very useful. Using some of the IP integration mechanisms described above, we have been able to integrate popular soft processors such as MicroBlaze, by using the Xilinx tools to define the processor and its I/O. The I/O then appears as connection to the rest of the LabVIEW FPGA code for setup in accelerated computing or directly attached to I/O. The processor can be programmed today in C, and resulting compiled code can be overlaid onto an existing bit file without recompiling the rest of the fabric.

4.3.11 Consistent Debugging Techniques and Simulation

In addition to consistent models of computation, libraries and patterns, it is important to have a consistent debugging and simulation flow in between targets as well. In the LabVIEW framework a developer can start development on the host, and can complete all logic or algorithm development there, then move it to the real-time processor in case they need to directly access I/O, and eventually to LabVIEW FPGA. Once in LabVIEW FPGA they can see a cycle accurate simulation of the application, and only then would they need to target the actual FPGA, so that the compilation time is minimized.

4.3.12 High-Level Language Features

Software engineering principles like object-oriented design and meta-programming are very useful to domain experts as well, especially as their code base grows incrementally or when they are starting on large projects to begin with. These techniques are also available in graphical form with LabVIEW and they extend to LabVIEW FPGA as well.

4.4 Use Case

To demonstrate some of the framework features and recommended techniques listed in Sects. 4.2 and 4.3, we describe in this section how we developed a device that connects to a regular pair of analog speakers and enables them to play music from iOS devices over the air. To this effect we used the LabVIEW framework and a RIO board to develop an AirPlay-based [Wik15] Wireless Music System (WMS).[1]

Since we wanted this design to be available as reference for students, we selected as a base interface board the NI myRIO embedded hardware device [Nat15d] that introduces students to industry proven RIO technology and allows them to quickly design real, complex engineering systems. In our case, it provides the following platform elements needed for our implementation: (1) Wi-Fi access for over the air music delivery; (2) a Digital to analog converter (DAC), connected to audio quality analog output (for analog speakers); and (3) a Xilinx Zynq 7010 all-programmable SoC, implementing the basic RIO architecture elements: a dual-core ARM instruction processor to run the basic protocols, and high-level soft real-time processing; a reconfigurable fabric to interface to the DAC, provide low-level timing, and hard real-time processing.

[1]We kindly acknowledge the collaboration with our colleague Trung N. Tran on this project.

There are a couple of free implementations of the AirPlay protocol [Wik15] for ARM processors running Linux, which we were able to leverage. The main application is developed using the LabVIEW Real-Time Module (Figs. 4.4, 4.5, 4.6), to run on a Linux-based real-time operating system, called NI Linux Real-Time, on the instruction processor. It spawns a process running ShairPort, which implements basic protocols and decoding for the music transmitted from iOS devices. ShairPort was compiled from C code available online [Wik15]. To advertise AirPlay service ShairPort searches for an installed mDNS service (binary opkg) and connects to it directly.

The ShairPort process provides the data in lossless format to the main LabVIEW Real-Time processing loop via UNIX pipes. The main loop implements a simple equalizer using a 3-channel filter bank (Figs. 4.7 and 4.8), using the built-in libraries. It then passes the data to a Host-FPGA DMA First in, first out (FIFO) queue, so that we can implement on the FPGA the only part that really requires very precise timing, namely the 44 kHz interface to the analog out port (Figs. 4.9 and 4.10). Notice that as domain experts we do not need to worry about the details of the interface to DAC, but just provide high-level data to the I/O interface. Similarly, we described the timing as a simple loop timer that gets the specific timing requirements from the RT processor via a run-time parameter.

The prototype is shown in Fig. 4.11, with the myRIO device attached to the speakers, streaming music from an iPhone iOS Device, which has correctly identified the myRIO device as an AirPlay WMS. This use case shows how a software programmer will find familiar elements in this extended LabVIEW flow, and will be eased into more hardware centric elements as needed/beneficial for specific requirements, in particular interfacing to real-world I/O at several levels.

4.5 Conclusions

In this chapter we have presented a graphical programming framework, LabVIEW, in particular the LabVIEW FPGA Module, and associated language and libraries, as well as programming techniques and patterns that we have found useful in making FPGAs accessible to scientists and engineers who are typically domain expert software programmers. We focused on a single RIO module as target, i.e. one instruction processor, attached fabric and I/O, and touched briefly on some of the challenges of networking RIO subsystems. We refer the reader to [Nat15b] and [Nat15c] for further details on the topics covered here, additional descriptions of related tools, as well as use and programming of networks of RIO modules, and more advanced topics, supported by many examples and more complex customer use cases.

Although it is beyond the scope of this chapter, we would like to point the reader to a recent specialized version of LabVIEW, the LabVIEW Communications System Design Suite [Nat15a]. It focuses on making FPGAs accessible to communications system designers and algorithm engineers. It is a single, cohesive environment that

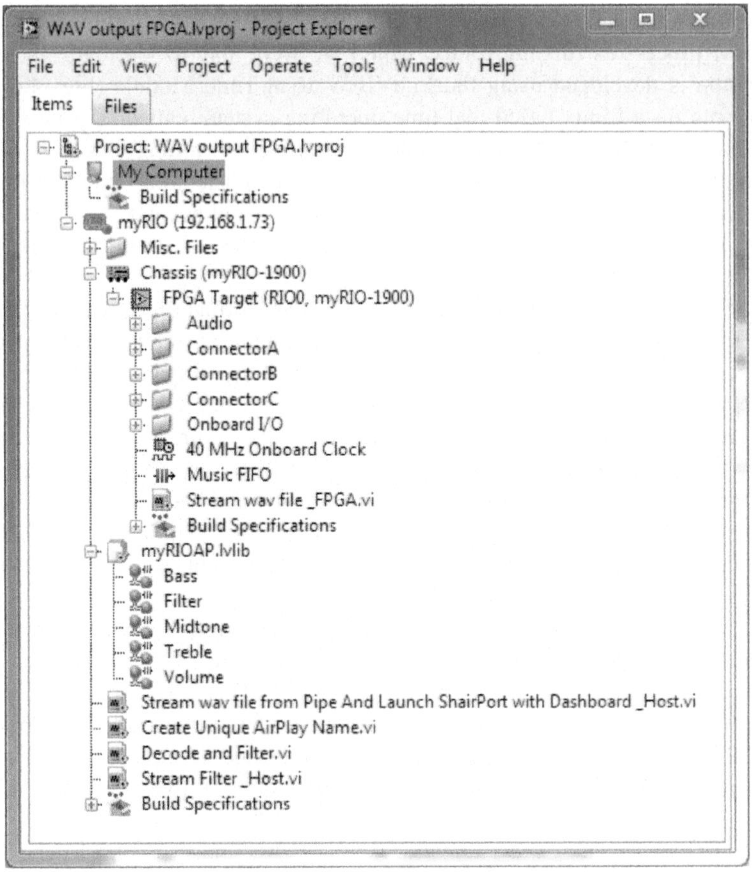

Fig. 4.4 LabVIEW project explorer view

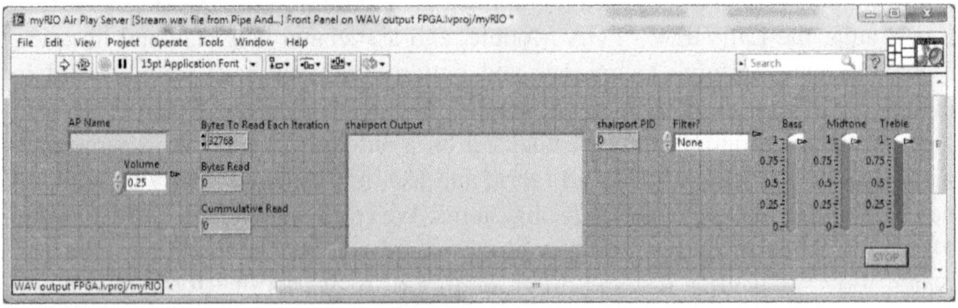

Fig. 4.5 LabVIEW real-time top-level App front panel

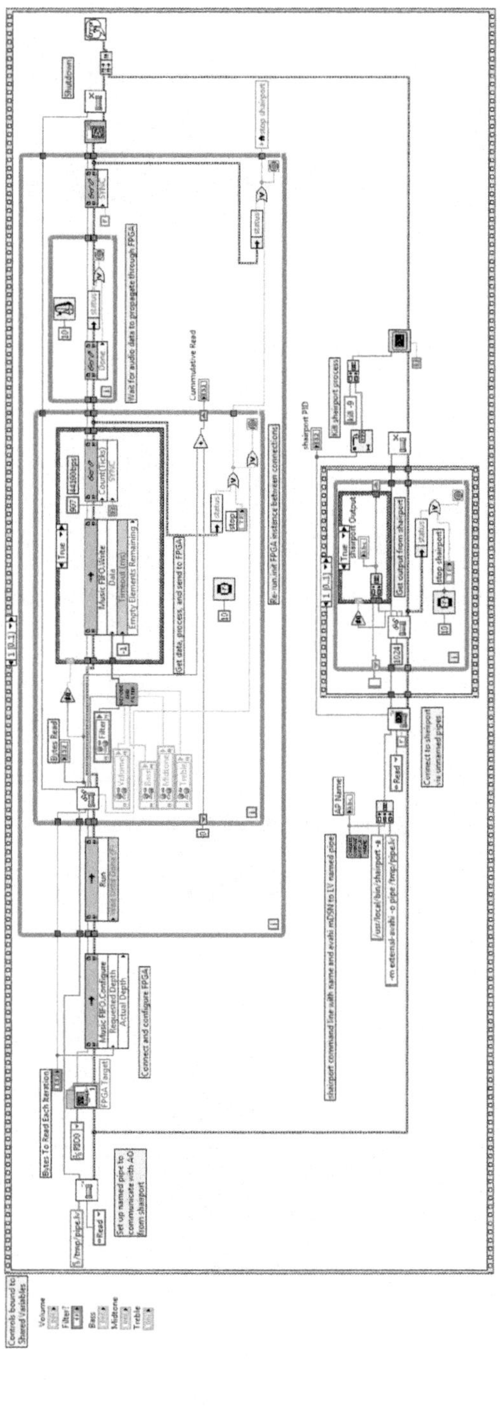

Fig. 4.6 LabVIEW real-time top-level App block diagram

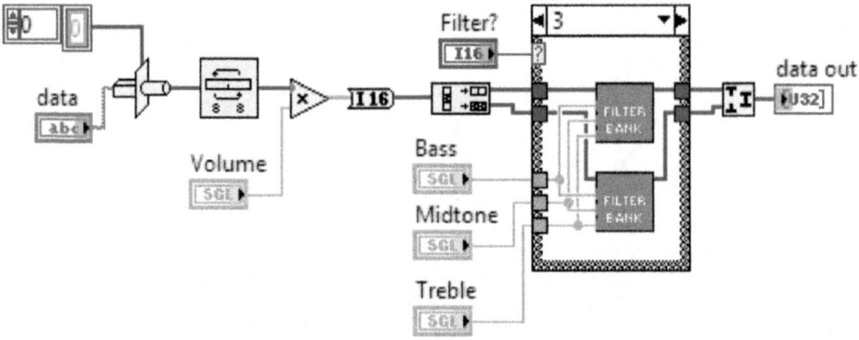

Fig. 4.7 LabVIEW real-time decoding block diagram

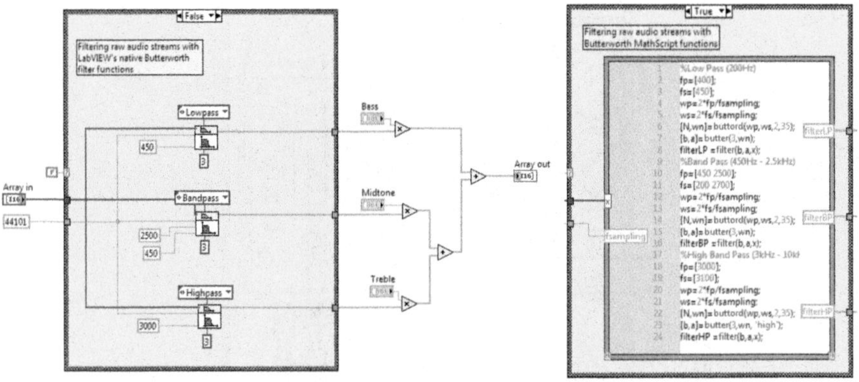

Fig. 4.8 LabVIEW real-time processing block diagram (graphical/textual)

Fig. 4.9 LabVIEW FPGA DAC timing front panel

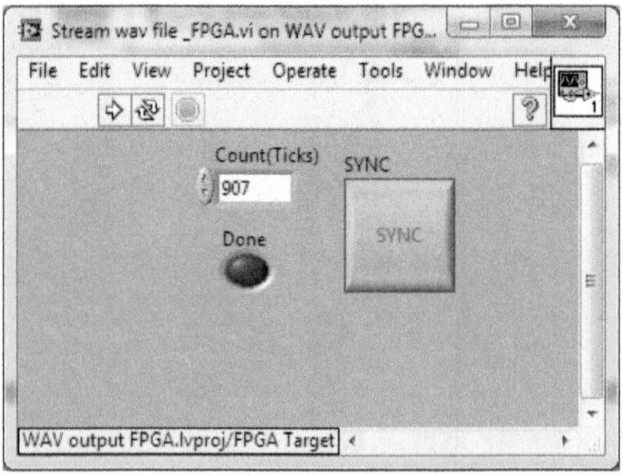

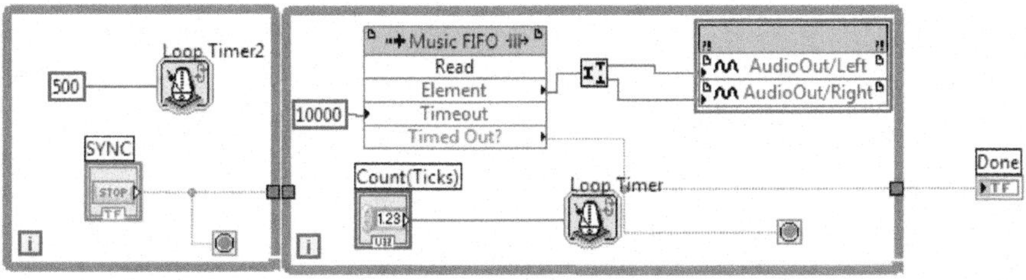

Fig. 4.10 LabVIEW FPGA DAC timing block diagram

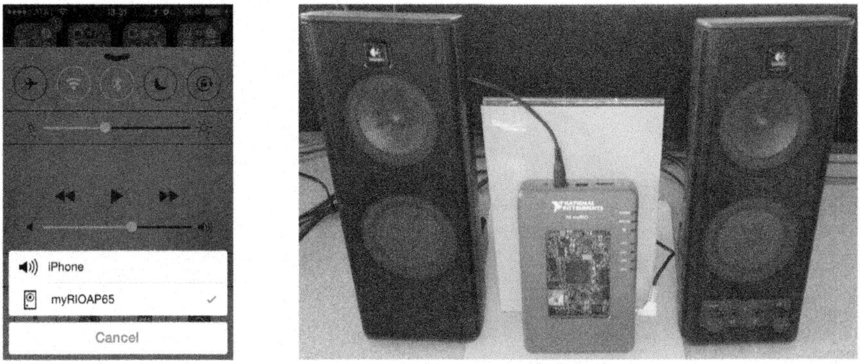

Fig. 4.11 myRIO WMS prototype

enables those domain experts to program both the processor and FPGA, while scaling across multiple Software-defined radio (SDR) systems so that one can define and manage an entire prototype solution with a single design tool, and rapidly deploy new algorithms to hardware. Furthermore, it specializes the G language with multi-rate dataflow that is familiar to communications algorithm engineers. It also provides specialized tools, e.g. for conversion from a floating point design to a more efficient fixed point representation, and also specialized IP, such as application frameworks that plug into LabVIEW Communications to provide a LTE and 802.11 software implementations.

Chapter 5
Spatial Programming with OpenSPL

Tobias Becker, Oskar Mencer, and Georgi Gaydadjiev

In this chapter we present OpenSPL, a novel programming language that enables designers to describe their computational structures in space and benefit from parallelism at multiple levels. We start with our motivation why spatial programming is currently among the most promising approaches for building future computing systems in Sect. 5.1. In Sect. 5.2 we introduce the basic principles behind OpenSPL and exemplify them with few simple examples targeting the first commercial offering of a Spatial Computer system by Maxeler Technologies. We validate the potential of Spatial Computers in Sect. 5.3 and conclude in Sect. 5.4.

5.1 The Case for Spatial Programming

With recent developments in semiconductor technology, the computing industry has experienced a dramatic shift in paradigms. Clock frequency is no longer the driving force behind improvements in performance. While the number of transistors keeps increasing, we can no longer afford the power budgets[1] to run all of them and at higher clock rates. Performance is now obtained through increasing amounts of parallelism, such as multi-threaded multi-core implementations. A disadvantage of this approach is that it still relies on the fundamentals of sequential computing, i.e. the high-level program and the compiled machine code describes an algorithms as

[1]The sum of power consumption and heat dissipation.

T. Becker (✉) • O. Mencer • G. Gaydadjiev
Maxeler Technologies, London, UK
e-mail: tbecker@maxeler.com

D. Koch et al. (eds.), *FPGAs for Software Programmers*,
DOI 10.1007/978-3-319-26408-0_5

a sequence of elementary steps.[2] These steps are then carried out in a sequential manner with all of the related limitations and challenges concerning (instruction) parallelism. Sequential programming models are very universal and convenient, however, inherently slow and inefficient for problems with inherent parallelism. Many micro-architectural and architectural innovations such as out-of-order execution, VLIW, SIMD, advanced branch prediction and multi-threading have been applied to gain additional parallelism. The fundamental limitations of sequential processing are still present despite all of these highly sophisticated techniques.

A far more efficient approach is to describe the program as a spatial structure that is then translated into a fundamentally parallel, highly customised implementation. Computing in space is emerging as an attractive alternative to programming sequential computing elements, such as microprocessors. Custom accelerators utilising parallelism have been explored for many large scale applications, but they usually suffer from poor designer productivity due to complex design methods. However, the novel software methods presented in this chapter give developers access to the creation of spatial accelerators with competitive performance without requiring any hardware design experience.

Spatial computation refers to the concept of laying out the computational structure in as a monolithic parallel datapath where all operations are carried out concurrently. A simple analogy for spatial computation is a factory floor with an assembly line. A mass production problem where the same steps are performed over and over again on each individual product, is most efficiently solved by a production pipeline. Every station along the pipeline is customised to perform one assemble step (data operation) and products (final results) are gradually completed while different parts (data) arrive just in time needed. The final results are finished products (processed data at the output) that leave the assembly line in a constant stream and without any observable control overhead. The same principle applied to computation results in large scale data flow structures where data is simply streamed in at the right moment and location, avoiding large caches and complex control.

The Open Spatial Programming Language (OpenSPL) has been developed to enable domain experts to generate such spatial computational structures for application-specific challenges. The key innovation lies in bridging the previously separate hardware and software domains, and providing a simple programming model for the spatial part on the computation. With OpenSPL, developers can now create large systems following the traditional software development life-cycle (SDLC) principles for both ultra high throughput computing and ultra low-latency communication while also reducing their carbon footprint by means of minimising energy consumption, cooling and overall data center space. OpenSPL achieves this by providing access to parallelism at all levels from algorithm, trough system architecture down to the individual bits and hence facilitating designers in their effort to produce the most optimal workload specific instance in space.

---

[2]Referred to as *instructions*.

5.2 OpenSPL Execution and Programming Models

OpenSPL targets machines consisting of a Spatial Computing Substrate (SCS) that is instantiated in a specific reconfigurable hardware technology containing flexible arithmetic units, customisable interconnect and a set of configurable memories. To facilitate simplicity of programming, easy automation and to boost designers productivity, OpenSPL relies on the Static Data-flow execution model. This represents a convenient high-level abstraction that system designers and scientists can grasp and master while in addition simplifying the compilation tools. The SCS arithmetic units are organised as a very large, deeply pipelined execution graph involving all operations (or the maximal number that can fit on a given fabric) constituting the targeted function laid down in the 2-dimensional (2D) space. The different operations are placed in 2D space and interconnected in such a way that all partial results travel the least possible distances between any two subsequent operations. All operations at all levels (including transfers along the data paths) operate in a globally synchronous fashion. The concept of unit clock ensures high degree of predictability and a specific behaviour characterised with the ability to consume and produce a single set of data elements at each clock tick once the steady state operation is reached.[3] Under such conditions the overall system throughput is bounded only by the available pin bandwidth[4] used to transfer data in and out of the SCS chip. Programming in OpenSPL relies on describing spatial compute kernels that clearly separate data- from control-flows of the targeted algorithm. Stated differently, the designer is expected to explicitly describe the computational algorithm as a static computation structure in 2D space while the data elements will simply flow through. This makes the most natural SCS variables to be *data streams*. Data-flow kernels mapped on the SCS are highly efficient as they do not require shared memory, complex multi-level caches, a program counter, or branch prediction. All the above-mentioned highly sophisticated units usually require substantial amounts of area (and power) in modern processors, and their absence results in significant area savings that on its turn frees more area for additional computational structures. In addition, the implicit deep pipelining operation allows spatial kernels to operate at clock frequencies significantly lower than modern silicon structures while still delivering higher throughput. To summarise, an implementation performing spatial computation can provide significantly higher performance as compared to general purpose processors by using the same number of transistors and silicon area even at lower clock frequencies, and all of the above with lower power consumption. Furthermore, OpenSPL naturally supports high degree of customisation for each SCS implementation to best support the specific algorithmic structure and the precise numerical demands of the application at all intermediate phases and stages.

[3]When all, typically thousands, pipeline stages operate fully on large number of partial results.

[4]The aggregated metric of the number of available pins and their individual bandwidths.

Last but not least, please note that Computing in Space is not intended to fully replace the conventional processors based on the von Neuman program execution principles. Quite the contrary, the SCS based accelerators will be used to offload and speedup only the data intensive parts of the targeted applications. Traditional, general purpose processors will be used to run the operating system and the overall user application (ideally only its control-flow dominated parts) that fully benefit from the SCS acceleration potential. In addition, different SCS configurations are handled on demand and as such there will be always at least one control processor responsible for reconfiguring the SCS part of the system and notifying completion of its execution. Such a hybrid approach provides the designers with a flexible system architecture that allows them to partition, balance and customise their algorithm's implementation in order to gain the best possible and ideally achieve the maximal performance point of operation on the system under consideration.

To summarise, in order to build a computing system implementing the OpenSPL basic principles as described above, the following components are required:

- Spatial (customisable) Computing Substrate constituted by:
  - Reconfigurable compute (arithmetic) fabric;
  - Customisable memory system (typically divided in at least two parts: Large (off-chip) Memory and Fast (on-chip) Memory);
  - Flexible IO and Networking support.
- Spatial Compiler (including SCS specific floorplanning, place and route tools);
- Operating System or runtime OS extension;
- Support for customisable Data Orchestration[5];
- (optional) Domain specific languages and libraries.

Some of the important concepts described above are represented by the motivating example of a simple 3-elements moving average OpenSPL program depicted in Fig. 5.1. This figure shows the relationships between the individual OpenSPL instructions and their implementations on the SCS surface. As it can be observed the designer is expected to explicitly define all input and output data streams; and that the Spatial Compiler automatically resolves dependencies among operations, e.g., the division uses the results of the sum operation.

OpenSPL can be used to efficiently program any system that supports the basic principles of spatial computing. Several good examples for such SCS-based architectures are Data-Flow Engines (DFEs) from Maxeler [FPM12], Automata Processor Engines (APEs) from Micron [DBG$^+$14] and even any future Quantum Processing Units (QPUs). It should be noted that different target architectures such as the samples above would have quite different substrate characteristics used to achieve spatial computation. Different algorithmic optimisations and techniques will be used to improve system performance and reduce energy expenditure. The main goal of OpenSPL is not to provide direct performance portability across these

[5]Two options exist: implicit support by the SCS and explicit management by the programmer.

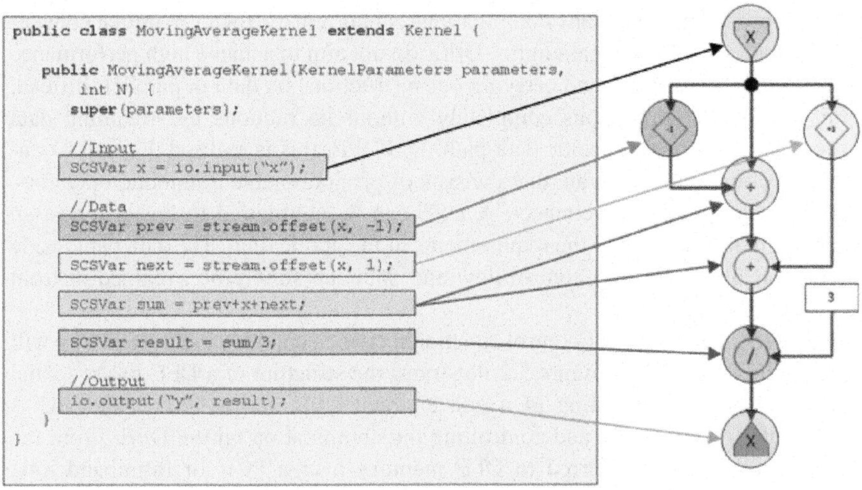
public class MovingAverageKernel extends Kernel {
public MovingAverageKernel(KernelParameters parameters,
int N) {
super(parameters);
//Input
SCSVar x = io.input("x");
//Data
SCSVar prev = stream.offset(x, -1);
SCSVar next = stream.offset(x, 1);
SCSVar sum = prev+x+next;
SCSVar result = sum/3;
//Output
io.output("y", result);
}
}
X
+
+
3
/
X

Fig. 5.1 Moving average over three data elements described in OpenSPL

different architectures (or between versions of the same platform) as this would be unrealistic in the majority of the cases. Instead, OpenSPL aims to establish key principles of spatial computations and to enable efficient application design for such architectures. The language itself and its available libraries and runtime software support have to be customised to a particular target architecture.

From the above mentioned SCS architectures DFEs deserve a special attention since they are the only SCS platform used in commercial products to date. Current DFE generations rely on the largest reconfigurable devices available on the market (also known as Field Programmable Gate Arrays—FPGAs) to enable custom implementations of the arithmetic structures as described by the OpenSPL kernels. The OpenSPL programming language principles are applied to create DFE-specific extensions to Java, resulting in a meta-language *MaxJ* that conveniently describes kernel structure generation. To automatically translate the MaxJ kernel descriptions to equivalent hardware description a dedicated tool named *MaxCompiler* is used. MaxCompiler produces a VHDL description and a fast cycle accurate simulation model of the kernel being compiled. The VHDL description of the kernel is passed to the vendor place and route toolchain to produce the raw bitstream needed to configure the DFE's reconfigurable device. A Linux based run time called *MaxelerOS* is used to manage the DFEs. The *SLiC API* (Simple Live CPU Application Program Interface) ensures a glue-less, subroutine call like, integration between the two sub-parts of the application (the code running on the CPU and the DFE code). In most DFE based systems designers are expected to explicitly describe all data movements in and out of each DFE in the system (all data stream directions and their individual patterns) in a separate Java file called the *Manager*.

In the following we focus on how to target DFE-based spatial computers with OpenSPL. A DFE is a spatial computing substrate that enables very high performance by running many computations concurrently. However, unlike conventional parallel computing paradigms, DFEs do not aim to achieve high performance by providing many cores and carrying our instructions on data in parallel. Instead, DFEs perform computations completely without instructions by streaming data through an application-specific data-path. In a DFE, this is realised through a configurable computing substrate that consists of programmable arithmetic operators, memories, logic and interconnect. A DFE can be configured to create a custom data-path that is tailored to the requirements of the application. The data path inside the DFE is also free from control flow and data can simply be streamed in from surrounding memory.

However, some form of control functionality is usually necessary and this will be preformed by a CPU. Figure 5.2 illustrates the structure of a DFE-based spatial computing system consisting of a conventional CPU and a DFE. The CPU is responsible for setting up and controlling the computation on the DFE. From the CPU, data will be transferred to DFE memory over a PCIs or Infiniband link. Once the data is located in DFE memory, it can be streamed through the spatial DFE implementation with very high efficiency and performance. Accordingly, the compute model involves splitting the application into a CPU part for control, and OpenSPL for spatial data-flow computing. CPU programming can be performed in any common programming language such as C or C++, while DFE programming is done in OpenSPL. OpenSPL is based on the Java language and provides a number of classes to express data-flow principles.

In the following we cover some of the key principles of OpenSPL. OpenSPL is meta language that describes data-flow computing; it uses Java syntax but is in principle different from regular Java programming (or any other imperative

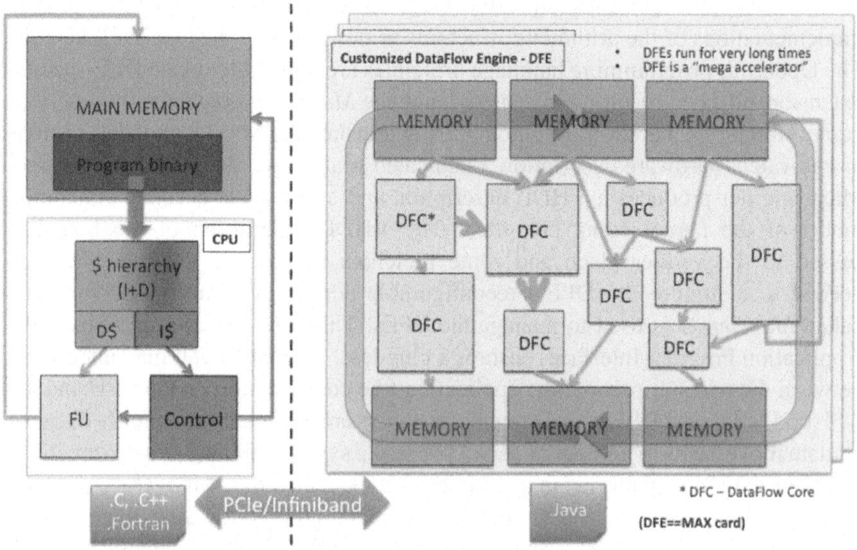

Fig. 5.2 Structure of a system with a spatial DFE accelerator and a standard CPU

Listing 5.1 C code of a simple computation inside a loop

```
for (i = 0; i < numDataElements; i++)  {
    float x = input[i];
    float y = x * x + 30;
    output[i] = y;
}
```

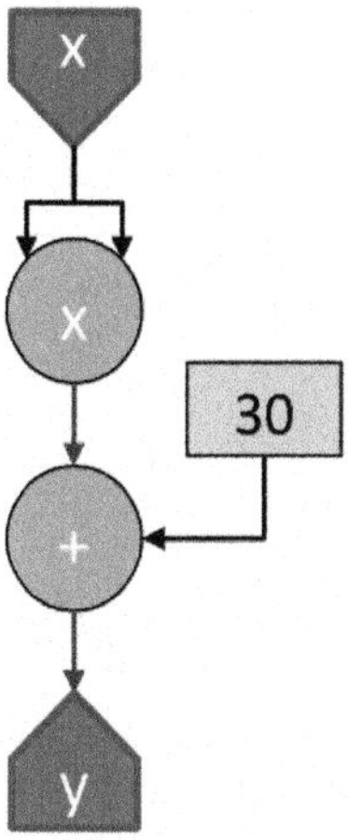

Fig. 5.3 A data-flow implementation for the computation inside the loop body

programming paradigm). The most important conceptual principle in OpenSPL is that we describe a fixed spatial data-flow structure that can perform computations by simply streaming through data, and not a sequence of instructions to be preformed by a traditional processor. The fixed nature of the data-flow implementation means that (a) no instructions are needed, (b) complex control mechanisms of a processor are avoided, (b) many operations can be performed concurrently in the pipeline, and (c) the computation is performed by simply streaming in data from memory, producing a result with every clock cycle.

To illustrate these principles we show how a simple loop computation can be transformed into a data-flow description in OpenSPL. Listing 5.1 shows C code that perform a simple calculation inside a loop body. The calculation is repeated in a number of iterations over an input data set. Figure 5.3 illustrates a data-flow implementation for the same computation. The computation inside the loop body can be performed by a fixed data path that contains both a multiplier and an adder. It is one of the key features of spatial computing to have several operators present at the same time and operating concurrently, instead of using a time-shared ALU inside a processor. Another key principle is the absence of control, instructions and caching. The data path is fixed and the computation is performed by streaming data from memory directly into the data path. The computation will be fast and efficient, and its throughput will be completely predictable.

Listing 5.2 An OpenSPL description that generates the data-flow implementation shown in Fig. 5.3

```
1 class Loop extends Kernel {
2     Loop() {
3         DFEVar x = io.input("x", dfeFloat(8,24));
4         DFEVar y = x * x + 30;
5         io.output("output", y, dfeFloat(8,24));
6     }
7 }
```

Listing 5.3 C code of a nested loop with dependency

```
for (i = 0; i < numDataElements; i++)  {
   float d = input[i];
   float v = 2.91 - 2.0*d;
   for (iteration = 0; iteration < 4; iteration++)
      v = v * (2.0 - d * v);
   output[i] = v;
}
```

Listing 5.2 presents an OpenSPL description that can generate the data-path shown in Fig. 5.3. In OpenSPL, a data-path is called a kernel and our OpenSPL descriptions begins by extending the `kernel` class (line 1). We then define a constructor for the `Loop` class (line 2). The code inside the body of the program will then generate the spatial data-flow description. It is important to remember that OpenSPL generates compute structures and but does not execute at run time. To create an input to the data-flow graph, we use the method `io.input`. This method takes two inputs: The first is a string representing the name that will later be used to connect the data-flow graph to memory. The second input specifies the data type for the stream.

The C-code shown in Listing 5.3 gives an example of an example C-application with two nested loops. We observe that the outer loop performs one iteration over each data element, while the inner loop describes a computation with a cyclic dependency of v from one loop iteration to another. This example can be effectively transformed into a data-flow description as illustrated in Listing 5.4. The outer loop is replaced by streaming inputs and outputs. The inner loop is described with using similar `for` loop, however it will result in an unrolled, spatial implementation as illustrated in Fig. 5.4. Note that the inner-loop dependency is resolved by the resulting data-path as it is acyclic, and v is reimplemented for each loop iteration, and connected to the result from the previous iteration (in space). This results in a long pipeline where we can simply stream into new input data d at each clock cycle, while the data will be gradually processed as it moves through the resulting pipeline.

As demonstrated by this simple example, OpenSPL differs from traditional imperative programming in various ways, which we shall describe in the following. First, control and data flow are separated. Control flow is described in C and targets

Listing 5.4 An OpenSPL implementation of a loop with dependency

```
1 class Loop extends Kernel {
2     Loop() {
3         DFEVar d = io.input("d", dfeFloat(8,24));
4         DFEVar two = constant.var(dfeFloat(8,24), 2.0);
5         DFEVar v = constant.var(dfeFloat(8,24), 2.91) - two * d;
6         for (int iteration = 0; iteration < 4; iteration += 1) {
7             v = v*(two - d * v);
8         }
9         io.output("output", v, dfeFloat(8,24));
10     }
11 }
```

the CPU; it is used as a host program to set up the computation and copy data to SCS memory. Data flow is described in OpenSPL, as shown in the example above.

All statements in OpenSPL describe the structure of a data-flow graph, and not sequential computational steps. The OpenSPL statement in line 7 of Listing 5.4 results in two separate multipliers and a subtractor (see Fig. 5.4). Unlike a time-shared multiplier on a CPU, the multiplications are now performed in parallel. Additionally, the `for` loop in line 6 of Listing 5.4 can be unrolled and all loop iterations can also be implemented as separate units that operate concurrently. Figure 5.4 illustrates a fully parallel implementation of the sample code.

The OpenSPL program is "run" only once to generate a binary configuration file for the Spatial Computing Substrate (SCS). This SCS binary contains a spatial implementation of the program, and it is linked with the host application in C. The C-code can be changed or invoke various DFE binaries without re-running the OpenSPL program. Only a change to the data-flow model requires to re-compile the OpenSPL program (by "running" its OpenSPL description).

This above principle has important implications on how OpenSPL programs behave. First, we clearly distinguish between variables that are evaluated at compile time and variables that contain run-time information. Regular Java variables such as `int` are evaluated only once when the program is run and result in run-time constants. Likewise, Java loops and conditionals describe how to generate a data-flow description and they do not make run-time decisions. For example, the `for` loop in line 6 of Listing 5.4 will not be evaluated at run-time to carry out four iterations. Instead, it will replicate the content of the loop body four times, which can be used for an unrolled and parallel implementation as shown in Fig. 5.4. Similarly, an `if` statement allows us to select different computations based on an input parameter, but no run-time switching will be performed. If various cases of need to be selected dynamically, then separate SCS implementations would be generated, and the switching between different cases would be performed by the host application that handles control.

OpenSPL provides the data type `DFEVar` for all variables that handle run-time data. For example, the variables `d` and `v` in Listing 5.4 handle run-time data that is streamed through the DFE. Since OpenSPL describes the structure of a data-flow implementation, it is not possible to evaluate run-time data of a `DFEVar` in a conditional. Instead, OpenSPL uses the ternary operator to switch between two data

Fig. 5.4 Spatial implementation of the loop. The inner loop is unrolled into a data path

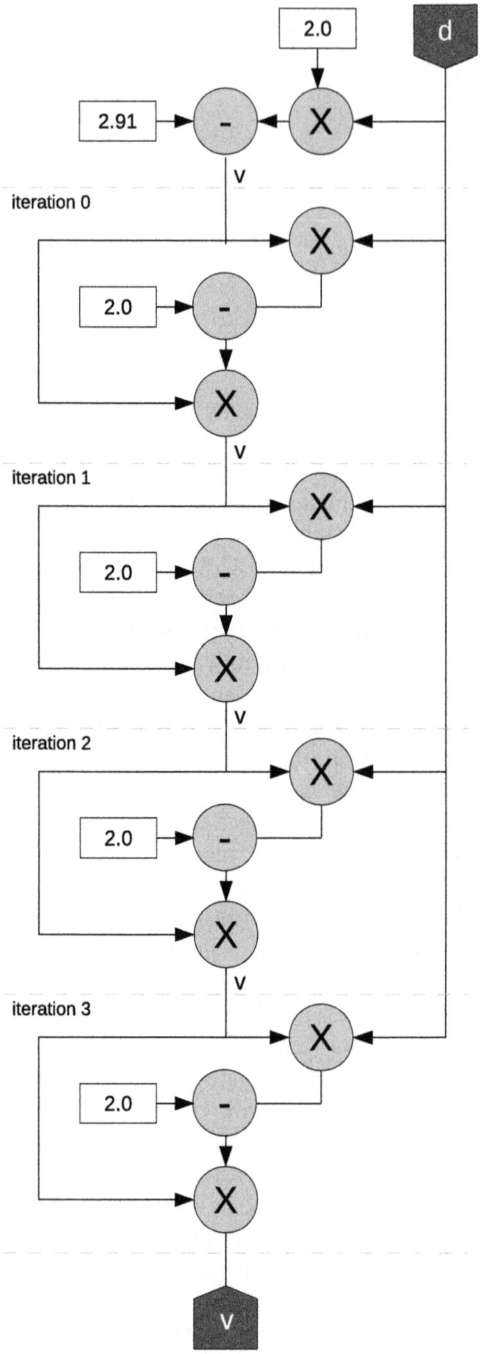

streams based on run-time data, resulting in a multiplexer in the data-flow graph. It is also not possible to read a `DFEVar` into a variable of type `int` as the values of a `DFEVar` will only be known at run time. However, it is possible to read a regular Java variable of type `int` into a variable of type `DFEVar`. This will result in a run-time constant since a variable of type `int` is evaluated at compile time. Another way to declare a constant is illustrated in lines 4 and 5 of Listing 5.4. The method `constant` allows to create a stream of constant values of the same data type as the `DFEVar` and no type casting is necessary.

The inputs and outputs of an OpenSPL program are so-called streaming inputs and streaming outputs. Lines 3 and 9 in Listing 5.4 show how the methods `io.input` and `io.output` are used to describe the inputs and outputs to the computation. The multiple operations within the OpenSPL program constitute a so-called kernel. Typically, streaming inputs and outputs replace the outer loop of a computation that iterates over data. Instead, we copy the input data to memory surrounding the DFE, stream it through the DFE, and collect the results in output memory. This happens in a fully concurrent and pipelined fashion, i.e. once a data element has passed through the first pipeline stage of a kernel, we can stream in the next data element, while the previous partial result moves down the pipeline.

OpenSPL supports customisable data types. Spatial implementations of computations can greatly benefit from tailoring the used data types to the requirements of the computation. For instance, floating point data types usually require more area in the spatial computing substrate than a fixed point representation. Likewise, smaller floating point formats require less area than large ones. In Listing 5.4, a floating point format with 8-bit exponent and 24-bit mantissa is used because it corresponds to the original single precision floating point format in used in the original C-code (Listing 5.3). As part of further optimisations, a user could attempt to reduce the range or precision of the floating point format, or switch to fixed point.

5.3 Application Examples

In this section we present the results of three highly relevant applications after their implementation on spatial computing substrate. The results clearly show the competitive advantage of the approach discussed in this chapter and implemented by Maxeler Technologies in their computing systems.

5.3.1 Seismic Modelling in the Oil and Gas Industry

Geological modelling and computation are taking an important role in the search for oil and gas. Improving the amount of oil and gas that can be recovered from an existing reservoir as well as discovering new reservoirs requires creating an image of the subsurface earth. Scientist typically conduct an acoustic experiment on the earth's surface and record the reflections from the different subsurface layers with tens of thousands of sensors. The resulting datasets can be several hundred Terabytes in size, and to compute the data, thousands of compute nodes are typically used for weeks at a time. Hence, the computational effort is substantial and increasingly complex geological models represent a significant high-performance computing (HPC) challenge.

Reverse Time Migration (RTM) is an application that is frequently used in geoscience and it can produce detailed subsurface images in complex geologies. The concept behind RTM is relatively simple. It involves computing the forward propagation of an acoustic wave from the source to the receivers based on an initial known earth model. At the same time, the actual received output is propagated backwards to the source. The forward and backward wavefields are cross-correlated with the goal to refine the earth model (i.e. to find an earth model where the computed propagation is coherent with the measured output). The computation involves 4D wavefields based on the acoustic wave equation:

$$\frac{\partial^2 u}{\partial t^2} = v^2 \left(\frac{\partial^2}{\partial x^2} + \frac{\partial^2}{\partial y^2} + \frac{\partial^2}{\partial z^2} \right) \tag{5.1}$$

To solve the wave propagation problem numerically, a finite difference stencil operator is applied to approximate the derivatives. The computational complexity arises from performing the cross-correlation on a 4D problem (four nested loops), the overall problem size and the memory access patterns. An optimised streaming architecture has been developed to map the computation to data-flow kernels on Maxeler DFEs [LCP+11]. The optimisations include re-arranging the data-access patterns to maximise throughput, running multiple parallel pipelines per chip, using a customised number format, and optimising computation's spatial and time resolution.

Table 5.1 shows a comparison of RTM performance on a 1U CPU server with two 8-core Xeon CPUs and a 1U Maxeler MPC-X system with 8xMAX3 DFEs. The Maxeler dataflow system delivers a 36× improvement in computational performance within the same chassis form factor. The Maxeler system is also 15× more energy efficient mainly due to the lower clock rate of the dataflow structure.

Table 5.1 Comparison of RTM performance on a CPU and DFE-node

Platform	Conventional 1U CPU-node 2x8 Xeon cores (Sandy bridge)	Maxeler 1U MPC-X node 8xMAX3 DFEs
Performance (points/s)	0.38G	13.6G
Speedup	1×	36×
Power (W)	377	900
Energy efficiency (points/Ws)	1×	15×

5.3.2 *Financial Analytics*

The use of complex mathematical models has become a key factor in the global financial markets, and especially derivatives rely on such models. This has led to the development of increasingly complex derivative instruments, but the process of valuing and managing the risk of complex portfolios has grown to a point where the computational requirements represent a significant computational challenge. Many banks typically operate large data-centres with thousands of CPU cores to perform the daily calculation of value and risk. However, the power inefficiency and overall scalability limitations of CPUs are becoming an increasingly pressing concern.

In finance, a derivative describes a contract that derives its value from an underlying entity, such as a stock or a bond. Examples of such derivatives are stock options where the underlying can be bought for a pre-arranged price at some time in the future. For example, an American Option gives the owner of that option the right but not the obligation to buy or sell stock at a fixed price at any time in the future up to a certain point of expiry. Calculating the price for such an option is based on stochastic models that can be solved with partial differential equations (PDEs). However, closed form solutions often do not exist and complex numerical solvers have to used. An American option for instance could be priced by using a 2D finite difference approach. More complex options such as Bermudan swaptions are also available. A Bermudan swaption is an option to enter into an interest rate swap on any one of a number of predetermined dates. A typical approach to this problem involves using a high-dimensional Monte Carlo model.

Another common type of derivative are credit derivatives. For example, in a credit default swap (CDS) the risk of default in a single underlying credit is exchanged for a regular payment. A CDS contract can be seen as a form of insurance on a bond defaulting. Financial institutions face the challenge of having to calculate the risk and value for hundreds of thousands of credit derivatives each day. A standard single name CDS requires very relatively little time to calculate. However, tranched credit derivatives on a pool of credits such as collateralised default obligations (CDOs) are substantially more complex. It has been shown that in a Monte Carlo approach, a single Maxeler DFE can provide a 270× speed-up over a single Xeon core when pricing CDO tranches [WSRM12].

Maxeler provides a Financial Analytics library that covers a range of financial valuation and risk functionalities that are frequently required in practice. Each module realises a core analytics component, such as curve bootstrapping or Monte Carlo path generation. To support flexible hardware/software co-processing and to enable ease of integration with existing systems, each module is available as both a CPU and DFE library component. Table 5.2 provides a comparison of different instruments priced per second for a range of instrument types supported in Risk Analytics. As it can be seen, a single 1U MPC-X node can replace between 19 and 221 conventional CPU-based units. The power efficiency advantage due to the data-flow nature of the implementation also ranges between one an two orders of magnitude.

Table 5.2 Comparison of CPU and DFE-node performance (instruments priced per second) for various instruments

Instrument	Conventional 1U CPU-node	Maxeler 1U MPC-X node	Comparison
European swaptions	848,000	35,544,000	42×
American options	38,400,000	720,000,000	19×
European options	32,000,000	7,080,000,000	221×
Bermudan swaptions	296	6,666	23×
CDS	432,000	13,904,000	32×
CDS bootstrap	14,000	872,000	62×

5.3.3 Atmospheric Modelling

Atmospheric modelling is an essential application in weather prediction and the study of climate change. The mesoscale atmosphere can be modelled by 3D Euler equations, which are partial differential equations. The Euler equations describe fluid dynamics and in the case of atmospheric modelling, they describe a compressible flow. The core of the computation is stencil-based. Due to the size of practical problems, the computation is very demanding and often carried out on large supercomputers. Parallel computation of the problem requires domain decompositioning, where the entire problem is split into smaller domains that can be computed in parallel. A halo exchange is required to update the elements on the domain boundary. Hence, practical implementations often face constraints in from of memory access patterns and data communications as well as data representation.

An efficient DFE solver for the 3D Euler equations has been developed that achieves efficient resource allocation and data reuse [GFY$^+$14]. The algorithm has been implemented in a streaming model suitable for DFE implementation. The memory access operations are orders such the data exchange is minimised, and data elements can be reused within the kernel. Furthermore, optimised mixed precision is used to satisfy the given relative error requirements while minimising hardware resources. The resulting implementation maps a complex Euler kernel into a single DFE which can perform 956 floating point operations per cycle.

The evaluation compares a CPU reference implementation against a hybrid CPU-DFE version and an hybrid many-core version. The CPU baseline targets a server with two 12-core Intel E5-2697 (Ivy Bridge) CPUs and uses OpenMP multi-threading and vectorization. The hybrid many-core version (CPU-MIC) adds three Intel Xeon Phi 60-core 5120d cards to the two baseline CPUs. The hybrid CPU-DFE is composed of a server with two 6-core Intel E5650 CPUs and a Maxeler MPC-X system with 8 MAX4 DFE cards each of which have 48 GB of memory. The CPU-DFE version is 18.5 times faster and 8.3 times more power efficient than the two 12-core CPU baseline, and is 6.2 times faster and 5.2 times more power efficient than the hybrid CPU-MIC system with three Intel Xeon Phi cards. The results are summarised in Table 5.3.

Table 5.3 Performance and power efficiency for CPU, CPU-MIC and CPU-DFE

	Performance (points/s)	Speedup	Power (W)	Efficiency (points/Ws)	Power efficiency
CPU 24-core	154k	1×	427	0.36k	1×
CPU-MIC	474k	3×	815	0.58k	1.6×
CPU-DFE	2.85M	18.5×	950	3k	8.3×

5.4 Conclusion

We introduced OpenSPL, a novel programming language built upon the concepts of dataflow execution. We introduced the basic principles behind the OpenSPL language and explained some of the most important concepts by using motivational examples based on the first commercial offering of computer systems embracing the principles by Maxeler Technologies. We used three commercial applications to validate our claims that OpenSPL designers can efficiently master parallelism at multiple levels by describing their computational structures in space. The significant speedups reported, e.g., between 19× and 221×, are in addition accompanied by non negligible reduction in power and energy consumption. We strongly believe that building large scale computer systems can greatly benefit from the techniques defined and supported by OpenSPL.

Chapter 6
OpenCL

Deshanand Singh and Peter Yiannacouras

Software programmers are unable to take advantage of the higher performance and lower power achievable from FPGA acceleration because of the additional hardware design skills needed to develop FPGA applications. A major component of this skill gap is in design entry, where FPGA design requires register transfer level (RTL) circuit descriptions instead of the sequential programming software programmers are proficient in. C-to-gates compiler technology can alleviate much of this complexity by automatically generating RTL implementations from sequential C input descriptions. However these tools often suffer from two common problems: (1) they often involve adopting new programming paradigms, languages, or semantics that are often tool-specific; and (2) they fall short of making FPGAs accessible to software programmers since the RTL output of any C-to-gates tool is unusable by a software developer. A hardware designer is needed to integrate the RTL into a complete system, optimize its communication with I/Os, close timing on it over several iterations through FPGA CAD software, and then bring up the design in hardware. This distances the software programmer from the end system, creating long design iterations and ambiguity with system-level functionality and performance.

The Altera OpenCL SDK addresses both these problems, first by targeting the OpenCL programming language which is an open standard for heterogeneous programming already widely used for Graphics Processor Unit (GPU) programming. This means software programmers can use the Altera OpenCL SDK to effortlessly retarget their skills as well as their previously developed OpenCL applications for FPGAs. Second, the Altera OpenCL SDK unites C-to-gates high level synthesis technology, the OpenCL heterogeneous programming standard, and complex FPGA CAD features to deliver a complete development-to-deployment

P. Yiannacouras (✉) • D. Singh
Altera Corporation, Toronto, ON, Canada
e-mail: pyiannac@altera.com

D. Koch et al. (eds.), *FPGAs for Software Programmers*,
DOI 10.1007/978-3-319-26408-0_6

solution for software programmers while requiring no FPGA-specific knowledge nor skill, nor interaction with an FPGA designer. Moreover, the OpenCL SDK provides standard software development tools such as in-system hardware profiling, CPU emulation requiring no FPGA hardware, and also optimization reports, all of which enable the software programmer to develop and optimize their design without knowledge of the FPGA internals.

6.1 Compile and Execute via FPGA Design Automation

A software development environment for FPGAs must hide all facets of FPGA application development including HDL design, system integration, the CAD software, timing closure, bitstream programming, hardware installation, and even deployment. Each of these have been automated/abstracted as described below:

Design Entry The core of the development environment is its programming language and compiler technology. The Altera OpenCL SDK uses OpenCL as its programming language input requiring no tool-specific language constructs or attributes. The OpenCL description is `C`-based and hence very easy to use by software programmers even without prior OpenCL knowledge (the programming model is further described in Sect. 6.2). The Altera OpenCL SDK is a fully 1.0 conformant OpenCL implementation allowing the Altera OpenCL Compiler (`aoc`) to accept OpenCL 1.0 kernel code and automatically generate a highly-pipelined custom datapath in HDL for each *kernel*. It can do so using the parallelism explicitly described by the user in OpenCL, or by inferring parallelism from single threaded code via loop pipelining.

Partitioning the System Design Problem via BSPs A complete FPGA system consists of some kernel computation connected to the I/O subsystem logic necessary for that kernel to communicate to the outside world. For example the kernel requires an I/O connection with the CPU serving as the OpenCL host, as well as off-chip memory such as DRAM, SRAM, or stacked RAM for storage. In addition the kernel may need to connect to external I/O such as an Ethernet or video stream. This I/O subsystem must be pre-designed by an FPGA designer who fully understands the I/O interfaces of the platform as well as the FPGA I/O architecture needed to support them. This subsystem design is packaged into an OpenCL BSP (board support package) and distributed by platform designers. This platform distribution model decouples the `aoc` compiler from any specific hardware/system setup. System designers or FPGA card manufacturers can develop and distribute BSPs alongside their hardware systems. OpenCL application develops merely install the Altera OpenCL SDK, the BSP, and the physical hardware without ever needing to access the hardware design itself nor the CAD tools necessary to interact with them. Thus, to the software programmer, the BSP is just a set of files that accompany the FPGA hardware that was purchased.

Post-Fit I/O Subsystem The I/O subsystem that is packaged into an OpenCL BSP is fully preserved post-placement-and-routing in a post-fit netlist. During OpenCL compilation, this post-fit netlist is imported without modification ensuring that every OpenCL compile is coupled with I/O subsystem logic that is placed and routed exactly as it was when the BSP developer verified its functionality in their own lab. This means the nodes, wires, and timing that implement the I/O subsystem are completely preserved.

Installation and Deployment by IT An OpenCL platform is designed to be purchased, installed and deployed by an organization's IT department. No FPGA-specific knowledge or skill set is required to do this. Users can install the Altera OpenCL SDK, acquire a physical FPGA platform, install the BSP for that platform, and verify the setup with built in diagnostic utilities. Many platforms are already optimized for deployment, for example the Stratix V PCIe platforms requires only a PCIe connection and no additional cables. Once inserted into a PCIe slot, software programmers can compile and execute OpenCL programs without any further physical interaction with the platform.

In-System Compile and Execute Software programmers expect a development environment where the compiler output can be immediately and correctly executed in hardware. The RTL output from high level synthesis tools requires several additional steps to execute in hardware: (1) the RTL must be timing analyzed to determine safe operating clock frequencies; (2) the physical clocks/resets must be supplied to the design; (3) the design must be integrated with the interfaces provided by the I/O subsystem; (4) the integrated system must be synthesized through FPGA CAD tools; (5) the produced bitstream must be programmed onto the FPGA. All of these steps are performed automatically by the Altera OpenCL SDK. The `aoc` compiler inserts all adaptation and arbitration logic necessary to communicate with the I/O subsystem interfaces. It will run just a single pass through the FPGA CAD software, and in that will automatically timing analyze the design and configure the user clock supplied by all BSPs to the maximum operating frequency the design can run at. The OpenCL host runtime will then automatically program the FPGA device with the produced FPGA bitstream as needed by the user's application. The end result of this is a push-button compile and execute flow that guarantees timing functionality while running the user's application at its maximum operating frequency. In addition to behaving identically to a CPU software programming environment, this empowers software programmers by providing them direct access to the end-system allowing them to run and profile their applications directly on the end-system in order to guide subsequent optimizations.

Software Productivity Tools: Emulating, Reporting, Profiling Two artifacts of FPGA design can affect the productivity of a software programmer: (1) the long FPGA compilation times; and (2) the custom architecture automatically generated by `aoc`. The first one is mitigated by including support for CPU emulation of OpenCL in the Altera OpenCL SDK. This allows programmers to run their OpenCL code on their development PC, enabling them to compile, run, and repeat in minutes.

Successful emulation of an application ensures that the OpenCL code is legal and functional without having to endure long FPGA compiles. The second artifact affects the performance of the system. In order to optimize the system performance, a software programmer requires knowledge of the system bottlenecks and how those relate to the OpenCL code. The profiler included with the Altera OpenCL SDK analyzes the performance in hardware and annotates the user's OpenCL source code pointing to specific instructions where, for example, memory access patterns are inefficient or the cache miss rate is high, etc. Programmers can use this to reason about how restructuring the access pattern could alleviate the bottlenecks. In addition, the `aoc` compiler generates at compile time a set of reports that highlight lines of code where certain optimizations were (or could not be) applied, as well as a breakdown of FPGA resource used by each line of the OpenCL C code. These tools provide software programmers the means to develop and optimize the automatically generated custom FPGA pipeline without understanding any of the internal register transfer level interactions of the system. It also ensures software programmers never need to utilize FPGA CAD design software.

6.2 Programming Model

OpenCL provides a heterogeneous computing model consisting of a host processor connected to multiple acceleration devices. A user's application is divided into two parts: (1) the host code that executes on the host and is written in plain C using the OpenCL host API; and (2) the device code that executes on the accelerators that is written in OpenCL Kernel C. OpenCL Kernel C is derived from C99 and includes several language extensions. These extensions provide the user the capability to explicitly describe the parallelism in the code in terms of "work-items" (or threads). In addition the user can explicitly target various levels of memory hierarchy. Data tagged as `local` will be stored in memory accessible only by the work-items in a designated "workgroup". The Altera OpenCL compiler will typically use on-chip distributed RAM blocks for these. Data tagged as `global` is accessible by the host and all threads in the system, subject to the consistency provided by the OpenCL memory model. OpenCL Kernel C also includes built in support for math, vector, synchronization, and other operations. The complete OpenCL programming model is specified by the Khronos group who distribute the full specification [Khr08].

To date the Altera OpenCL SDK is conformant to the 1.0 OpenCL specification. However the SDK also provides its own vendor extensions to target features specific to FPGA architecture. These extensions are listed below:

6.2.1 *Channels*

FPGAs can provide fine-grained task-level parallelism by streaming data produced by one component directly to a downstream component for immediate consumption. The Altera channels extension allows OpenCL application developers to do this between OpenCL kernels. In addition, channels can be used to access external I/O such as a video or network packet stream. Channels can also be used via the new Pipes feature in the OpenCL 2.0 specification.

6.2.2 *Heterogeneous Memory*

FPGA systems often make use of a variety of memory technologies simultaneously. The main data storage is often several gigabytes of DRAM such as DDR4, while small random access data can be stored in SRAM such as QDRII. To enable software developers to take advantage of this the SDK allows: (1) an OpenCL platform to provide multiple memory systems, each tagged with a string such as "QDR"; (2) the OpenCL developer to query the BSP for a list of memory strings; (3) the developer can tag data in their OpenCL kernel code with these strings to control which data resides in the various memory technologies available.

6.3 Application Development Example: Gzip

As an example, we describe the OpenCL implementation of hardware data compression using gzip on FPGAs motivated by its potential for use both in data centers [Cra98] and in communication networks [Pap03, Sum08]. In both scenarios, fast data compression is required to improve the overall system speed in the face of slower I/O either to disk or network. Gzip implements the DEFLATE algorithm for compression, which consists of two parts, LZ77 compression and Huffman encoding. In this example we describe only the LZ77 component.

6.3.1 *LZ77 Compression*

This compression algorithm replaces repeat occurrences of bits with a reference to their previous occurrence [ZL77]. Consider Fig. 6.1 for instance. The algorithm traverses the sentence serially, one character at a time in this case, and looks for repetitions. When " word " is found the second time, it is replaced with a pointer to the previous occurrence of it. This pointer consists of a marker @, match length and

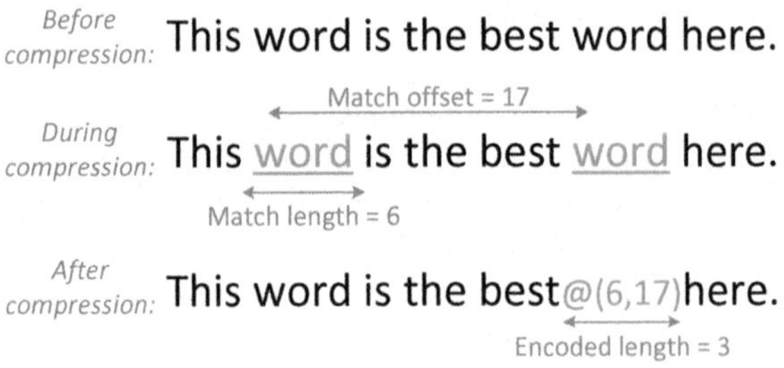
Before compression:
This word is the best word here.
Match offset = 17
During compression:
This word is the best word here.
Match length = 6
After compression:
This word is the best@(6,17)here.
Encoded length = 3

Fig. 6.1 Example of LZ77 compression

match offset. The match length is the length of the match being replaced—in this case it is 6 bytes including spaces. The match offset, is the distance to the previous occurrence of the word.

In LZ77, we only replace our matched word with a pointer if it results in compression. That means that the (marker/length/offset) must be smaller than the word being replaced—words of length 3 bytes are therefore never replaced for instance.

6.4 LZ77 Implementation and Optimizations

In this section we discuss implementation details and show how OpenCL code translates to hardware. The OpenCL implementation is based on a Verilog implementation by IBM [MJA13]. Additionally, we describe performance and compression ratio optimizations that we performed.

6.4.1 System Architecture

Figure 6.2 shows the overall system architecture; the host portion of the implementation selects a byte to represent the marker and a suitable Huffman tree; it also sends to the FPGA the Huffman tree along with the data to be compressed. The host modifies the Huffman tree only if the "update_tree" flag is set; a new tree improves compression quality when the new data being compressed contains a different set of bytes.

Our implementation uses an x86 based system as the host processor connected over PCIe Gen2 x8 to a Stratix V A7 C2 FPGA with local DDR3-1600. This implementation is extendable to standalone FPGA systems such as the case where data streams in and out of the FPGA through Ethernet cables. In these environments, new generations of FPGAs with embedded hard processors [Alt13a] can easily replace the functionality of the x86 host.

System Architecture

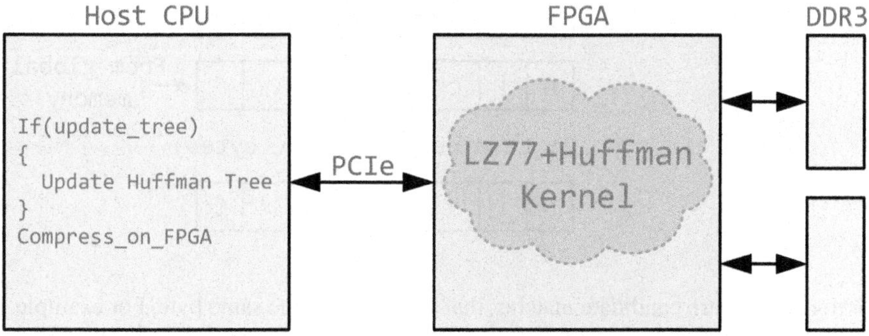

Fig. 6.2 OpenCL system architecture

Listing 6.1 Load 16 bytes per cycle

```
//shift current window
#pragma unroll
for(char i = 0; i < VEC; i++)
  current_window[i] = current_window[VEC+i];

//load in new data
#pragma unroll
for(char i = 0; i < VEC; i++)
  current_window[VEC+i] = input[inpos+i];
```

6.4.2 *Shift in Data and Compute Hash*

The first step is to load new data from global (DDR3) memory—the loaded data is then stored in on-chip registers. Listing 6.1 shows the details of how this is done. The on-chip register array "current_window" buffers data that is currently being processed. First, we shift the second half of "current_window" into its first half, then load VEC bytes from the global memory buffer "input". Figure 6.3 illustrates "current_window" before and after loading in new data. This will then allow us to process VEC substrings each cycle, by extracting portions of the "current_window" (each of length LEN) as shown. The parameter LEN determines the maximum match length possible in LZ77 compression.

Next, we want to search the previous text for substrings that partially or fully match the "current_window" substrings. Previous text is buffered in history dictionaries in local memory (in OpenCL terminology) or FPGA block RAM (BRAM). To lookup these dictionaries for possible matches, we use a hash value corresponding to each of the "current_window" substrings. A very simple hash function just uses the first byte of the substring thus guaranteeing that the

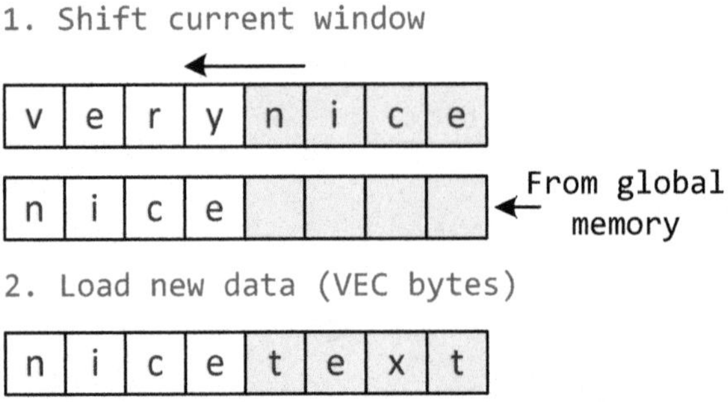

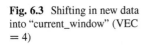

Fig. 6.3 Shifting in new data into "current_window" (VEC = 4)

dictionaries return candidate matches that all start with the same byte. For example, if the "current_window" substring is "nice", hash[nice] = n (remember that each letter is just an 8-bit ASCII number).

This simple hash function has two shortcomings: first, it creates an 8-bit value, that means it can only index 256 addresses while the dictionary BRAMs (M20Ks on Stratix-V FPGAs) have a depth of 1024 words. Second, it only contains information about the first byte (out of VEC bytes) in the current substring and hence the candidate matches returned from the dictionaries only resemble the "current_window" substrings in the first letter. To overcome these two shortcomings we experimented with the hash function and found that the following function improves absolute compression ratio by ~7 % on average compared to the aforementioned simple hash:

$$\begin{aligned} hash[i] &= (current\_window[i] \ll 2) \\ &\quad xor\,(current\_window[i+1] \ll 1) \\ &\quad xor\,(current\_window[i+2]) \\ &\quad xor\,(current\_window[i+3]) \end{aligned}$$

The improved hash function XOR's the first 4 bytes of the current substring, with the first byte shifted left by two bits, and the second byte shifted left by one bit. This creates a 10-bit hash value that is able to index the full depth of the M20K BRAM, and incorporates information about the first 4 bytes as well as their ordering. In testing different hashing functions, the emulator that came with the Altera OpenCL compiler was very useful as it allows testing the algorithm fairly quickly on the CPU.

6.4.3 Dictionary Lookup and Update

Using the hash value computed for each substring, we lookup candidate matches in history dictionaries. These history dictionaries buffer some of the previous text on-chip, and are implemented as a hierarchy of BRAMs—this local memory hierarchy

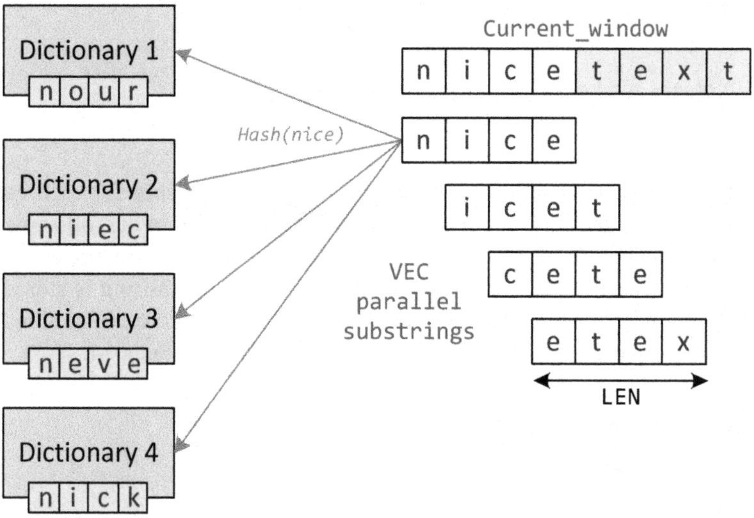

Fig. 6.4 Each substring looks up candidate matches in VEC dictionaries. Each dictionary returns a candidate substring from history that resembles the current substring (VEC = 4)

is key to the high throughput of our LZ77 implementation. To look for matches quickly, there are VEC dictionaries, each buffers some of the previous text. Using the hash value, each substring looks for a candidate match in each of the dictionaries; resulting in VEC candidate matches for each current substring.

The example in Fig. 6.4 shows that the word "nice" finds "nour", "nice", "neve" and "nick" as candidate matches—each of those comes from a different dictionary and they all occurred in the text before our current substring "nice". This lookup is repeated for the other substrings "icet", "cete" and "etex" as described in Listing 6.2—"pragma unroll" on the outer loop tells the compiler to replicate the read ports on each dictionary BRAM so that there are exactly as many read (or write) ports on the physical BRAM to provide conflict-free accesses. In our case, each dictionary has VEC read ports and one write port. The inner loop in Listing 6.2 specifies the width of the read ports. In this case "pragma unroll" tells the compiler to coalesce the memory accesses into one wide access of LEN bytes per read/write port, and the generated on-chip memory supports that width to be able to load/store each substring in one access. In our implementation with VEC=16 and LEN=16, this local memory topology can load 64 16-byte substrings and store 16 16-byte substrings each cycle.

Listing 6.2 Lookup candidate matches in history dictionaries

```
//loop over VEC current window substrings
#pragma unroll
for(char i = 0; i < VEC; i++)
  //load LEN bytes
  #pragma unroll
  for(char j = 0; j < LEN; j++)
  {
```

```
    compare_window[j][0][i] = dict_0[hash[i]][j];
    compare_window[j][1][i] = dict_1[hash[i]][j];
    ...
    compare_window[j][15][i] = dict_15[hash[i]][j];
  }
}
```

The result of dictionary lookup is a set of candidate matches for each current substring stored in an array "compare_window"—this is used in the following step to look for matches for LZ77 compression. After dictionary lookup we update the dictionaries with the current substrings such that each substring is stored in a different dictionary. In Fig. 6.4 for instance, "nice" will be stored in dictionary 1, "icet" in dictionary 2, "cete" in dictionary 3 and "etex" in dictionary 4.

6.4.4 Match Search and Reduction

In this step, each "current_window" substring is compared to its candidate matches in "compare_window" and a match length is computed for each one as illustrated in Fig. 6.5. The "length" array in Fig. 6.5 and Listing 6.3 contains the number of matching bytes from the start of each current and compare windows. The largest value is then chosen and stored in the "bestlength" array—there is now one bestlength value for each "current_window" substring.

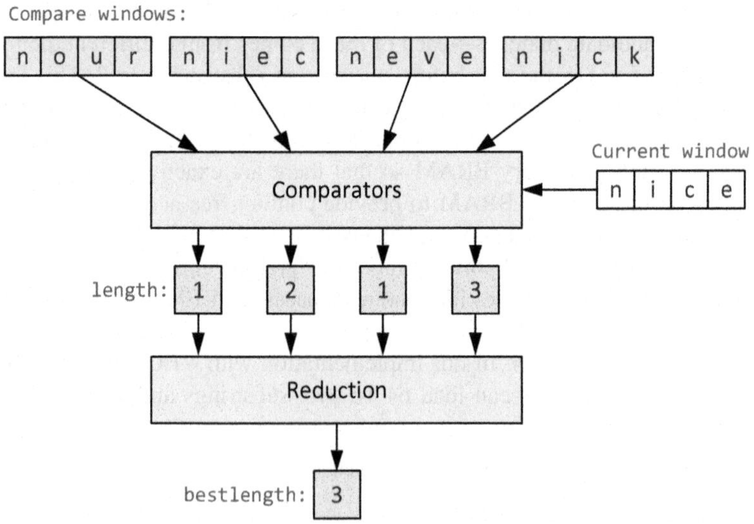

Fig. 6.5 Each "current_window" substring is compared to its candidate matches in "compare_window" and the best match length is selected (VEC=4)

Listing 6.3 Compare each "current_window" substring to its candidate matches in "compare_window" and find the length of each match

```
//loop over each comparison window
#pragma unroll
for(char i = 0; i < VEC; i++)
{
 //loop over each current window
 #pragma unroll
 for(char j = 0; j < VEC; j++)
 {
  //compare current/comparison windows
  #pragma unroll
  for(char k = 0; k < LEN; k++)
  {
   if(current_window[j+k] == compare_window[k][i][j]
      && !done[j])
     length[j]++;
   else
     done[j] = 1;
  }
 }
 //update bestlength
 #pragma unroll
 for(char m = 0; m < VEC; m++)
  if(length[m] > bestlength[m])
   bestlength[m] = length[m];
}
```

Bad Coding Style The code in Listing 6.3 consists of three nested loops; the innermost loop does a byte-by-byte comparison to find the match length. Listing 6.4 shows functionally equivalent code that works better on a CPU but does not translate into efficient hardware. The reason is that the while-loop bounds cannot be determined statically so the compiler will not be able to determine how many replicas of the comparison hardware it needs to create. The compiler issues a warning and only one replica of the hardware is created—the loop iterations share this hardware and execute serially on it causing a big decrease in throughput.

Listing 6.4 Typical C-code targeting CPU processors does not necessarily compile into efficient hardware

```
//compare current/comparison windows
#pragma unroll
while(current_window[j+k]==compare_window[k][i][j])
  length[j]++;
```

Area Optimization Listing 6.5 demonstrates a subtle area optimization. The if-statement in Listing 6.3 gets translated by the compiler into a long vine of adders and multiplexers as shown in Fig. 6.6. This is because we need to know both the value of "length" and the condition of the if-statement before finding the new value of "length". Listing 6.5 removes this problem by using the OR-operator instead of addition to store the match length. Since the ordering of the OR operations doesn't matter, the compiler can create a balanced tree to group the operations on "length_bool" together. This reduces area for two reasons: first, it creates a shallower pipeline depth, meaning fewer registers and FIFOs will be required to

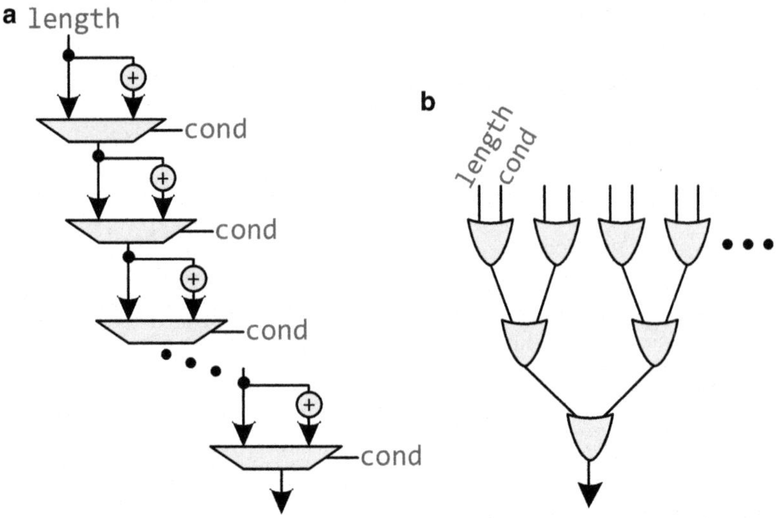

Fig. 6.6 The code in Listing 6.3 produces a vine of adders/multiplexers as shown in (**a**), while using an OR-operator in Listing 6.5 allows the compiler to create a balanced tree of gates that uses lower FPGA resources (**b**)

balance the kernel pipeline. Second, shifters and OR gates require less resources than adders and multiplexers.

Listing 6.5 An area efficient implementation of the innermost comparison loop in match search

```
//compare current/comparison windows
#pragma unroll
for(char k = 0; k < LEN; k++)
{
 if(current_window[j+k] == compare_window[k][i][j])
  length_bool[i][j] |= 1 << k;
}
```

The resulting "length_bool" now contains an array of ones instead of an actual number—if "length" was 3 for instance, "length_bool" will be 0111 where the ones indicate which bytes were equal between the current and compare window substrings. We leverage this one-zero encoding of "length_bool" (instead of the binary encoding of "length") in selecting the best length which results in further area savings. Overall, this area optimization reduces total logic utilization by ~31k logic elements, or 5 % of the Stratix-V A7 FPGA device.

6.4.5 Match Filtering

The previous step creates a "bestlength" array of length VEC; each entry corresponds to one of the "current_window" substrings. The match filtering step now

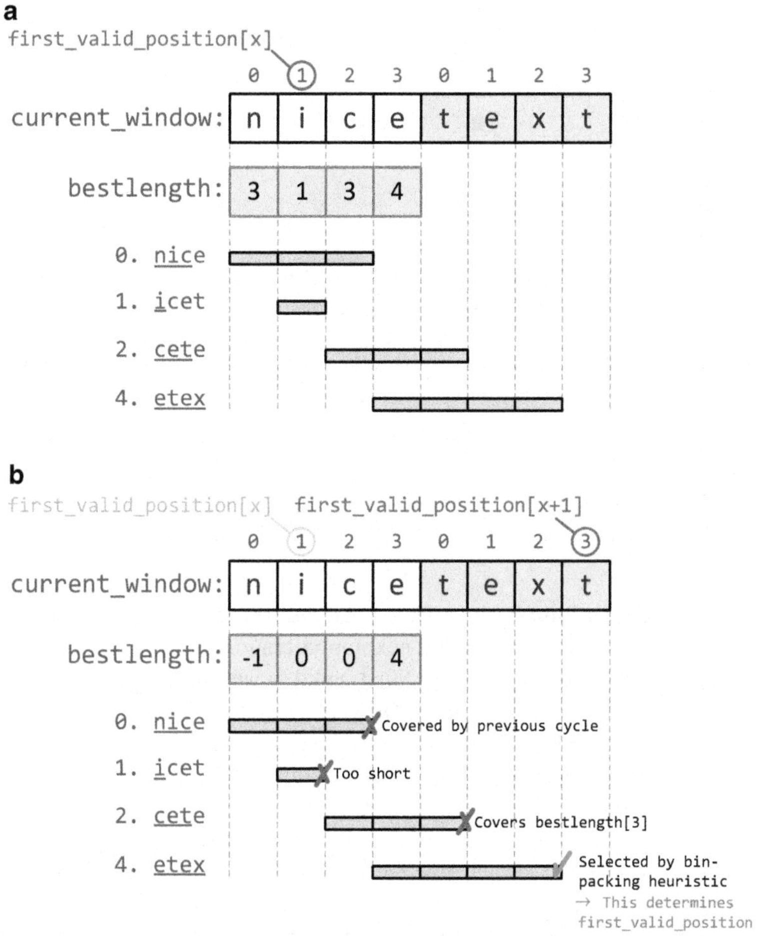

Fig. 6.7 Bestlength array (**a**) before, and (**b**) after filtering (VEC=4)

picks a valid subset of the "bestlength" array such that it maximizes compression; it consists of four steps:

1. Remove "bestlength" matches that are longer when encoded than the original. In Fig. 6.7 "bestlength[1]" is removed because its LZ-encoding will consist of 3 bytes at least (for marker,length,offset).
2. Remove "bestlength" matches covered by the previous step. In Fig. 6.7 "bestlength[0]" is removed because the 'n' was already part of a code in the previous loop iteration.
3. Select non-overlapping matches from the remaining ones in "bestlength". We implement a bin-packing heuristic for this step; the one we choose to implement is "last-fit"; this selects the last match first ("bestlength[3]") then removes any "bestlength" that covers it ("bestlength[2]").

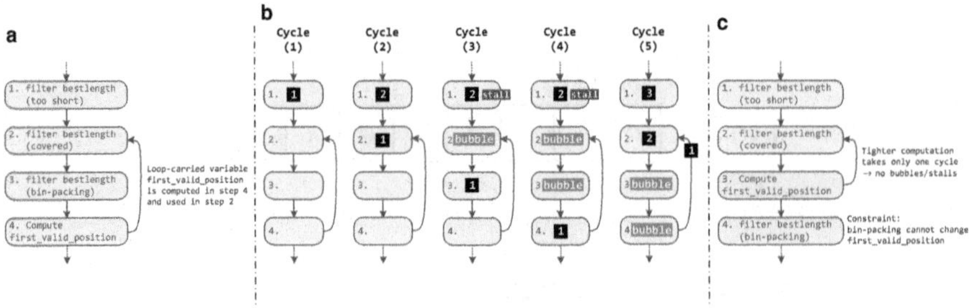

Fig. 6.8 Loop-carried dependencies may cause a kernel pipeline to stall thus reducing throughput. By optimizing the loop-carried computation a high-throughput pipeline can be created. (**a**) Unoptimized loop-carried computation. (**b**) Execution timeline of unoptimized loop-carried computation. (**c**) Optimized loop-carried computation

4. Compute the "first_valid_position" for the next step. This depends on the "reach" of the last used match—in the example in Fig. 6.7 the last match covers bytes 0, 1 and 2 in the following cycle so the "first_valid_position" in the following cycle is 3 as shown.

Loop-Carried Computation A variable is **loop-carried** whenever it is computed in loop iteration x and only used in the next loop iteration $x + 1$. In our application, one of the loop-carried variables is "first_valid_position". In hardware this is implemented as a feedback connection between a later stage in the pipeline to an earlier stage in the pipeline as shown in Fig. 6.8a—"first_valid_position" is fed-back from stage 4 to stage 2 in match filtering. If this feedback path takes more than one cycle, this loop-carried computation causes the kernel pipeline to stall until it is completed.

Figure 6.8b shows the kernel pipeline executing over time assuming that the "first_valid_position" computation takes three cycles. Loop iteration 2 is stalled in the first step until "first_valid_position" from loop iteration 1 is computed in the fourth step; this causes pipeline bubbles as illustrated in Fig. 6.8b. This also means that we can only start new loop iterations every three cycles—the *initiation interval* (II) of the loop is 3. For any FPGA design, our target should be to optimize this loop-carried computation such that we get an initiation interval equal to 1; this avoids pipeline bubbles—a design with II=1 has triple the throughput of a design with II=3. For a stallable pipeline such as the one generated by OpenCL; the loop-carried computations can be thought of as the critical path of the design.

In our application, the computation in Fig. 6.8a resulted in II=6; the Altera OpenCL compiler optimization report informs the user of the loop's II and points to the problematic variable in the user's source code—in this case it pointed to "first_valid_position. To optimize the computation, we take the bin-packing heuristic that filters "bestlength" off of the critical path by moving it after "first_valid_position" computation as shown in Fig. 6.8c. This leads to an II=1 as desired, meaning we can process a new loop iteration every cycle, instead of every six cycles, effectively increasing throughput six-fold. However, we now have a design constraint on the bin-packing heuristic; it cannot alter the value of "first_valid_position" to maintain correctness.

6.5 Results and Comparison

A complete gzip application is implemented using the LZ77 algorithm described above, as well as the Huffman encoding. The full implementation is described in [AHS14]. To highlights its advantages, the generated system from the Altera SDK is compared to the following CPU, Verilog FPGA, and ASIC implementations:

- **CPU Implementations**—Intel [GGF+11] demonstrated a highly optimized implementation of DEFLATE running on a Core™ i5 650 processor at 3.20 GHz. The results showed that a single core could sustain a data rate of 2.7 Gigabits/s with an average compression ratio of 2.18×.
- **FPGA Implementations**—Hardware implementations of FPGA-based DEFLATE algorithms have been previously studied in the literature [RBK07, TMK10, ESK07, MJA13]. The best known results have been reported in a recent work by IBM researchers [MJA13]. Their implementation on a Altera StratixV A7 FPGA is able to sustain a throughput of 3 Gigabits/s [Hof13]. Note that the IBM implementation is the highest performance hardware implementation that has been publicly reported.
- **ASIC Implementations**—Several companies have created dedicated ASIC devices [AHA14, Exa13, Ino12, San12, Int13] that address the needs of high performance lossless compression. The AHA3642 [AHA14] provides the best reported results: 20 Gigabits/s compression throughput while providing a 3.6× compression ratio. Intel's 89xx series of communication chipsets also offers similar levels of performance [Int13].

While ASICs provide incredible efficiency, the algorithms are completely fixed and cannot be changed after the chip is produced. Verilog FPGA implementations are akin to assembly language for hardware. Such a design is time-consuming, difficult to verify, and hence also difficult to modify. With the OpenCL SDK a user can develop a high level description that is easy to modify, with similar productivity to CPU development, while unlocking the acceleration/efficiency benefits of FPGA technology.

Figure 6.9 compares our achieved throughput with the best known commercial implementations of Gzip on FPGAs [MJA13] ASICs [AHA14], and CPUs [GGF+11]. As the plot shows, our implementation is only about 10 % slower than the best known FPGA implementation and 10 % faster than the fastest commercial ASIC implementation. However, note that the ASIC implementation reports an average compression ratio of 3.6× on the Canterbury corpus [Can15], whereas our FPGA implementation achieves 2.43× on the same benchmark set. This is attributed to both the expert knowledge of the industrial vendor that we are comparing to, as well as the higher area budget available to ASICs.

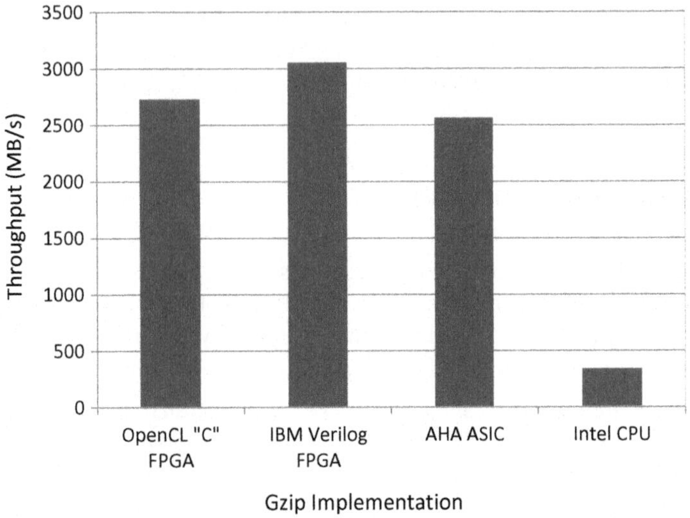

Fig. 6.9 Comparison against commercial implementations of Gzip on FPGA, ASIC and CPU

To evaluate the Altera OpenCL compiler against traditional RTL design, we compare our OpenCL implementation of DEFLATE to the Verilog implementation of IBM on which our design architecture is based [MJA13]. The entire gzip application was developed in 1 man-month, although the development time of the Verilog implementation is not known, we expect it to be substantially longer.

The OpenCL generated circuit has a performance of 2.66 GB/s running at 179 MHz. This is only 10 % slower than the 2.98 GB/s at 200 MHz achieved in Verilog. In addition, the OpenCL logic occupies 51 % of the logic and 68 % of the RAM on the SV A7 device. Based on the chip image in [MJA13], we expect the Verilog implementation to occupy 45 % logic and 45 % RAM. This yields an OpenCL overhead of 6 % logic, and 23 % RAM.

Compared to Verilog, performance only drops by 10 %, and even though area is increased by 25 % we believe OpenCL makes a compelling alternative to hardware design. Similarly to how standard-cell ASIC design flow is typically used instead of full-custom layout for microelectronic circuits, we believe that hardware designers will migrate to the use of high-level languages like OpenCL for most designs. With OpenCL, this kernel was coded in 1 week and optimized in the following 3 weeks. The OpenCL SDK allows a programmer to develop this full application without any hardware skills nor tools, it is near identical to writing software code. In addition the C-like code allows for better readability, quicker algorithmic changes, and portability across platforms over a register transfer level description. Moreover, the emulator and profiler make OpenCL development easy to test and optimize as described in the following sections.

6.6 Development Environment

The Altera OpenCL SDK provides a complete set of software design tools: (1) OpenCL emulation on a CPU; (2) reports on area and performance; (3) compile and execute development environment; and (4) in-system profiling tools for analyzing throughput bottlenecks and memory behaviour. Each of these are described in more detail below.

6.6.1 Emulation Environment

The emulation environment executes complete OpenCL applications on a CPU. Since this doesn't involve any FPGA compilation, the turnaround time is less than a minute. CPU emulation allows a user to validate their Altera OpenCL code while supporting Altera specific extensions such as channels and heterogeneous memory. Using this flow, a software programmer can quickly iterate and converge on legal and correct OpenCL code.

6.6.2 Area and Optimization Reporting

The `aoc` compiler emits valuable reports describing how much of the total FPGA area is consumed by the device, as well as which lines of code were optimized (or why they were not). Users can use this to iterate on their design without long FPGA compile times. Based on the area estimate, a user may opt to scale up or scale down their design. The optimization report would also guide the user to employ code transformations or attributes for breaking dependencies that force the `aoc` compiler to be conservative and hence enable it to generate more optimized hardware.

6.6.3 Hardware Execution Environment: Compile and Execute

The differentiating feature of the Altera OpenCL SDK is the push button compile and execute development environment which converts an OpenCL-written software program to an FPGA programming file implementing an optimized datapath clocked at its maximum operating frequency alongside pre-verified post-fit I/O subsystem logic. This enables software programmers to directly execute the compiled output in hardware allowing them to realize and optimize real end-system level metrics such as throughput and latency. However doing so requires insight into system-level behaviour and bottlenecks. Users can employ the hardware profiler delivered with the Altera openCL SDK to query various components of the system-level execution.

6.6.4 Hardware Profiling Environment

FPGA systems are typically optimized based on RTL simulation, logic analyzer output, and deep knowledge of the cycle to cycle behaviour of each component in the system. Software programmers need a similarly effective optimization path but without having to learn or interact with hardware design tools. The Altera OpenCL SDK provides an in-system profiling capability. User's can simply pass the `-g` flag to the `aoc` compiler and instrumentation hardware will be automatically inserted into the design. When the design is executed in hardware, the host runtime will automatically query the instrumentation hardware and assemble a range of insightful measurements that characterize the current system behaviour. This is distilled and presented to the user via the profiler GUI which annotates the user source code highlighting bad memory access patterns, frequent stalling, high cache miss rates, etc. In addition the GUI will present the user with a system-level timeline of kernels that were running and memory transfers that were in-flight.

6.7 Summary

The Altera OpenCL SDK empowers software programmers by unlocking FPGA acceleration technology with a full software-like compile-and-execute development flow augmented with a range of productivity tools for debugging and profiling. Programming is done in a standard C-based heterogeneous programming model with no need for tool specific attributes or semantics. Software programmers can go directly to in-system hardware realizations without any FPGA design knowledge and without an FPGA designer in the loop. This truly enables FPGA use by software programmers, enabling not only design entry, but also design iterations towards optimizing and solving real world system level problems.

Chapter 7
Big Data and HPC Acceleration with Vivado HLS

Moritz Schmid, Christian Schmitt, Frank Hannig, Gorker Alp Malazgirt, Nehir Sonmez, Arda Yurdakul, and Adrian Cristal

7.1 Introduction

Recent years have seen a new generation of HLS tools, which do not only allow to generate hardware architectures from hardware behavioral models, but perform synthesis starting from algorithms specified in HLLs. One of the reasons for this development is the ever growing popularity of reconfigurable logic, which aims at providing the performance and energy efficiency of integrated circuits at a flexibility that is very close to software. Despite FPGAs being a very attractive hardware target for the acceleration of computations, the programming still often requires knowledge of hardware structures and the associated Computer Aided Design (CAD) tools for implementation. Being able to use a familiar language for algorithm specification makes developing for an FPGA platform as target more approachable to algorithm designers.

There has been a lively discussion going on, about whether programming languages, such as C or C++ are suitable languages for the specification of hardware implementations [Edw06]. Often mentioned is the fact that these languages are intended to specify programs that are executed in a sequential manner on a microprocessor and lack proper mechanisms to describe the behavior of parallel hardware.

M. Schmid (✉) • C. Schmitt • F. Hannig
Friedrich-Alexander-Universität Erlangen-Nürnberg (FAU), Erlangen, Germany
e-mail: moritz.schmid@fau.de; frank.hannig@fau.de

G.A. Malazgirt • A. Yurdakul
Bogazici University, Istanbul, Turkey

N. Sonmez • A. Cristal
Barcelona Supercomputing Center, Barcelona, Spain

Centro Superior de Investigaciones Cientificas (IIIA-CSIC), Barcelona, Spain

Universitat Politecnica de Catalunya, Barcelona, Spain

D. Koch et al. (eds.), *FPGAs for Software Programmers*,
DOI 10.1007/978-3-319-26408-0_7

Also, arbitrary precision arithmetic, one of the key advantages of hardware, as well as data streaming and pipelining are not supported by such languages. Moreover, these languages feature capabilities that are difficult to realize in hardware, such as dynamic memory allocation, recursion, and polymorphism.

A very successful approach to allow HLS from C-based languages is (a) to make only a few restrictions to the host language and (b) provide extensions to support the modeling of hardware features. Tools following this approach can often synthesize almost any purpose into hardware. Since the compiler requires extensive user-controlled guidance, detailed knowledge of the underlying hardware target is yet essential to achieve a high quality of the synthesis results, such as high throughput, low resource requirements, or efficient use of the target platforms resources.

One of the many commercial tools that follow this approach is Vivado HLS from Xilinx as part of Vivado Design Suite. It is based on the acquisition of AutoESL Design Technologies, a spin-off from UCLA whose HLS tool AutoPilot [ZFJ+08] became the basis of Vivado HLS. It allows design entry in C, C++, and SystemC specifically for Xilinx FPGAs [Xil15e]. The tool is aimed at synthesizing individual functions into IP Cores in either VHDL, Verilog, or SystemC (although SystemC support has been dropped since the most recent version). Xilinx provides libraries for challenging problems and a partial port of the Open Source Computer Vision (OpenCV) [PBKE12] framework to aid designers. Moreover, the tool can make use of the extensive library of Xilinx IP cores, for example to support floating-point arithmetic. Specific focus has been laid on interface synthesis, which in addition to general purpose communication also supports ARM's AXI4 standard to integrate generated hardware modules with the FPGA fabric.

7.2 Background on Vivado HLS

Over the past decades, the FPGA community has seen the move to ever higher abstraction levels. Each new abstraction layer aims at hiding design complexity, which offers increased productivity at the cost of less visibility of the challenges associated with the lower abstraction level. The current move from behavioral to functional design specification brings the benefit of accelerating the overall design cycle including expedited design simulation, automatic correct-by-construction RTL creation and easy design space exploration. C-based design entry is a very popular mechanism to create functional specifications and many engineers are familiar with either C or C++. Most of the language constructs are supported, except for recursion and those involving dynamic allocation of memory, which is known to be challenging on hardware.

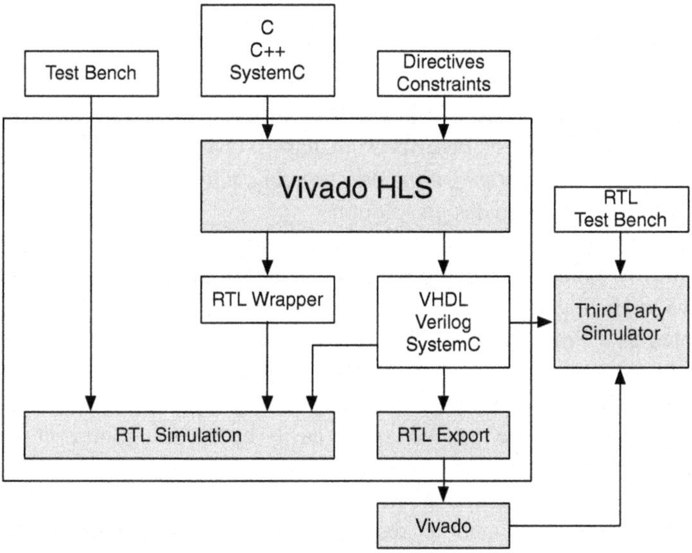

Fig. 7.1 Design flow in Vivado HLS

7.2.1 The Vivado HLS Design Flow

Figure 7.1 shows an overview of the design flow in Vivado HLS. Design entry can be made using either C, C++, or SystemC. Additionally, it is possible and recommended to specify a functional, self-checking test bench, the design can be verified against before the actual synthesis. Moreover, the functional test bench is used to craft the RTL simulation for the synthesized design. RTL sources are available for use outside of Vivado HLS after the synthesis, although it is also possible and often useful to use the RTL export function of the tools, which can make the synthesis results available in the Vivado IP core library, as well as generate an IP core for System Generator or Xilinx EDK. The contained RTL simulation can be used for design validation, however, not to find where errors might stem from. A third party external RTL simulator can be run on the generated output of RTL synthesis in conjunction with a custom RTL test bench. Moreover, there exists an almost one-click simulation setup. Exporting bundles the RTL source files together with a script to import these into a Vivado synthesis project. This script can be edited to also include a custom RTL test bench and setup an external simulator. In Vivado it is then only a single click to start the simulation.

High-Level Synthesis

Synthesis of a functional specification in C to an RTL description for hardware implementation involves not only the synthesis of the specified algorithm but also the synthesis of an interface to the hardware module. The combination of both can lead to a plethora of different designs. The individual steps to be taken during high-level synthesis can be categorized into control and data path *extraction*, *allocation* of suitable resources, *binding* of the elements of the control and data path to the resources, and *scheduling*, as described in Chap. 2. Over the course of these steps, design *constraints* and *optimizations* can severely influence the available *design space* and thus the achieved design solution.

The first step in HLS is to extract control information from the functional specification by expecting the control flow of the program, i.e., conditionals and loops. Vivado HLS achieves this by extracting a control flow graph, which is then implemented as a finite state machine in hardware. For the initial extraction, it is sufficient to assume single cycle instructions. In further steps, the control logic will most likely have to be adapted according to allocated resources and interfaces. Data path extraction can be achieved rather easily by unrolling loops in the design specification and evaluating conditionals.

Scheduling and binding are the two most essential steps during HLS. After instructions have been bound to resources for execution, a schedule can be determined to plan when which instruction is executed. The order might also be reversed by first determining a schedule for execution, for example, by specifying very strict timing constraints, and performing allocation and binding according to the schedule. For this, Vivado HLS always requires the user to specify a basic set of constraints, consisting of the targeted hardware platform and clock period. It is up to the user to constrain the design further using so-called *directives*. Amongst others, directives can be used to specify a certain interface, how loops are to be handled, and which resources are to be allocated for certain instructions.

Unsupported C Language Constructs

Vivado HLS is able to synthesize a large subset of the C modeling standard. However, there are several language constructs, that are not supported for high-level synthesis. One requirement is that the C function must contain the entire functionality and must not rely on any calls to system functions or the underlying operating system. Moreover, C constructs must use a fixed, bounded size and the implementation must be unambiguous. Therefore, Vivado HLS cannot synthesize any system calls, such as printf(), time(), sleep(), etc. Some of these functions are supported to simulate the design with the functional test bench, however, they are ignored for synthesis. As memory cannot be allocated dynamically during runtime, all memory requirements must either be outside of the scope of the function or their fixed bounds must be known at compile time. In general, pointer casting is only supported for native C types. Furthermore, the C++ STL is not supported.

Arbitrary Precision Data Types

C-based native data types are only on 8-bit boundaries. RTL on the other side can support arbitrary lengths. HLS includes a mechanism to support arbitrary precision data types. This is crucial for area optimization, as it is suboptimal to be forced to use a 32-bit multiplier in a design, where a 17-bit operator would be sufficient. HLS provides such data types for C and C++, as well as it supports the data types of SystemC. If arbitrary precision data types are to be used in C functions outside of Vivado, the Vivado compiler `apcc` must be used to compile the source code. Although standard compilers, such as GNU Compiler Collection (gcc) can compile the sources, the produced executable will not be able to handle the compilation directives. These restrictions do not apply to C++, as the custom data types are implemented as a class, which can be compiled correctly by standard compilers. For C++ and SystemC, Vivado supports fixed-point data types and provides an appropriate library. Arbitrary fixed-point precision can be defined by specifying the overall word-length, the amount of bits for the integer part, as well as instructions to control rounding behavior. In addition, HLS supports the synthesis of single and double precision floating-point data types.

Interfaces

Interfaces for RTL designs synthesized by Vivado HLS are generated from the arguments of the target function. The `interface` directive can be added to enforce a certain interface protocol, however, supported protocols greatly depend on the type of the argument. Vivado differentiates between C function interfaces and block interfaces. C function interfaces define how data is passed to the synthesized function, whereas block level interfaces apply an interface to the generated IP core, which controls, for example, when to start execution or when the block is ready to accept new data. Function interfaces are applied to individual function arguments, block-level interfaces are applied to the function itself.

7.2.2 Optimization and Synthesis Guidance

Vivado HLS provides the user with several options to optimize the design in terms of resource usage and performance. Optimizations can be categorized into function optimization, meaning to optimize the execution of functions in relation to each other and optimization of the elements of a function, such as loop constructs and storage elements, for example, arrays. Optimizations are specified as *directives* and can be entered either as *pragmas* directly in the source code or by using a directives script file. Both options have their reasoning, as pragmas might be used for directives which always apply, for example, loop unrolling, and directives which might change, such as pipelining with a certain initiation interval, can be easily grouped together in the script file.

Performance Metrics

Vivado HLS uses two metrics to describe design performance, namely the *latency* and the *initiation interval*. In general, latency specifies how many clock cycles it takes to complete all of the computations of a module. If the algorithm contains loops, the *loop latency* specifies the latency of a single iteration of the loop and the *Initiation Interval* (II) determines after how many cycles a new iteration can be started. If the iterations of the loop are executed in a sequential manner, meaning that an iteration has to finish all computations before the next iteration can start, the Initiation Interval (II) is equal to the loop latency. In this case, the total latency, or as named in Vivado, the loop interval, lat_{total} can be computed by multiplying the iteration latency lat_{it} by the number n of iterations to be performed, also known as the *trip count*. An additional latency must be added for entering and exiting the loop, as well as performing tasks outside of the loop, which we account for as lat_{extra}. In other words,

$$lat_{total} = (n \cdot lat_{it}) + lat_{extra}.$$

For large values of n, the extra latency may become negligible. However, loops in algorithms can often be parallelized in order to decrease the overall latency. In this case the II might become significantly smaller than the iteration latency which may decrease the overall latency significantly,

$$lat_{total} = (n \cdot II) + lat_{iter} + lat_{extra}.$$

The II can thus also be seen as a measure of throughput of the generated RTL design. The antagonist of parallelization and throughput optimization are area constraints. To be able to execute in parallel, operators must either be replicated or at least pipelined. Both increase the hardware resource requirements of the generated RTL. A brute force optimization strategy is to unroll all loops of the algorithm and provide dedicated operators and data storage for each iteration. In consequence, this will also demand a high amount of resources. Fortunately, HLS enables strategies to speed up the execution of an algorithm while keeping the resource demands at a minimum. In addition to demanding a certain clock frequency for the design, users can specify latencies across functions, loops, and regions. Moreover the amount of resources can be constrained and data dependencies can be relaxed, for example, by permitting read before write.

Function Optimization

There are several possibilities to guide the synthesis of subfunctions within a function. An optimization also known from traditional C and C++ programming is function *inlining*, which copies the function code to the location where the function is called. The concept is ideally suited for small functions and for those, which are shared amongst several other functions. In terms of Vivado HLS, using the `inline` directive on a function puts it within the context of its encapsulating function, thus, enabling optimization during synthesis. Another useful directive is function `instantiation`, which can be used on functions with a complex control set to split these up into smaller simple functions and instantiate these directly where they apply. Moreover, function synthesis can be guided by constraining the minimum and maximum number of cycles for the function latency, using the `latency` directive, as well as by choosing a specific interface through the `interface` directive.

Array Optimizations

As memory dimensions must be fixed, arrays are the preferred structure for data storage and may thus have great influence on area and performance. As a consequence, there are several optimizations available that can be applied to arrays. As arrays are typically implemented using a Dual-Port-RAMs (DPRAMs), most directives are concerned about how the C structure is implemented in hardware. The `resource` directive makes it possible to select a specific DPRAM implementation in RTL, such as, for example, one using asynchronous read but synchronous write, or a traditional true dual-ported implementation. Furthermore, small arrays may rather be implemented using registers or a larger multi-dimensional array might be split up and implemented using several DPRAMs. For these cases, the `partition` directive can be applied to control the implementation. Also, the contents of an array can be reorganized using the `reshape` directive. This is useful if, for example, multiple consecutive values should be read from an array. The array can be *reshaped* and the contents can be read in parallel using a larger word-width. Finally to use hardware resources more efficiently, multiple small arrays can be mapped together into a single DPRAM component using the `map` directive.

Loop and Function Optimizations for Parallelization

A crucial part of the optimization of functions is concerned about loops and sequential function calls to exploit *temporal parallelism* by overlapping their execution, also known as *pipelining*. Pipelining a loop nest on a specific level requires that all lower-lever loops are completely unrolled. Unrolling a loop can be instructed by the `unroll` directive and aids in exploiting *instruction-level*, or *logical parallelism*, by generating dedicated hardware for each operation in the loop, which results in the highest hardware requirements, but also the highest performance. Completely unrolling is advisable only for the innermost loops of a loop nest, such as a convolution kernel. Moreover, unrolling requires the loop bounds to be known at compile time. Higher-level loops can be pipelined by the `pipeline` directive. In contrast to unrolling, the operators will be shared across the iterations of the loop, resulting in substantial speedup compared to sequential execution at a very reasonable increase in hardware cost. To ideally prepare loops for optimization, loops should be specified *perfectly* instead of *imperfectly*. Consider the imperfectly specified loop construct to implement a shift register in Listing 7.1. To *perfectly* describe the loop, it should be rewritten as shown in Listing 7.2. Perfectly specified loops may be optimized better than imperfectly specified loops, especially when Vivado HLS analyzes loop-carried dependencies. However, it may also occur that Vivado HLS determines dependencies within a loop (*intra*) and between loops (*inter*) too conservatively and prevents the level of optimization that would actually be possible. For these occurrences, Vivado HLS provides the dependency directive to actively set dependencies to `false` if they incorrectly prevent loop optimization. Moreover, if loop bounds are dependent on a function argument, Vivado HLS is currently unable to determine the trip count, that is, the number of iterations of the loop. In such cases, HLS can be informed about the maximum number of iterations using the `tripcount` directive to specify a lower and an upper bound on a loop. In addition, separate sequential loops can be merged to reduce overall latency and improve operator sharing and nested loops can be collapsed into a single loop using the `flatten` directive.

Listing 7.1 Shift register implementation using an imperfectly specified loop

```
int A[7]; // this is the shift register
int new_data; // the image data
// implement the shifting as a loop
for(int i = 0; i < 6; i++){
  A[i] = A[i+1];
}
A[6] = new_data;
```

Listing 7.2 Perfectly specified loop in shift register implementation

```
for(i = 0; i < 7; i++){
  if(i == 6)
    A[i] = new_data;
  else
    A[i] = A[i+1];
}
```

If a top-level function contains separate sequential loops or sequentially called subfunctions, such as for the description of an algorithm on the application level as a sequence of operation level steps, the loops and calls can be pipelined and paralellized by using the `dataflow` directive. Dataflow creates a pipelined architecture, thereby synthesizing each called function and executed loop into an individual module. It also optimizes the communication between the functions and can improve the design throughput, but it leaves the specification and generation of storage elements between the pipelined modules to the designer.

Chaining Functions Using Data Streaming

As a top-level function may include several sub-functions, Vivado HLS provides the `stream` class to provide streaming-based data communication. Stream objects can be synthesized as simple registers, however, they can also be targeted towards FIFO buffers, if they need to retain a greater number of data items. Functions interconnected by streams are usually scheduled to execute sequentially. If concurrent execution is desired, the `dataflow` directive can be specified in the top-level function. An essential requirement for pipelining the execution across multiple sub functions is that for each stream, there is only a single producer and a single consumer. Choosing the right amount of data items to buffer to avoid dead locks in concurrent execution is the responsibility of the designer.

7.3 Case Studies

Due to high performance requirements and more and more limited energy budgets, FPGAs are very interesting accelerators for data centers. To demonstrate how Vivado HLS may aid designers in productive development of highly efficient accelerators, we present case studies for two different application domains. Section 7.3.1 will illustrate the acceleration of data base queries and can achieve speedups of up to 140× compared to traditional, software-based solutions. Section 7.3.2 will demonstrate the development of a multigrid solver for scientific computing.

7.3.1 Query Processing Acceleration Using Vivado HLS

About three exabytes of data is created and stored in databases each day, and this number is doubling approximately every 40 months. Querying this enormous amount of data has been a challenge and new methods have been actively researched. In this section, we present the design of hardware accelerators to speed up database analytics for in-memory databases.

Many previous studies, in order to have faster query processing capabilities, have looked into accelerating database analytics in hardware: using ASICs [WLP+14], or using FPGAs statically [CO14, PPA+13, MTA10], or using dynamic reconfiguration [KT11, BBZT14]. In this section we look into how such accelerators can be designed using the Vivado HLS tool.

Traditional accelerator design requires writing complex RTL code that is prone to errors and difficult to debug. HLS uses high-level software implementations of algorithms. In this section, using Vivado HLS, we design hardware accelerators for data filtering, aggregation, sort, merge and join operations for a Virtex 7 FPGA. We describe the design, implementation, and the tradeoffs of employing each accelerator module. Later, we use these modules to simulate an in-memory

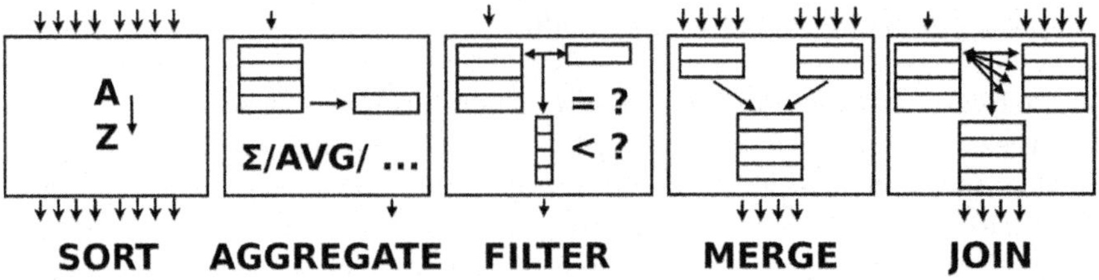

Fig. 7.2 Implemented database operations

database accelerator. We present performance and area results of simulating two TPC-H benchmarks and compare the results with a modern Database Management System (DBMS), PostgreSQL software implementation. We demonstrate more than two orders of magnitude speedup, which provides preliminary evidence that HLS is a very good match to design database accelerators [MSY+15].

Accelerator Implementation

In order to support full and complex database analytics in hardware, we have focused on accelerating data filtering, arithmetic, logic, sorting, aggregation and equi-join operations, as shown in Fig. 7.2. All the units are designed to work at 200 MHz on a Virtex 7 XC7VX690TFFG1761-2 FPGA.

- **Data Filtering, Arithmetic and Logic Operations:** Database filtering operations are relational operations that test numerical or logical relations between columns, numerical and/or boolean values. For this purpose, we have designed a pipelined, parametrizable width, n-way compute engine that takes rows as inputs, applies a filtering operation on the desired columns and produces an output bitmap. This bitmap determines the selected rows for further processing after the filtering operation. The main importance of filtering operations in a Structured Query Language (SQL) query is to filter out unwanted data from further processing, thus reducing the size of the input set. The most important design choices of filtering operations is selecting the correct parallelism for the maximum utilization of memory bandwidth. Similarly, we have designed pipelined, parametrizable, n-way arithmetic and logical compute engines. The arithmetic compute engine supports integer ADD, SUB, MULT and DIV operations, whereas the logical compute engine supports the logical AND, OR and NAND operations.
- **Aggregation Operations:** Aggregation operations compute a single value from a collection of values and we provide an n-way aggregator engine that supports MAX, MIN, COUNT, SUM and AVERAGE. It takes multiple values in the form of an array as the input and calculates a final aggregate value. We apply the binary fan-in method which is fully pipelined at each stage. The number of pipeline stages is dependent on the input array size such that $\log_2 \mathit{arraysize}$ stages should

be designed. The advantage of the binary fan-in method is that Vivado HLS can easily map equations to two-input comparators and support full pipelining.

- **Sort and Merge:** Ordering data is widely used in queries for data organization and presentation. Therefore, efficient sorting is highly sought in database analytics. For this reason, we have implemented a 64-way Bitonic sorting network [Bat68], which is highly efficient on FPGAs, providing high throughput, while using acceptable amounts of real estate.
  A Bitonic network, which is fed with unordered data from RAM, can only partially sort a large data set and is not adequate for high volumes. Therefore, partially sorted values are written back to RAM. This first pass creates arrays of partially sorted data and these arrays must be merged for global sorting. For this purpose, we designed a 4-deep, 2-way merger that can output 4 values each cycle, that is similar to the merge unit presented in [CO14]. It uses 14 comparators, and was written as a long case statement (16 possible output combinations for a 2×4 array). The depth of four of the merger unit has allowed to utilize the available bandwidth without consuming too much FPGA area, hence it is possible to use multiple instances given higher bandwidth specifications. Although larger mergers could be designed, the amount of comparisons would superlinearly increase.
- **Table Joins:** A join operator is used to combine fields from two tables by using values that are common to each. Joining two tables is a very time consuming operation, especially when dealing with decision support systems. Table joins account for more than 40 % of the total execution time while running TPC-H queries [WLP$^+$14].
  We opted for implementing the merge-join algorithm because of two reasons. First, we already have a highly-parallel sorting network which the merge-join requires. Second, the sorting network and merge unit can be highly parallelized through HLS. In contrast, a hash table might require managing collisions, and other complicated circuitry which are difficult to express in HLS. Another option, nested loop joins are more suitable for software implementations. For these reasons, we have designed a parallel merge-join block (for equi-joins) that can match and join a window of $n \times n$ rows on two given columns by comparing sorted rows of database tables in parallel.
  In order to perform a merge-join, we first need to sort and merge both the input tables by using the sort and merge modules. Later the join operates on the sorted tables. Similar to our merge unit, our join operator can join four rows from two separate tables. Our experiments have shown that increasing the number of rows in the merge unit presents different latencies and area consumption. In our design a 4×4 join operator has a 4-cycle latency.

All the accelerators were coded from scratch using C++ for Vivado HLS. The development process has been very rapid compared to RTL. The sorting network is also implemented as a large network of comparators and the merge block is coded as a long case block. Compared to RTL implementations, each C/C++ implementation of the accelerators have fewer lines of code. The major difference between the full software implementation of an accelerator and its Vivado HLS version are the directives, custom data widths and file Input/Output (IO). Vivado HLS directives manage compiler optimizations and synthesize the hardware accordingly. Similarly, custom data types allow fine-grained width control on variables on the synthesized hardware.

Hardware generation in Vivado HLS is controlled through optimization directives. For our hardware accelerators, we apply several directives. The `array_partition` directive is used to partition input data into registers. This has prevented data access bottlenecks due to DPRAMs. The `inline` directive is used to remove the function hierarchy of the C++ language. All loops in the design are unrolled using the `unroll` directive. The level of unrolling is calculated based on the memory bandwidth requirements. We used the `pipeline` directive in order to reduce the II to one, so that all accelerators can process data in every clock cycle.

Running Full, Complex SQL Queries

To test our engine, we ran example queries from TPC-H, a decision support benchmark, which illustrates decision support systems that examine large volumes of data, execute queries with a high degree of complexity, and give answers to critical business questions [Tra08]. In this work, we've run TPC-H queries 6 and 14 [Tra08] which involve filtering, join and aggregation operations. These queries present some different properties: (a) q6 can be fully run without writing intermediate results to memory, while (b) q14 introduces memory overheads, as well as performing a large join operation (of size 200K $\times$ 74K). One exception to a software run is that currently, we do not provide support for floating point numbers, and we only use fixed point arithmetic, however using Vivado HLS, it would be straightforward to provide floating point support. Therefore, we converted the dates into 32-bit timestamps and the often used type of `numeric(15,2)` was converted to fixed-point, similarly to [WLP$^+$14]. In order to run full TPC-H queries, we also needed to design some additional special units such as: multiply-accumulate, divider, or conditional outputs.

In a typical SQL Relational DBMS (RDBMS), a query execution planner does the job of establishing an ordered set of steps that are used to access data. Similarly, a planner in hardware can be devised to schedule operations to modules. The scheduling problem also affects if running multiple queries in flight can be supported or not. In this work, we manually establish the query plan, as we describe in the next subsection. We want to automatize this process in the future.

The input data is organized in memory in table columns. We assume that a given index `n` for column `a` points to the same row for column `b`, i.e. the tuple order is preserved across the columns. To have a realistic simulation environment, we constrained our memory accesses to a maximum of 512-bits each cycle/channel, and with a 16-cycle latency at 200 MHz, which matches the DDR3 specifications on our VC-709 FPGA board (featuring 2 channels of 4 GB DDR3, up to 120 Gbits/s per channel). Although this does not perfectly model DRAM properties, we believe that it can be a representative way of adding memory latency and bandwidth constraints to our model. We used both of the available DDR3 channels, whereas more channels can easily be converted into more performance by laying out the data columns in parallel and thus increasing bandwidth. We also assumed that for exploiting the maximum bandwidth, columns might be distributed into the two channels.

Depending on the query plan, we compose our accelerator modules together and write the intermediate results into memory as needed. A unified memory model enables our accelerator system to be customizable for each query and is capable of expressing complex queries. Combining all modules in a long pipeline would be another alternative design. However, it is not very clear how this design decision would affect performing iterations, such as in a merge-sort scenario.

Experimental Results

In this work, we have used Vivado HLS and ISE version 14.1. As a basis for comparison, we ran the same TPC-H queries on a popular DBMS, PostgreSQL 9.2 running on a 32-core Intel Xeon E5-2670 at 2.60 GHz, with 256 GB DDR3-1600, using 4 channels and delivering up to 51.5 GB/s. We built all indexes, and ran the benchmarks on a RAMdisk to get the best possible performance out of our server.

The SQL queries used are available in [Tra08]. Based on these queries, in Figs. 7.3 and 7.4, we present the query plans, where a sort sign also includes inherent merge steps. We ran TPC-H in the 1 GB scale, and used the two main tables that the queries operate on. The lineitem table has around 6M rows and contains 16 columns that start with a `l_`, whereas the part table (starting with a `p_`) has 200K rows. The numbered steps in the figures represent the execution order of the operations based on the query plan.

Query 6: The three filter operations required were designed to work in a 5-way Single Instruction Multiple Data (SIMD) fashion, after which the resulting bit-maps get AND-ed. If the result is a 1, it gets multiplied-aggregated and finally outputted. This query, compared to the DBMS run, gave the highest speedup because it fits our computational capacity well and completes in a single step since we never need to write any intermediate results to memory.

Query 14: In the first two steps, a 16-way filter operation reduces a 6M element table to approximately 74K rows and writes these into memory. Later, in preparation to the join operation, these columns are sorted on the key, `l_`partkey. Later, the 200K table is sorted on the `p_`partkey in step 4. Finally, these two tables get joined on the keys and two sums are kept, one for those rows that include `PROMO`, and another for all the rows. The `LIKE` keyword is implemented as part of the filter block with simple comparators. Finally, we divide the two aggregated results to get the result.

Table 7.1 shows the area usage and latencies for our generated hardware for each query. The most area consuming operations are the bitonic network due to its highly parallel nature and join unit. Other operations are negligible in size. Vivado maps multiplication and division operations to built-in DSP blocks. However, flip flop

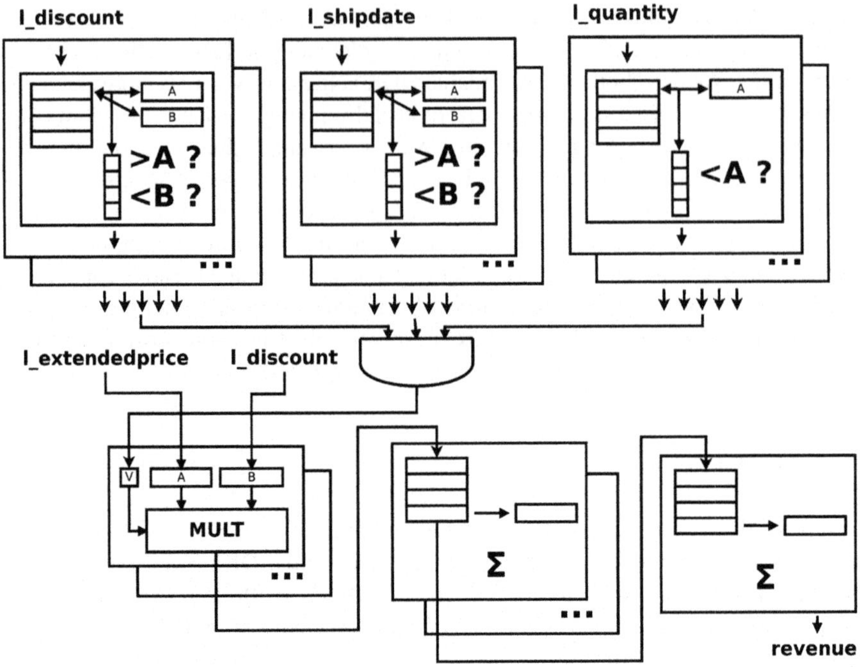

Fig. 7.3 Q6 query plan

utilization is increased when the `array_partition` directive is used aggressively. Vivado HLS has been very efficient while synthesizing Query 6, it uses the least amount of area and provides the highest speedup. Query 14 has a more complex control logic than Query 6 because the merge and join units are not completely data parallel, therefore the state machine has to check for the sequential cases.

Our performance results running these queries are presented in Fig. 7.5. It can be seen that our efficiency in performing n-way filtering and thus working with a reduced set of data to be sorted/joined has paid off, as we were able to achieve between 15–140× of speedup, compared to the DBMS software.

Conclusions

As our results demonstrate while simulating an in-memory database accelerator, the Vivado HLS tool can present high performance gains running complete database queries and is a promising way to address the big data explosion. We have shown between 15–140× speedup compared to Postgres software DBMS running selected TPC-H queries.

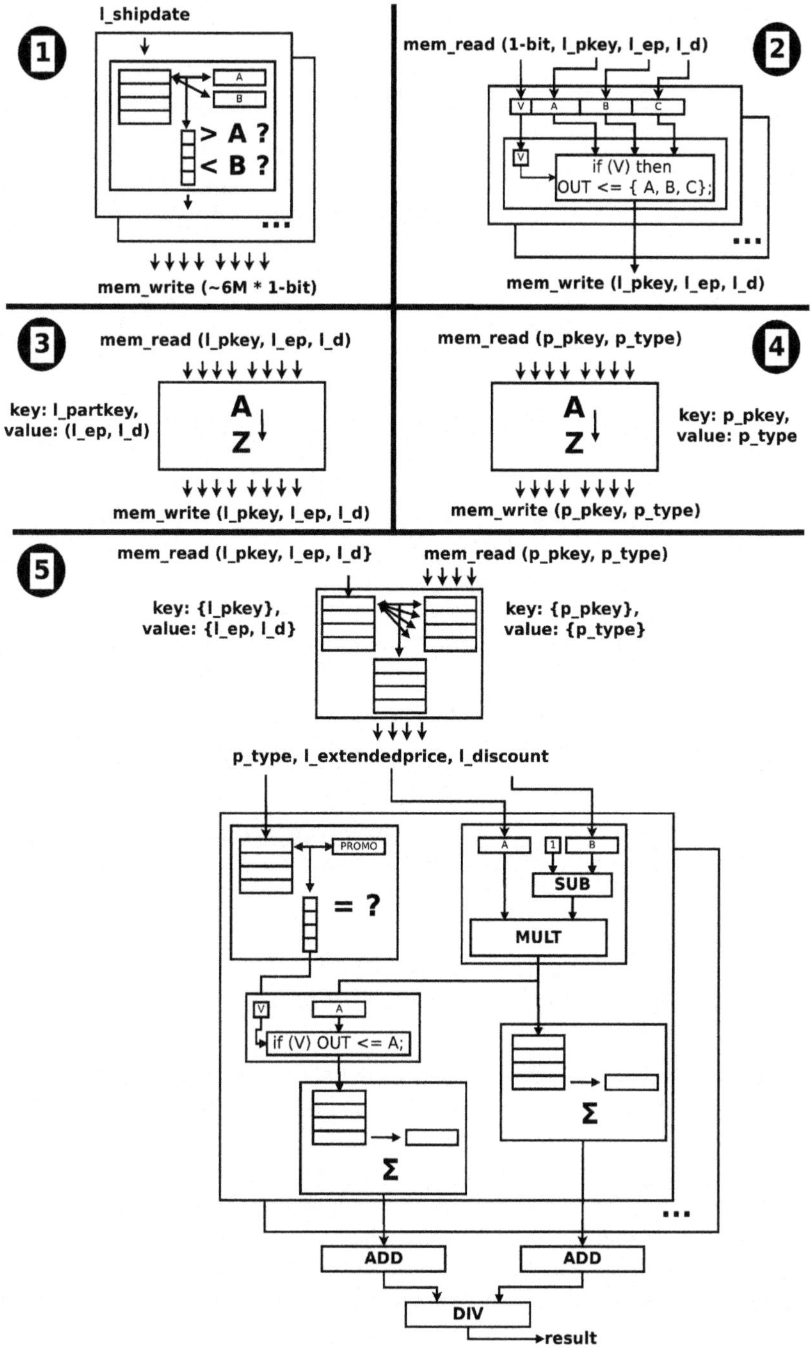

Fig. 7.4 Q14 query plan

Table 7.1 Hardware usage for Query 6 and Query 14

Query 6	LUT	FF	DSP	Latency
5 × 64-bit between	845			0
5 × 32-bit between	415			0
5 × 32-bit LT	195			0
5 × 64-bit Mult		19	80	18
5 × 128-bit sum	384	642		1
1 × 128-bit sum	128			
Q6 Total	1967	661	80	19

Query 14	LUT	FF	DSP	Latency
16× 32-bit between	2704			0
32-bit merge	2613			0
32-bit bitonic64	69216	20491		10
3 × 200-bit comp	1284			0
2 × 128-bit agg	267	265		1
2 × 128-bit agg	267	265		1
3 × 64-bit sub-mult	192	19	48	18
Join 32-bit	42088	10137		9
Q14 Total	118659	31187	48	39

Virtex 7 FPGA	433200	866400	2940	

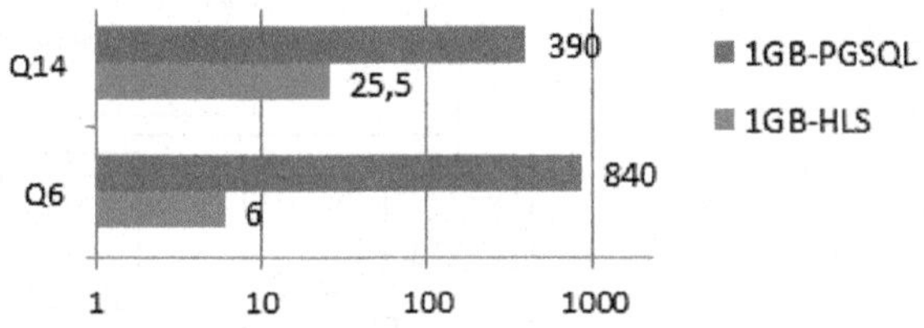

Fig. 7.5 Runtimes for two queries (in ms, log scale)

7.3.2 Multi-resolution in Scientific Computing

Multi-Resolution Analysis (MRA) is a mathematical method that is based on working on a problem at different resolutions. A few among numerous applications of MRA are signal detection, differential equation solving, information retrieval, computer vision, as well as signal and image processing. The algorithms used to solve problems in industry and scientific computing are becoming more and more complex and must deliver enough performance to process vast amounts of data often under rigid resource and energy constraints. Starting at full resolution (base), for each consecutive level a more coarse-grained representation of the data set is created, as shown in Fig. 7.6. On each level, the same computational operations can be applied, affecting a different relative region size, without modifying the filter kernel. The recursiveness in multi-resolution methods and the high degree of parallelism makes these an ideal target for data streaming-oriented FPGA-acceleration.

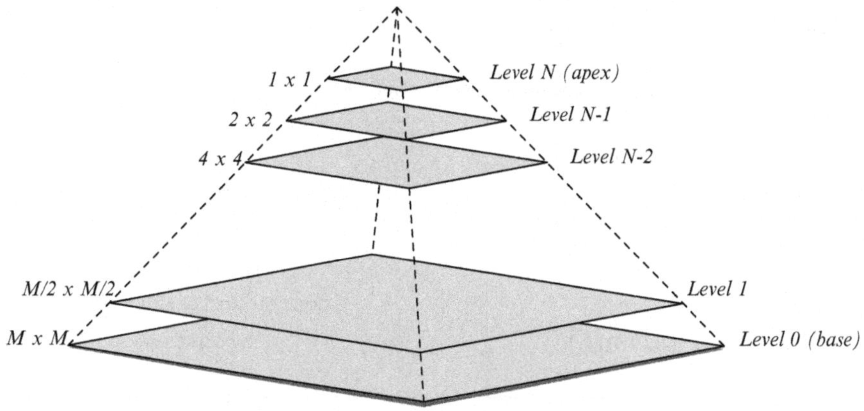

Fig. 7.6 Conceptual representation of multi-resolution data

In scientific computing, MRA is applied in so-called *multigrid methods* [ST82], which are a popular choice for the solution of large systems of linear equations that may stem from the discretization of Partial Differential Equations (PDEs) [GL91]. One of the most researched PDEs is Poisson's equation (refer to [DLYS13]) which is used for modeling diffusion processes, e.g., in the simulation of temperature distributions [Fre96].

The V-cycle, one variant of a multigrid method, is shown in Algorithm 7.1. In the pre- and post-smoothing steps, high-frequency components of the error are damped by smoothers such as the *Jacobi* or the *Gauss-Seidel* methods [Van94]. In the algorithm, ν_1 and $\nu_2 \in \mathbb{N}$ denote the number of applied smoothing steps. Low-frequency components are transformed into high-frequency components by restricting them to a coarser level, thus making them good targets for the smoother once more.

On the coarsest level, direct solving of the remaining linear system of equations is possible due to its low number of unknowns. However, it is also possible to apply a number of smoother iterations. In the case of a single unknown, one smoother iteration corresponds to directly solving the problem.

Such multigrid algorithms are essentially stencil computations, for which the conventional way of implementation is to allocate a continuous chunk of memory and apply the stencils by iterating sequentially over the memory, i.e., a multigrid algorithm is realized as a sequence of smoother, residual calculation, restriction and prolongation stencils. Usually, these kernels are executed linearly, where application of a new kernel starts only after completion of the previous one. As a consequence, to improve the overall performance, kernel execution times have to be reduced.

In contrast, FPGAs offer a massively parallel hardware architecture which can achieve the best results in combination with data streaming. The concept of implementing the multigrid algorithm as a sequence of stencils can be carried over to the FPGA architecture by converting the computational kernels into hardware modules

Algorithm 7.1: Recursive V-cycle to solve $A_h u_h = f_h$

if *coarsest level* **then**
  solve $A_h u_h = f_h$ exactly else by smoothing iterations;
else
  $\bar{u}_h^{(k)} = S_h^{\nu_1}\left(u_h^{(k)}, A_h, f_h\right)$; `{pre-smoothing}`
  $r_h = f_h - A_h \bar{u}_h^{(k)}$; `{compute residual}`
  $r_H = R r_h$; `{restrict residual}`
  $e_H = V_H(0, A_H, r_H, \nu_1, \nu_2)$; `{recursion}`
  $e_h = P e_H$; `{interpolate error}`
  $\tilde{u}_h^{(k)} = \bar{u}_h^{(k)} + e_h$; `{coarse grid correction}`
  $u_h^{(k+1)} = S_h^{\nu_2}\left(\tilde{u}_h^{(k)}, A_h, f_h\right)$; `{post-smoothing}`
end

and laying them out in parallel on the chip. The modules are then interconnected by data streams to form a pipeline, through which data is streamed from one entity to another. Once the pipeline is completely filled, all of the computations are carried out concurrently, providing a continuous output flow of results from a continuous delivery of input data.

A key concept in hardware development is to design a component once and replicate it as often as necessary. For multigrid algorithms, we can make use of this principle by designing one stage of the algorithm and replicate it to implement the recursion levels. An important fact to consider is that the lower stages always only have to process a fraction of the data of the next higher stage, e.g., a quarter in the case of 2D. Although this could be exploited in hardware by lowering the clock frequency of the lower stages, a much more sophisticated approach is to increase the pipeline interval. A high performance and be achieved if the top most level uses a pipeline interval of one, which means the architecture can accept new input data in every clock cycle. In consequence, it also produces results in every clock cycle, after a certain latency. To achieve this, however, a dedicated operator must be instantiated for each operation of the algorithm, and therefore the implementation requires a large amount of hardware resources. If the pipeline interval is increased on the lower levels, hardware operators can be shared among the operations, which leads to significantly lower resource requirements.

Figure 7.7 shows a structural representation of the FPGA implementation of the multigrid solver for the PDE. Here, *restriction* and *prolongation* are the building blocks for grid traversal, that is up- and downsampling the data sets. Downsampling in stream processing is implemented by simply dropping samples. To avoid aliasing, the data must be reduced in bandwidth, for example by Gaussian filtering. For upsampling, the sampling rate is increased by a factor of two and the new samples of value 0 between successive data elements are interpolated. As the resulting spectrum is a two-fold periodic repetition of the input spectrum and only the frequency components of the original signal are unique, the repetitions should be rejected by passing the resulting sequence through a lowpass filter with subsequent upscaling.

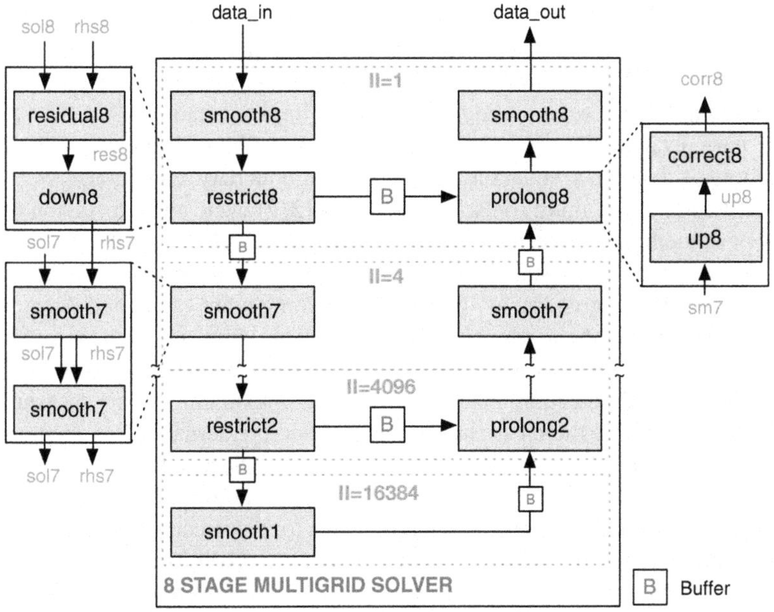

Fig. 7.7 Structural representation of the multigrid algorithm implementation

An alternative to using a Gaussian filter for the implementation is bilinear or bicubic interpolation [TBU00].

The *smoother* is implemented using the Jacobi over-relaxation (JOR) method as a stencil kernel, which is basically a local operator. Local operators process a local neighborhood, also often referred to as *window*, and therefore need to access an element of the data set more than once. Thus, handling streaming data on an FPGA requires a memory architecture to retain data for multiple accesses. Memory resources on modern FPGAs can be broadly categorized into Block RAMs (BRAMs) and Flip-Flops (FFs). An efficient memory architecture for streaming data uses a combination of *line buffers* for storage of complete lines and *memory windows* for the actual processing of local neighborhoods. A more elaborate treatise of local operators in Vivado HLS can for example be found in [SAHT14].

In addition to the actual implementation of each stage, the figure also shows the II for the stages, data streams, and indicates which connections require buffering (depicted as B). Although it is possible to instantiate buffers on every stream, there are actually only three cases where the interconnection requires buffering. These are (a) after downsamplers (part of restrict), (b) before upsamplers (part of prolong), (c) before nodes that combine data streams and have different path lengths. The necessity for buffering after downsampling and before upsampling is due to different II between the stages. The buffer requirement for combining nodes, such as the correction step of the prolongation, becomes evident from the structure of the accelerator. For example, the connection between restriction and prolongation requires a very large buffer, since prolongation must wait until data arrives after having traversed all of the lower stages. Other interconnections in the architecture can be set to simple registered handshake connections, which will lower the total amount of hardware resources required for buffer implementation.

We have evaluated considering a solution to a finite differences (FD) discretization of Poisson's equation with Dirichlet boundary conditions. We have implemented a typical multigrid solver using a $V(2,2)$ cycle and a recursion depth of 8. For smoothing, a weighted Jacobi with a pre-calculated optimal ω is used. The solution on the coarsest grid of 32×32 is approximated by multiple smoother steps. For the chosen grid size of 4096×4096 floating point values and 8 recursion levels, the buffers on the top four levels become very large and would overwhelm the amount of available resources, even on very large FPGAs. A solution to allow the fastest possible execution and keep within the maximum amount of available resources is to offload the most challenging buffers to external DDR3 memory. A drawback is that HLS tools, such as the here used Vivado HLS, do not support this from within the tool, but require an FPGA support design (for example, refer to [SSH+15]). We add input and output arguments for the streams to be externalized to the function definition and specify their type as AXI4-Stream (AXI4S). In this way, we obtain a high-performance interface to the FPGA fabric for each data connection and do not need to make extensive modifications to the actual accelerator source code. Table 7.2 shows evaluation results of the hardware synthesis from Vivado HLS, which give a rough approximation of the resource requirements of the design. Indeed, after externalizing the top four largest buffers for results and Right-Hand Side (RHS), the design can be fit onto the chip. Table 7.3 lists the Post Place and Route (PPnR) results after integrating the modified accelerator into the described FPGA support system and shows that the multigrid solver can be implemented on a Kintex 7 while achieving a maximum clock frequency of over 200 MHz. We have evaluated the PPnR hardware results on the Kintex 7 FPGA to measure the performance in terms of how many clock cycles it takes it takes to process a 4096×4096 grid of floating point values. In combination with the clock frequency of the design, this yields an accurate measurement of the performance. Contrasting to a software solution, the pipelining principle also applies here, thus, it is not necessary to wait until the result is ready, but we can start processing a new grid, as soon as all of the input values of the previous grid have been consumed.

Table 7.2 Vivado HLS resource estimates for the multigrid solver design comparing on-chip and external buffering

Resource	On-chip	External	Available
FF	256368	106311	407600
LUT	880314	177379	203800
BRAM	11812	323	445
DSP48	442	442	840

Table 7.3 PPnR resource requirements of the complete multigrid solver design

LUT	FFs	DSPs	BRAMs	Slices	F_{max} (MHz)
105951	135442	460	808	39147	202.34

Table 7.4 Comparison of the performance of the multigrid solver on different hardware targets

Target	Latency (ms)	Throughput (Vps)
Kintex 7	83.1	12.3
Intel i7	223.1	4.5

Table 7.5 HLS synthesis estimates for implementation of the multigrid solver using double precision arithmetic

FPGA	LUT	FFs	DSPs	BRAMs	F_{max} (MHz)
Kintex 7	140	43	111	124	232.0
Virtex 7	73	29	33	53	229.4

Values are given as percentage of available resource type

Data streaming also allows us to hide the communication time with a host that supplies the input data completely, as current high-speed serial interconnects, such as Peripheral Component Interconnect Express (PCIe) and others, can achieve data rates that exceed the throughput of the accelerator by far.

In order to compare the hardware accelerator to state-of-the-art approaches, we have used the DSL ExaSlang [SKH$^+$14] to generate highly optimized C++ code for execution on a single machine. The evaluation was done on a single core of an Intel i7-3770, which is clocked at 3.40 GHz and features L2 and L3 cache sizes of 1 MB, respectively 8 MB. For the CPU, AVX vectorization was enabled. Table 7.4 lists the performance results in terms of latency in milliseconds for processing a single iteration of the V-cycle and the throughput in terms of how many iterations of the V-cycle can be processed per second ([Vps]), on average. It is also worth mentioning that it is irrelevant to the performance of the accelerator, whether the input data uses only single-precision floating point or is implemented for double-precision input data. Although the chosen mid-range Kintex 7 FPGA cannot provide the necessary amount of logic and memory resources, switching to a larger FPGA, such as a member of the Virtex 7 family, here an XC7VX485T, can easily solve this issue. To underline this, we have changed the data type in the source code for HLS of the multigrid solver from single to double precision floating point arithmetic, for which the estimated resource requirements are listed in Table 7.5.

Conclusions

In this case study, we have presented an approach to map descriptions of multigrid algorithms to hardware designs for execution on FPGAs by using C-based High-Level Synthesis. We have verified our approach by synthesizing a multigrid-based solver for Poisson's equation for a Kintex 7 FPGA and comparing it to an Intel i7 CPU. The evaluation numbers show that employing FPGAs in High-Performance Computing (HPC) is a promising approach to increase computing power, while, at the same, to reduce the energy footprint of clusters.

7.4 Summary

In this chapter, we have provided an overview of the internals of Vivado HLS, as well as how its features can be used to specify hardware accelerators and optimize the synthesis results. Due to its versatility, Vivado HLS is suitable for many different application domains. We have shown two detailed examples in the form of case studies, where the first has illustrated the development of database query accelerators and the second has shown how MRA may be accelerated using HLS. In addition to providing highly efficient solutions, we moreover want to draw attention to the high productivity that is enabled using Vivado HLS, as for both solutions, the required lines of code are far less than for hand-crafted RTL solutions. Despite of all of these benefits it must also be mentioned that developing hardware accelerators using Vivado HLS requires a high degree of familiarity with RTL design and may still impose a steep learning curve for engineers with little to no prior FPGA design experience.

Acknowledgements This work was partly supported by the German Research Foundation (DFG), as part of the Priority Programme 1648 "Software for Exascale Computing" in project under contract TE 163/17-1. In addition, funding from the European Union's Seventh Framework Programme (FP7/2007-2013) under grant agreement No. 318633, the Turkish Ministry of Development under the TAM Project, number 2007K120610 and Severo Ochoa Mobility grant program support was received.

Chapter 8
Source-to-Source Optimization for HLS

Jason Cong, Muhuan Huang, Peichen Pan, Yuxin Wang, and Peng Zhang

This chapter describes the source code optimization techniques and automation tools for FPGA design with high-level synthesis (HLS) design flow. HLS has lifted the design abstraction from RTL to C/C++, but in practice extensive source code rewriting is often required to achieve a good design using HLS—especially when the design space is too large to determine the proper design options in advance. In addition, this code rewriting requires not only the knowledge of hardware microarchitecture design, but also familiarity with the coding style for the high-level synthesis tools. Automatic source-to-source transformation techniques have been applied in software compilation and optimization for a long time. They can also greatly benefit the FPGA accelerator design in a high-level synthesis design flow. In general, source-to-source optimization for FPGA will be much more complex and challenging than that for CPU software because of the much larger design space in microarchitecture choices combined with temporal/spatial resource allocation. The goal of source-to-source transformation is to reduce or eliminate the design abstraction gap between software/algorithm development and existing HLS design flows. This will enable the fully automated FPGA design flows for software developers, which is especially important for deploying FPGAs in data centers, so that many software developers can efficiently use FPGAs with minimal effort for acceleration.

This work was performed while the author J. Cong served as the Chief Scientific Advisor of Falcon Computing Solutions Inc.

J. Cong (✉)
Falcon Computing Solutions, Inc. Los Angeles, CA, USA

Computer Science Department, University of California, Los Angeles, CA, USA
e-mail: cong@falcon-computing.com

M. Huang • P. Pan • Y. Wang • P. Zhang
Falcon Computing Solutions, Inc. Los Angeles, CA, USA

D. Koch et al. (eds.), *FPGAs for Software Programmers*,
DOI 10.1007/978-3-319-26408-0_8

8.1 Motivations

As described in Chap. 2, high-level synthesis or behavioral synthesis can generate the RTL description from a high-level programming language such as C/C++. The automated algorithms are applied to determine and optimize the cycle-accurate behavior of the instruction-level operations like multipliers/adders and fine-grained controls such as branches and loops. These algorithms include operation scheduling, resource allocation, and binding. There are also design directives or pragmas provided by most of the HLS tools to give users the freedom to control and optimize the process of these three algorithms, and the ability to perform front-end preprocessing. However, extensive code rewriting on the programming language level is still required to achieve a good overall hardware design. We will start with an example FPGA design of a MPEG video decoder, as shown in Fig. 8.1. The decoder contains six submodules, and each of them can be described as a function in C/C++. We can implement each submodule using HLS efficiently, but there are many system-level decisions to make in order to achieve an overall good design using the limited resources in a specified FPGA platform. The following are some examples of system-level decisions.

- Scheduling of the submodules: How to change the execution order for parallelism and data locality? How to choose pipeline versus sequential execution, the portion of the system to pipeline, and the data granularity for the pipeline?
- Resource trade-offs between the submodules: Which design option should be selected for each submodule (each option has different performance/area results)? And what is the number of parallel duplications for each submodule?
- Storage allocation for data: For each array, whether to implement on-chip or off-chip? What is the minimal size for the memories after folding, and the address transformation for memory folding?

```
for (m = 0; m < N; m++){
    Parser(I → B, A);
    for (i = 0; i < 6; i++) {
        IQ(A → D); IT(D → E);
    }
    MC(B → C);
    Recon(C, E → O)
    Filter(C, E → O)
}
```

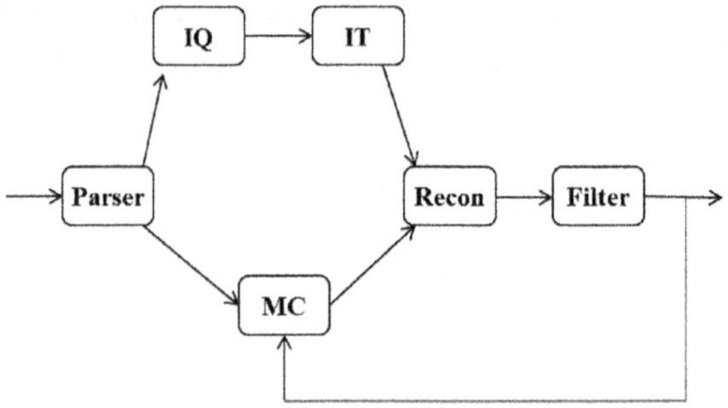

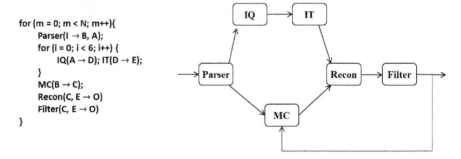

Fig. 8.1 Motivational example for system-level design

- Access efficiency for data: What are the bank number and the partition patterns in memory partitioning for on-chip access parallelism? How to determine prefetching buffers and address sequence optimization of the off-chip memory access?

These system-level decisions and optimizations will significantly impact the system implementation result, but they are not fully automated by the state-of-the-art HLS tools. Most of these choices are related to the low-level hardware design knowledge. Lacking such knowledge, it is a big challenge for software engineers to design for FPGA even with the state-of-art HLS tools. This system-level design challenge becomes more significant when the system scale is large, and even hardware engineers are not able to determine the best choices before the implementation. Therefore, the design abstraction of the source-to-source optimization is required to be higher than traditional HLS to increase the efficiency in design space exploration.

In fact, module-level design space exploration is also nontrivial. For example, the IDCT algorithm can be simply described as a matrix multiplication form in Fig. 8.2a, where in and out are input and output image blocks, and c is the transform coefficient matrix. Different microarchitectures may finally result in different design trade-offs in the implementations of the same functionality. Figure 8.2b–d show the simple pipelined versions with different execution orders of the for-loop iterations in the source code. Case (b) has an iterative data dependence on the accumulation of the partial sum $c[y, x]$, so it cannot be fully pipelined and has less throughput

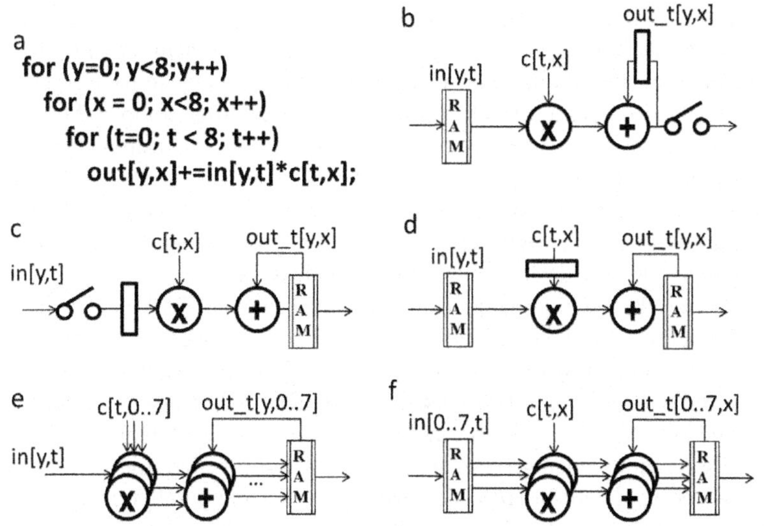

Fig. 8.2 Microarchitecture options for IDCT. (**a**) Original code of IDCT. (**b**) Single pipeline along t. (**c**) Single pipeline along x. (**d**) Single pipeline along y. (**e**) Parallel version of (**c**). (**f**) Parallel version of (**d**)

a
```
for (y=0; y<8;y++)
  for (x=0; x<8; x++)
    for (t=0; t<8; t++)
      out[y,x]+=in[y,t]*c[t,x];
```

b
```
for (t=0; t<8; t++)
  for (y=0; y<8;y++)
    for (x=0; x<8; x++)
      out[y,x]+=in[y,t]*c[t,x];
```

c
```
for (t=0; t<8; t++)
  for (y=0; y<8;y++)
    for (x=0; x<8; x++) {
      #pragma HLS pipeline
      #pragma HLS unroll
      out[y,x]+=in[y,t]*c[k,t];
    }
```

d
```
for (t=0; t<8; t++)
  for (y=0; y<8;y++)
    for (x=0; x<8; x++) {
#pragma HLS pipeline
#pragma HLS unroll
#pragma HLS partition out dim=2 factor=8
#pragma HLS partition c dim=2 factor=8
      out[y,x]+=in[y,t]*c[k,t];
    }
```

e
```
data_type out_t[8][8];
for (t=0; t<8; t++)
  for (y=0; y<8;y++) {
    data_type in_t = *in;
    for (x=0; x<8; x++) {
#pragma HLS pipeline
#pragma HLS unroll
#pragma HLS partition out_t dim=2 factor=8
#pragma HLS partition c dim=2 factor=8
      out_t[y,x]+=in_t*c[k,t];
      if (t==7) *out = out_t[y,x];
    }
  }
```

Fig. 8.3 Code rewrite for the implementation of Fig. 8.2e. (**a**) Original code. (**b**) Execution order change. (**c**) Micro-architecture pragmas. (**d**) Memory partitioning. (**e**) Data reuse and interface handling

than cases (c) and (d). But after the accumulation, case (b) does not need a SRAM buffer to store the partial sum. Cases (c) and (d) can both achieve full pipelining with an initial interval of one cycle. Considering that the image to be decoded is fed in through streaming, case (d) will need an additional buffer compared to case (c) in order to store the input data for reuse in the calculations that follow; but this also decouples the execution of upstream modules. We can then parallelize the designs to obtain cases (e) and (f). At the cost of the input buffer, case (f) does not need to partition the memory of *c*. In general, it is hard for software engineers to explore all these microarchitecture design options in an efficient way.

With a preselected microarchitecture such as the one in Fig. 8.2e in mind, the HLS coding is still not trivial. First, the hardware execution order can be represented by loop transformation as in Fig. 8.3b. Then, microarchitecture intention are conveyed to the HLS tools by inserting pragmas in the correct position of the source code—such as the pipelining and parallelism in Fig. 8.3c. However, in typical cases, these pragmas themselves cannot guarantee the generation of a pipelined or parallelized hardware in HLS. For example, to enable parallel accesses on the arrays c and out from duplicated multipliers and adders, these two arrays need to be partitioned. HLS provides pragmas to implement the array partition, but decisions on how to partition the array must be determined by designers, e.g., the selection of array dimensions and number of banks. If the wrong partition pragma parameters are specified, the parallelization or pipelining optimizations may be degraded or disabled. Figure 8.3d shows the additional array partitioning pragma required by loop pipeline and parallelization. Until now, we could generate the core data path with a parallelized pipeline structure, but it is not finished. Special care is also required for the module interface. For example, input and output are supposed to be FIFO channels. Therefore, not only does the access reference of these ports need to be updated into a FIFO form, but also the access order of the read and write sides for the FIFO channel must be matched (otherwise the system may be dead-locked). Data read from input FIFO *in* will be reused in the successive loop x iterations, so reuse buffer *in_t* is added before loop x. From Fig. 8.3a–e, we can see that implementing a specific microarchitecture efficiently in HLS is not a straightforward task.

In summary, the manual rewrite of source-level C/C++ code is still extensive when using the state-of-art HLS tools in order to fully consider system-level design trade-offs and optimizations, module-level microarchitecture refactoring, and efficient implementation of specific microarchitectures. Therefore, source-to-source optimization and automation is greatly needed to alleviate the design effort and difficulty in these areas and lessen or eliminate the roadblocks for the software designers, allowing them to use FPGA more easily and widely.

8.2 Merlin Compiler Overview

The Merlin Compiler is a source-to-source automation tool for FPGAs developed by Falcon Computing Solutions (FCS). It takes the C/C++ program as input and generates the optimized C/OpenCL program that can be used to generate the executable system by its own implementation flow, or other commercial OpenCL implementation tools like Xilinx SDAccel and Altera OpenCL SDK. The Merlin Compiler is dedicated to automating the code rewrite effort on all three aspects previously described, and providing an easy-to-use and efficient FPGA programming environment for software developers.

8.2.1 Overall Design Flow

The Merlin Compiler is targeted at enabling software developers to work on FPGAs with a fully automated compilation flow—not only with push-button implementation but also with intelligent optimizations. Compared to the existing FPGA design tools, it provides:

(a) Push-button compilation flow mapping of general C/C++ programs into full executable system designs

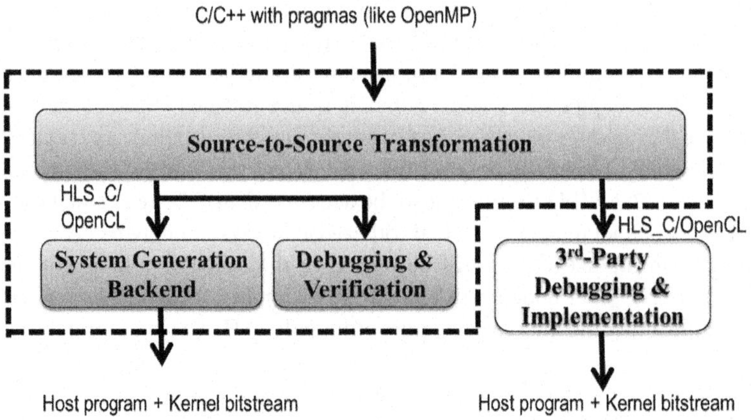

Fig. 8.4 Merlin Compiler overall flow

(b) Automated source-to-source code transformation for micro-architecture optimizations for different granularities, including both system and module levels.
(c) Capability to map the same input code to different platforms from different vendors with corresponding optimizations and coding styles.
(d) Integrated design environment containing system evaluation and debugging at different design abstraction layers.

The overall design flow can be summarized as shown in Fig. 8.4. The input to the design flow is the C/C++ program written by the software designers. A small set of pragmas is defined by the Merlin Compiler to enable users to specify the high-level design intention for the way they want the compiler to generate the design in FPGA. These pragmas are similar to the OpenMP pragmas used in parallel programming in multi-core CPUs, which are simple to understand and easy to use. For example, a user can specify a specific for-loop to be parallelized, and then the process of parallelizing the loop and all the hardware considerations to implement the parallelism efficiently will be handled by the source-to-source compiler.

The HLS synthesizable code is generated by the Merlin Compiler after applying the optimizations for the specified FPGA platforms. Then this generated code can be fed into the OpenCL implementation flows such as Xilinx SDAccel and Altera OpenCL SDK to generate the optimized design directly. The Merlin Compiler also generates the necessary scripts for these implementation tools. Additionally, it provides a built-in system backend and verification functionality to support a wider range of hardware platforms.

Figure 8.5 shows the source-to-source optimization flow. The input source code is first parsed by a C/C++ frontend infrastructure, and an abstract syntax tree (AST) is generated; this provides the interface for the following program analysis and code transformation. In the analysis stage, loop structure and array access information is parsed and recorded; in the modeling stage, a graph-based intermediate representation is established to hide the details for coarse-grained analysis and transformation.

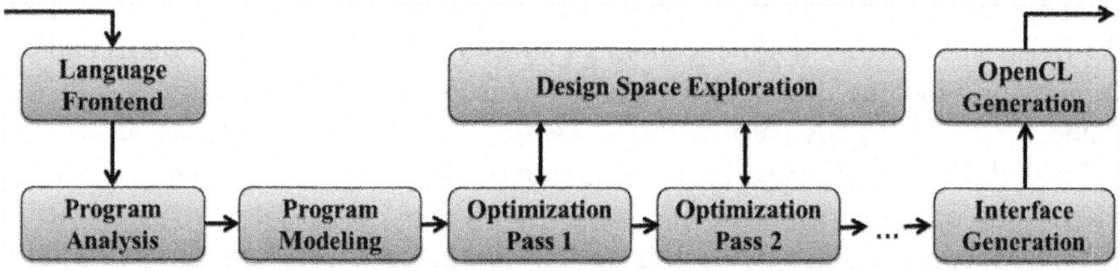

Fig. 8.5 Merlin Compiler source-to-source transformation flow

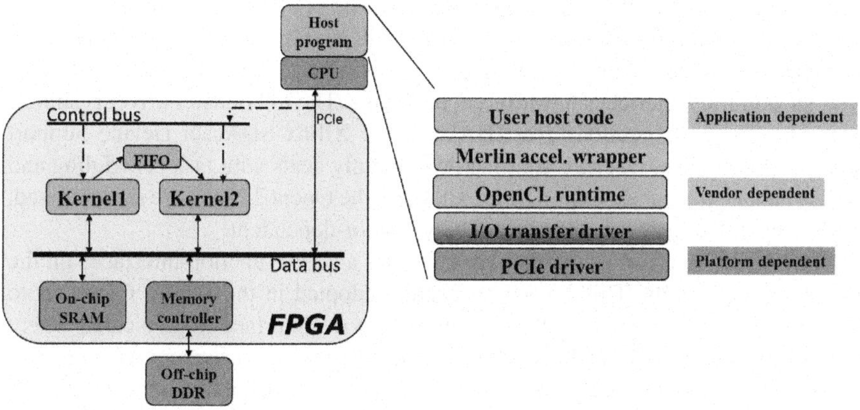

Fig. 8.6 Hardware and software architecture

After multiple optimization stages, the SW/HW interface is generated and finally wrapped into the OpenCL format.

Verification and debugging are always very crucial in a development environment. The Merlin Compiler integrates the automated verification in different design abstraction layers. First, the optimized C program before OpenCL generation is verified with the input C program in CPU execution. OpenCL is then executed by CPU emulation, and then RTL generated by HLS is verified via C+RTL co-simulation. Following this, the on-board hardware execution is performed. The execution in all these levels is invoked automatically via the testing scripts generated by the Merlin Compiler, and the comparison benchmark is established in a uniform way.

8.2.2 *Hardware and Software Architecture*

HLS tools only generates the RTL for the kernel modules. To build a complete executable system, there are several integration tasks for an OpenCL backend. These tasks include: peripheral IP integration and interconnection, bus device address map generation, driver generation, and OpenCL runtime for task queuing and scheduling. Figure 8.6 shows the hardware and software architecture for the Merlin Compiler. The Merlin Compiler contains a built-in OpenCL backend, which enables the portability of the designs between different FPGA device vendors. A CPU is connected to the FPGA card memory (DDR) via PCIe interface. And all kernel logics in the FPGA are connected via on-chip bus (typically AXI bus) to the shared on-chip and off-chip memories.

In the software architecture, a full stack of five layers is generated automatically. The bottom layer is the Linux kernel service to talk with PCIe hardware directly. This layer is typically platform-dependent according to the PCIe slot configuration of the server and the PCIe IP used in FPGA. The IO transfer driver is the layer that hides the platform information from OpenCL runtime. The low-level operation sequences are wrapped into standard single or burst data/control transfer functions with the common interfaces for different platforms. The I/O transfer driver is similar to the hardware abstraction layer (HAL) in the Xilinx SDAccel Device Support Archive (DSA) flow. OpenCL runtime layer mainly deals with task scheduling and buffer management at execution time. Although the OpenCL APIs are standardized, the implementation of the APIs is typically vendor-dependent.

The Merlin accelerator wrapper layer defines a set of function interfaces on the top of OpenCL APIs. This layer is internally adopted in the Merlin Compiler to wrap the detailed OpenCL APIs into pure software-interface library calls, where all the hardware details will be fully virtualized from the designers who use this library. In addition, it enables the common FPGA accelerators, generated once by the Merlin Compiler, to be re-integrated and reused across many applications under any pure software development environment easily and seamlessly. The Merlin accelerator wrapper layer also provides the possibility and flexibility to support the implementation backend tools not based on OpenCL. By separating the software stack into these layers, the portability and reusability of each layer is maximized.

8.2.3 Execution Model

The Merlin Compiler takes C/C++ as the input language to specify the whole design, and pragmas are used to pass the necessary information to the compiler to generate efficient system design, including both software and hardware. A pragma-based approach provides the interoperability for different C/C++ based design tools.

A simple and flexible execution model is adopted in the Merlin Compiler from the perspective of software programmers. The Merlin Compiler has a host/accelerator execution model that targets a typical CPU+FPGA platform. The main thread starts on CPU, and the computation-intensive part is offloaded to the accelerators in FPGA. FPGA accelerator is specified as a function call. Figure 8.7 gives an example code. By default, this function is running in a blocking way in a single thread execution mode, which means the CPU thread will wait for the accelerator function to finish before continuing execution, just like a normal function call in CPU.

The Merlin Compiler execution model is also compatible with other popular multithread execution environments such as OpenMP [Ope15b] and OpenACC [Ope15a], where each acceleration function is executed in the context of one thread in the CPU program. So in this way, the Merlin Compiler allows the flexible dynamic task scheduling between FPGA and CPU tasks, such as parallel CPU and FPGA execution, and parallel FPGA acceleration threads. When multiple FPGA acceleration threads are issued in the host program, they will share the FPGA device in different temporal intervals.

A runtime manager, as shown in Fig. 8.8, provided by the Merlin Compiler tool chain, allows the sharing of FPGA resources among different processes or threads in the CPU+FPGA node. The runtime manager performs a series of scheduling tasks to make the access to FPGA accelerator transparent to the programmers, including FPGA access authorization, FPGA access request queuing, priority management, task distribution among multiple FPGAs, and exception handling.

The Merlin Compiler assumes the distributed memory model used for the host/accelerator execution model. Both CPU and FPGA have their own memory space and, when accessed by FPGA accelerators, need to be transferred back and forth via PCIe connection. The data to be transferred is determined by the arguments of the accelerator function as shown in Fig. 8.9. There are three data transfer modes which correspond to destinations in FPGA: kernel registers, on-chip memory and off-chip memory. Scalar variables or expressions will be passed to the FPGA kernel directly. For arrays, the regions being accessed by the FPGA kernel will be transferred to off-chip memory by default, and these data will then be read back to the FPGA chip during the execution of the kernel. Data can also be streamed to the on-chip buffer for the kernel to use directly. A coarse-grained pipeline pragma at the top level of the acceleration function can enable this kind of data stream from host to FPGA kernels.

8.2.4 Programming Model

The Merlin Compiler provides a high-level abstraction of the optimizations of the microarchitecture of the accelerators. Programmers only need to focus on two high-level concepts in design optimization: parallelization and pipelining. By

Fig. 8.7 Defining an accelerator in host program

```
host() {
    ...
    for (i = 0; i < ...; i++)
    {
        #pragma ACCEL task
        kernel(a, b, c, d);  // a function to accelerate
    }
}
```

```
#pragma omp parallel
{ #pragma omp sections
{ #pragma omp section
  cpu_func(...);
#pragma omp section
#pragma omp for private (i)
  for (i = 0; i < N; i++) {
     #pragma ACCEL task
     fpga_func(...);
}}}
```

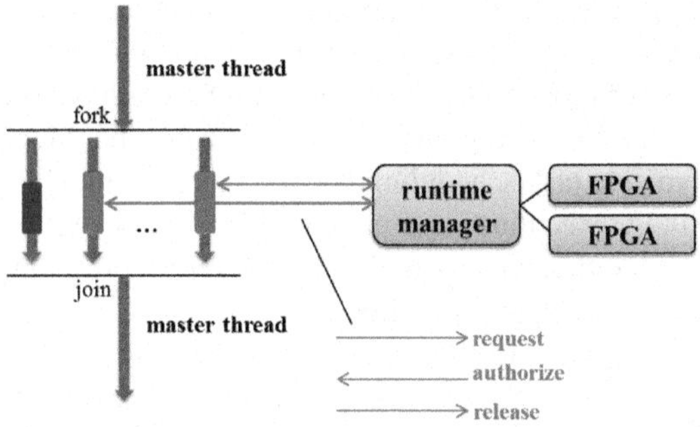

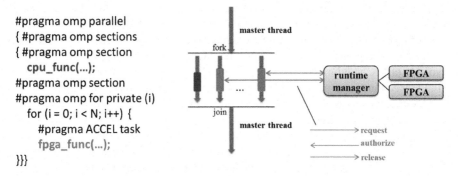

Fig. 8.8 Execution model in a multithread, multi-FPGA environment

```
host() {
    ...
    int a[1024];
    int b[1024];
    float c;
    int d[1024];

    for (i = 0; i < ...; i++)
    {
        #pragma ACCEL task
        kernel1(a);        // streaming

        #pragma ACCEL task
        kernel2(b[0], c, d); // normal
    }
}
```

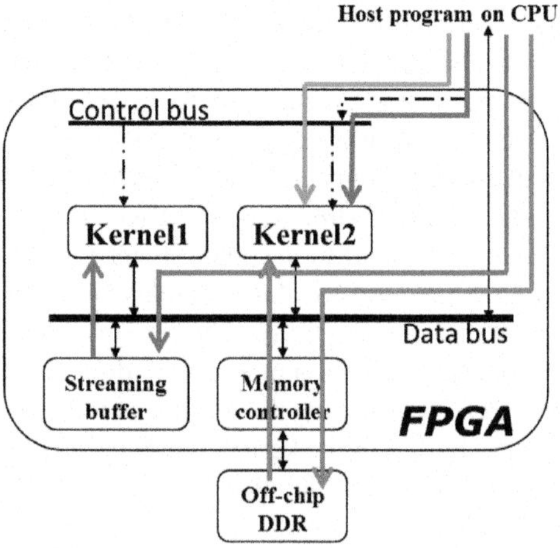

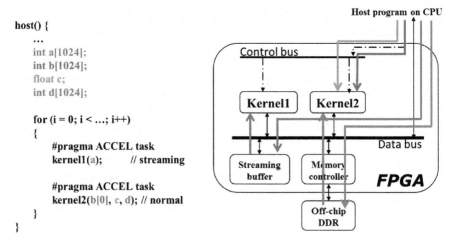

Fig. 8.9 Data communication between CPU and FPGA

considering the granularity of the optimizations as well, we have the four quadrants shown in Fig. 8.10. These four quadrants are related to four well-known concepts in software development: (1) Coarse-grained parallelism corresponds to task-level parallelism, e.g., parallel threads running at multi-core CPU. (2) Coarse-grained pipeline is related to dataflow streaming, where tasks are executed in parallel, driven by the availability of input data and vacancy of output buffer. (3) Fine-grained parallelism is similar to the concept of single-instruction-multiple-data (SIMD), where hardware operators and data-paths are duplicated to perform the operations of consecutive data elements. (4) Fine-grained pipeline corresponds to instruction pipelining, a basic microarchitecture technique adopted in the design of CPUs, where the execution of multiple instructions can be partially overlapped.

Each of the four quadrants has various microarchitecture-related details in FPGA accelerator design. For example, in coarse-grained parallelism, distributing tasks into parallel computation units is not trivial; and memory coalescing plays

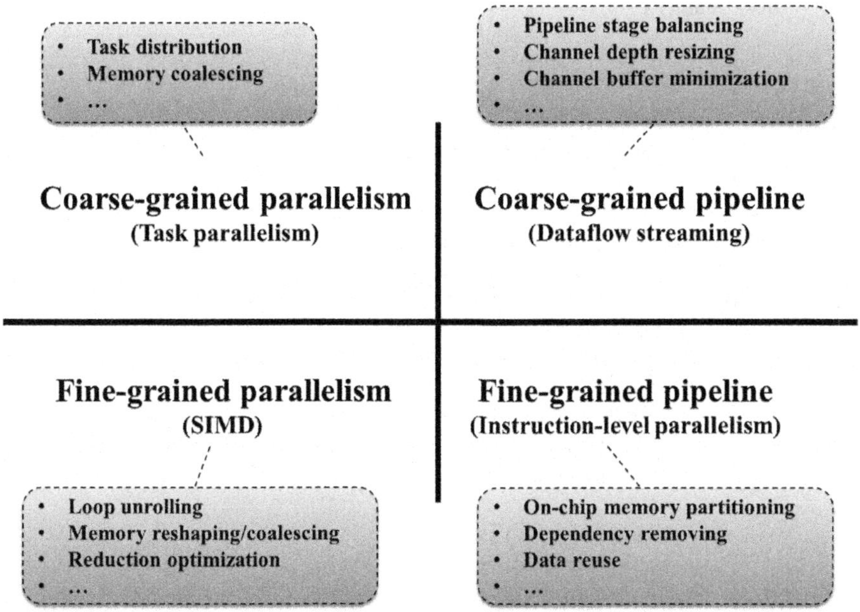

Fig. 8.10 Four quadrants of optimizations

a significant role in improving the off-chip bandwidth efficiency. In the Merlin Compiler programming model, those hardware details are hidden and abstracted into these four quadrants, and source-to-source automation will implement the corresponding optimizations efficiently.

The Merlin Compiler only relies on the following four simple pragmas from users to specify their intention on the FPGA acceleration design.

```
#pragma ACCEL task
#pragma ACCEL pipeline
#pragma ACCEL parallel [factor=N]
#pragma ACCEL auto
```

The `task` pragma defines the function to be accelerated, which specifies the interface between software and hardware in the design. By inserting `pipeline` and `parallel` pragmas in the different locations of the program, the high-level intention of the four optimization quadrants is conveyed to the Merlin Compiler. And finally, the `auto` pragma allows the design be optimized in a fully automated way, where programmers even do not need to provide the high-level pipeline and parallel pragmas. In this case, the loop transformation will be invoked to find a good code structure, and in the meantime, an optimized scheme to apply pipelining/parallelization on the code structure is also explored automatically. The position of the `auto` pragma in the source code determines the scope of automatic

```
host_main() {
    ...
    for (i = 0; i < ...; i++)
    {
        #pragma ACCEL task
        kernel(a, b, c, d); // SW/HW interface
    }
}
void kernel(int *a, int *b, int *c, int d)
{
    for (i = 0; i < N; i++)    { // coarse-grained parallelism
        #pragma ACCEL parallel factor=4
        for (j =0; j < M; j++) { // task-level pipelining
            #pragma ACCEL pipeline
            for (k = 0; k < M; k++) { // fine-grained parallel/pipeline
                #pragma ACCEL pipeline_parallel factor=3
                c[...] = a[...]*b[...]+d+...;
            }
            Loop1: for (...) { ... }
            Sub_func2(...);
    } }
}
```

Fig. 8.11 Example code for Merlin Compiler

```
void kernel(int *a, int *b, int *c, int d)
{
    #pragma ACCEL auto
    for (i = 0; i < N; i++)    {
        for (k = 0; k < M; k++) {
            for (j =0; j < M; j++) {
                c[...] = a[...]*b[...]+d+...;
            }
            Loop1: for (...) { ... }
            Sub_func2(...);
    } }
}
```

```
void kernel(int *a, int *b, int *c, int d)
{
    for (i = 0; i < N; i++)    {
        #pragma ACCEL parallel factor=4
        for (j =0; j < M; j++) {
            for (k = 0; k < M; k++) {
                #pragma ACCEL pipeline
                c[...] = a[...]*b[...]+d+...;
            }
            Loop1: for (...) { ... }
            Sub_func2(...);
    } }
}
```

Fig. 8.12 Fully automatic optimizations

optimization. The `pipeline` and `parallel` pragma can also be used in the `auto` scope to give hints to the compiler (Figs. 8.11 and 8.12).

8.3 System Optimization and Automation

The Merlin Compiler provides the abstraction of the four quadrants for hardware design optimizations. This section introduces some details of the optimizations performed in the source-to-source automation flow.

8.3.1 Fine-Grained Pipelining and Parallelization

Fine-grained source-to-source loop optimizations are applied at instruction-level when pipeline pragmas are inserted in the innermost loop. *Fine-grained pipelining* enables the pipelined execution of the instructions in the loop iteration. Initiation interval (II) can measure the throughput of the pipeline, which is a fixed clock cycle period required between the issues of the two consecutive loop iterations [CJLZ09]. It is highly affected by the computation resources, memory port limitation, intra-loop iteration dependency and inter-loop iteration dependency. An example of fine-grained loop pipelining is shown in Fig. 8.13b. *Fine-grained parallelization* allows simultaneous execution of the instructions in continuous loop iterations. The

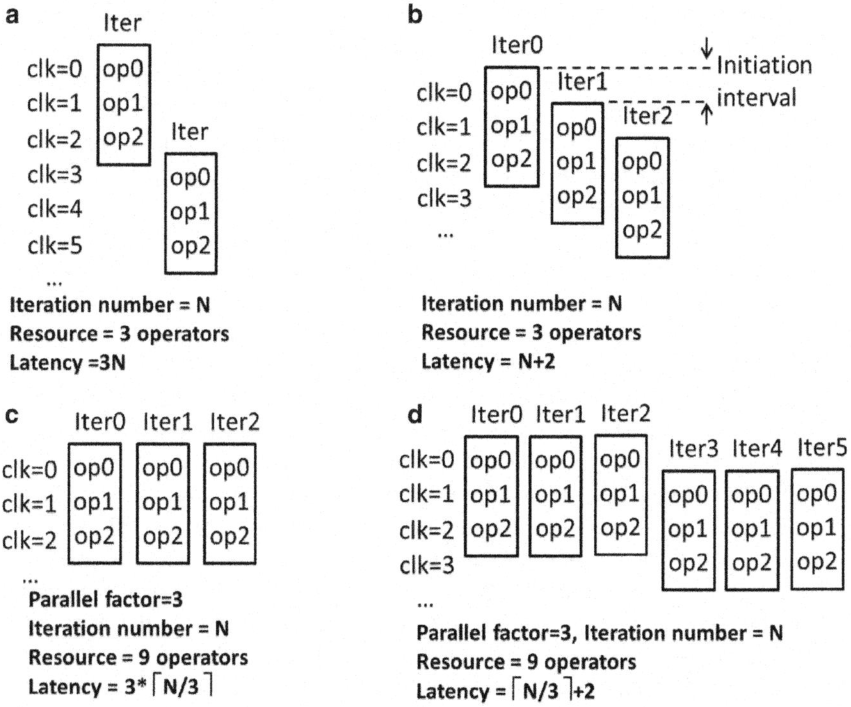

Fig. 8.13 Fine-grained loop pipelining and parallelization: (**a**) original execution sequence, (**b**) pipelined sequence, (**c**) parallelized sequence, (**d**) pipelined sequence after parallelization

number of parallelized loop iterations is defined by a user-given parallel factor. The source-to-source transformation duplicates the computation resources to enable the parallelized execution. An example of fine-grained loop parallelization is shown in Fig. 8.13c. Given a parallel factor 3, three continuous loop iterations execute in parallel.

The features of pipelining the parallelization can be combined together, which is called fine-grained pipeline-parallelization. As shown in Fig. 8.13d, the loop iteration is first duplicated for parallelized execution, and each duplication has a pipeline at the instruction level.

Comparing these three optimizations in Fig. 8.13b–d, the fine-grained pipeline is more area-efficient than fine-grained parallelization, but fine-grained parallelization can reduce more latency—especially when the parallel factor is large. When there are enough available resources and high performance is required, fine-grained pipeline-parallelization can be a good choice since it combines the advantages of both optimizations.

Fine-grained pipelining and parallelization have similar requirements on data access and computation parallelism. There are three major issues that can impact parallelism in the innermost loop: loop-carried data dependency, memory port limitation, and small loop trip count.

Loop-carried dependency limits the freedom of operations in scheduling. Assuming that such a dependency exists between op1 in iteration i and op0 in iteration $i + 1$ in Fig. 8.13a, all of the three kinds of fine-grained loop optimizations cannot be applied, as iteration $i + 1$ must execute after iteration i.

Ideally, if memory port number is not limited, all the data elements required in an execution of the loop iteration can be accessed in one clock cycle. However typical block RAMs in FPGAs have a limited number of ports to feed data to the highly parallelized execution units. Access conflict may happen when multiple accesses to one memory bank occur simultaneously.

Even when there is no loop-carried dependency and access conflict, a small loop trip count can reduce the benefits gained from using fine-grained pipelining/parallelization, and it sometimes has a negative effect on the design. If the iteration number is $N = 2$, the latency of the original design in Fig. 8.13b is 6 and the latency after fine-grained pipelining is 4. There is only a 2-clock-cycle benefit from pipelining, but it will generate a large amount of control logic, registers or even RAM memories. A trade-off must be introduced under this circumstance.

For loop-carried data dependence, we can apply loop transformation to move the dependence to the outer loops. For memory port limitation, automated memory partitioning is very helpful (discussed later in this section). For a small loop trip count, loop flatten can be applied to merge the nested loop into a flattened one.

One important technique in fine-grained pipelining and parallelization is memory partitioning because it improves memory bandwidth by allocating the array into several non-overlapping banks. Each bank is implemented with a separate memory block so that simultaneous accesses to different parts of the data elements of the same array can be realized. In the case shown in Fig. 8.14a, if the on-chip memory has only one port, fetching all the data elements from array "a" takes five cycles in

a
```
for (i =0; i < M; i++)
    for (j =0; j < N; j++)
        c_out[i][j] = a[i][j]+a[i][j-1]+a[i-1][j]+a[i+1][j]+a[i][j+1];
```

b
```
for (i =0; i < M; i++)
    for (j =0; j < N; j++)
        c_out[i][j] = a[i][j]+a[i][j-1]+a[i-1][j]+a[i+1][j]+a[i][j+1];
                        ↕        ↕          ↕          ↕          ↕
                      Bank0    Bank1      Bank2      Bank3      Bank4
```

Fig. 8.14 Example for memory partitioning: (**a**) original code (**b**) ideal bank mapping of data elements for access parallelism

a loop iteration. But if the data elements can be accessed from five different banks, as shown in Fig. 8.14b, five data elements can be accessed in only one cycle.

State-of-the-art high-level synthesis tools, such as Xilinx Vivado HLS (see Chap. 7), provide the capability to perform memory partitioning, but this requires users to determine the detailed parameters, such as the partitioning scheme (block or cyclic), the target partitioning dimension of the array, and the number of banks to use. All of these factors are hard to set for a programmer without much HLS experience.

The source-to-source optimization flow will analyze the access pattern of the loop iteration. The partitioning parameters are determined automatically by access conflict analysis during compiling time. Different memory partitioning schemes can be used, such as cyclic partitioning, block partitioning, and block-cyclic partitioning. We depict these in Fig. 8.15b–d, respectively [WZCC12], where each square denotes a data element in the array.

In addition to memory partitioning, other optimization strategies for performance improvement, such as loop fusion and loop permutation, will be introduced in Sect. 8.3.4.

8.3.2 Coarse-Grained Pipelining

When we add the pipeline pragmas at outer loops, the coarse-grained pipeline is enabled in the source-to-source optimization flow. The hardware mechanism and implementation details of the coarse-grained pipelines are different from those of fine-grained pipelines.

Coarse-grained pipelining enables the pipelined execution of tasks in an outer loop. Coarse-grained code segments (called pipeline nodes) within a common outer loop are pipelined, which means these segments are running in parallel at the

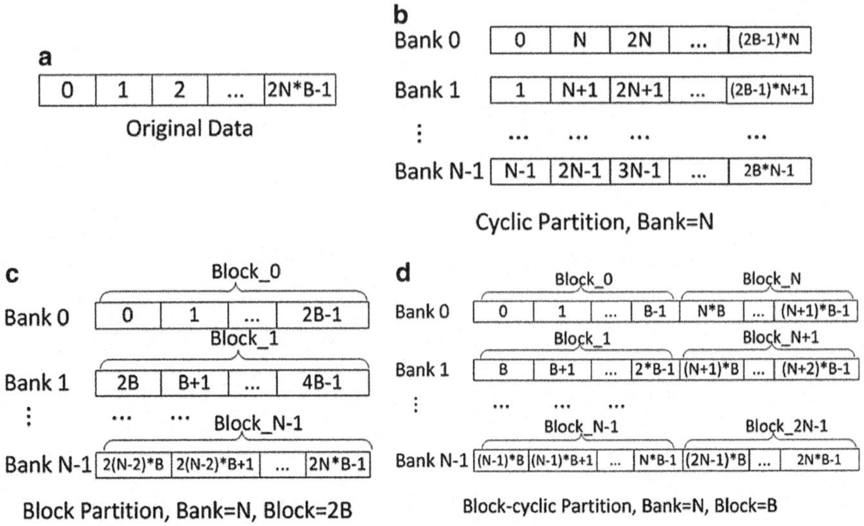

Fig. 8.15 Memory partitioning schemes: (**a**) original data, (**b**) cyclic partitioning, (**c**) block partitioning, (**d**) block-cyclic partitioning [WLC14]

same or successive iterations of the outer loop. These pipeline nodes can be for-loops, function calls, basic blocks and several adjacent statements in the outer loop. In contrast to the fine-grained pipeline, the latency of the code segment is normally larger than fine-grained instruction and cannot be determined easily during the compilation time. In some cases, special hardware is required to synchronize the execution of the pipelining nodes and more complex microarchitecture consideration is required—such as pipelining stage assignment, channel depth balancing, and buffer size optimization. Coarse-grained pipelining can be applied in multiple loops in a loop nest, and in this case, a hierarchical coarse-grained pipeline is implemented.

As shown in Fig. 8.16a, the statements in the loop are packed into two tasks: task0 receives off-chip data and task1 performs computation. The pipeline is executed in task-level as shown in Fig. 8.16b. Similar to fine-grained loop pipelining, the throughput of the pipeline can be measured by the initiation interval. Figure 8.16c gives an example of the hierarchical coarse-grained pipelining.

Task-level pipelining is implemented in some modern HLS tools, such as Vivado HLS (see Chap. 7). It provides a dataflow pipelining option to allow the parallelized execution of a sequence of the tasks in a pipelined way. However, it has many limitations when applying to the for-loops, such as the requirement of single producer and single consumer for the channel data.

Our flow will automatically partition a set of consecutive statements into multiple tasks according to both the intrinsic structure of the loop and the dependency graph. An existing loop nest or a function call in the loop can be recognized as a task, or grouped with other instructions in a task if strong feedback data dependency

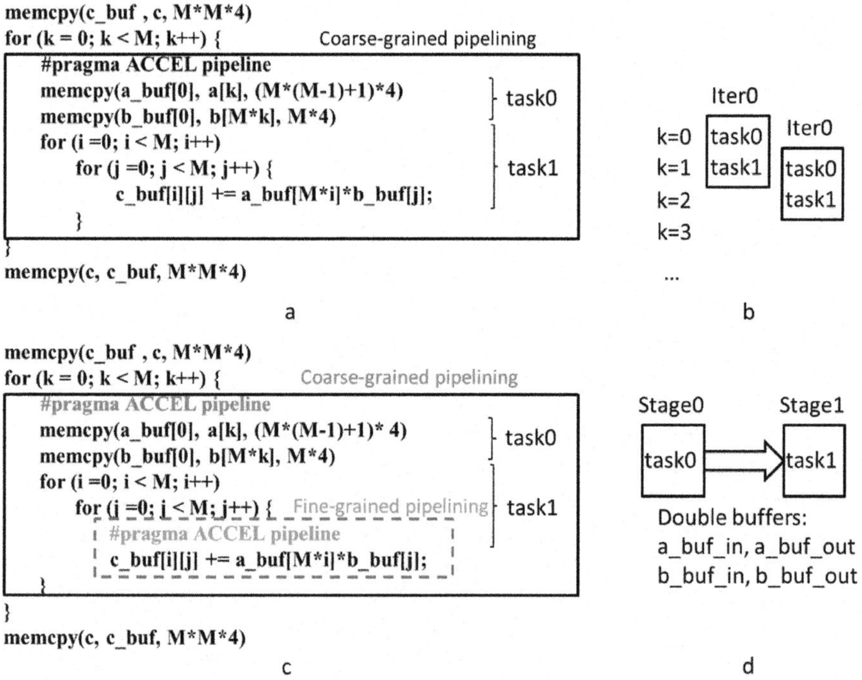

Fig. 8.16 Example (matrix multiplication) for coarse-grained pipeline: (**a**) task partitioning, (**b**) coarse-grained pipelining in task-level, (**c**) coarse-grained pipelining with fine-grained pipelining in a task, (**d**) double buffer in the data channel

exists between instructions. According to task-level dependency analysis, tasks are scheduled into different pipeline stages. Data channels between two stages are generated to ensure the correct data transfer.

A data channel between two adjacent pipeline stages can be implemented in two ways: double buffer or data streaming (see Chap. 7). In double-buffer (or so-called ping-pong buffer), the array is buffered in two RAM memories for both producer and consumer, as shown in Fig. 8.16d. This method is used in traditional software pipelines and ensures that the whole package of elements is transferred; however, sometimes it is too conservative and memory-resource consuming. Our flow applies accurate access range analysis based on the polyhedral model [Pou10] to reduce the buffer size. Bandwidth and storage resource are saved without transferring and storing the unnecessary data.

An alternative way to implement the data channel is data streaming, where the data element is directly transferred without using any double-sized RAM resource for storage. It requires the data producing and consuming relationship to be sequential. Compared to double buffer, although it saves the storage resources, it cannot handle the case when a data producing and consuming relationship is not purely sequential, or the access pattern is unpredictable during compiling time.

Our flow will choose a manner for implementing the data channel according to the characteristic of the data transfer pattern.

8.3.3 Coarse-Grained Parallelization

Like task-level parallelism vs. instruction-level parallelism, coarse-grained parallelism has different implementation details compared to fine-grained parallelism. In fine-grained parallelization, parallel low-level operations generally have the fixed and small latency in terms of the number of cycles, so the synchronized execution model is applied among all the parallel computation units. And this brings two-fold benefits: (1) the overhead of task scheduling and dispatching is largely saved, and (2) the synchronized data access pattern can improve the efficiency of the data access to on-chip and off-chip memories.

However, at the coarse-grained level, in general it is not explicitly necessary to synchronize the execution of individual computation units using HLS, because (1) this synchronization is complex and costly to implement, especially when the number of parallelized units is large; (2) this synchronization degrades the overall performance because of less overall utilization of the hardware, especially when the execution time varies among the tasks; (3) coarse-grained parallel units typically have parallel shared (off-chip) memory accesses, and these accesses are intrinsically asynchronous in the shared bus; (4) intended synchronization inside a kernel (such as a coding style in CUDA and OpenCL) can be converted into an equivalent form with separated loops in HLS code, where the separating point in HLS represents the synchronization implicitly.

Figure 8.17 demonstrates the microarchitecture of the coarse-grained parallelism. In this case, we assume function `foo` has internal for-loop and subfunction calls, so coarse-grained parallelization is applied. Hardware units (tasks) of function `foo` are duplicated into a fixed number according to the resource utilization in FPGA. The execution of each kernel unit is driven by the iteration token from the task dispatcher. The centralized task dispatcher deals with the execution control related to the coarse-grained for-loop with the parallel pragmas. If for-loop does not end, the dispatcher will check the status of the task units and find an available one that

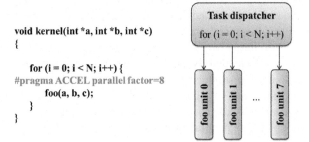

```
void kernel(int *a, int *b, int *c)
{
    for (i = 0; i < N; i++) {
#pragma ACCEL parallel factor=8
        foo(a, b, c);
    }
}
```

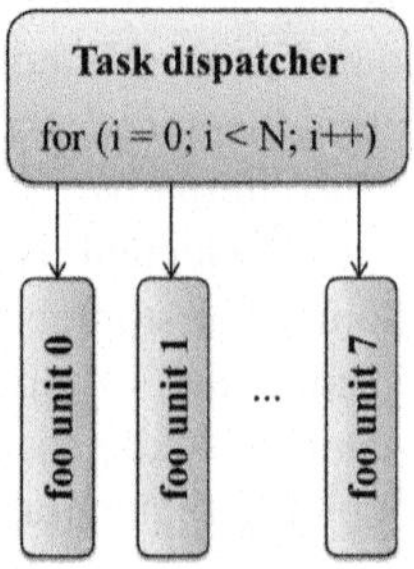

Fig. 8.17 Coarse-grained parallelism

is idle or has finished the previous task. Then the dispatcher sends the current loop iteration number as a token to the available task unit, and the unit starts execution. After the execution is finished, the task unit returns to the idle status and waits for the next token from the dispatcher.

An efficient coarse-grained parallelization implementation requires a good partitioning of the tasks—making the task instances as independent as possible. The Merlin Compiler will detect the dependency and add constraint logics in the task dispatcher to maintain the precedence relations, which may result in low utilization. The basic approaches to improving the parallelism include selecting the correct loop nest and loop level to parallelize, and performing loop transformation to exploit parallel loops. Accesses to the shared on-chip memory among the task units are parallelized if the compiler can find a feasible memory partitioning scheme to isolate the accesses within each task unit. Accesses to the shared off-chip memory are, in general, directly parallelized if there is no data dependency. The access ports from all the parallel units to AXI bus can be either in parallel or bundled into one port according to the FPGA resource utilization and DDR bandwidth utilization.

8.3.4 Automated Optimizations in the Merlin Compiler

In addition to the four major optimizations triggered by explicit pragmas, the Merlin Compiler also contains various implicit optimizations which are performed implicitly along with the pragmas to help improve the results of the pipeline and parallelization.

Off-Chip Memory Optimizations Off-chip memory bandwidth is one of the key bottlenecks for system performance. The Merlin Compiler does not provide specific pragmas for off-chip memory efficiency. The off-chip optimizations are triggered by pipeline and parallel pragmas, and the implementation details are hidden from the tool users.

Memory Burst Generation To better utilize the bandwidth of off-chip memory bandwidth, data transfer is separated into address and data transactions as shown in

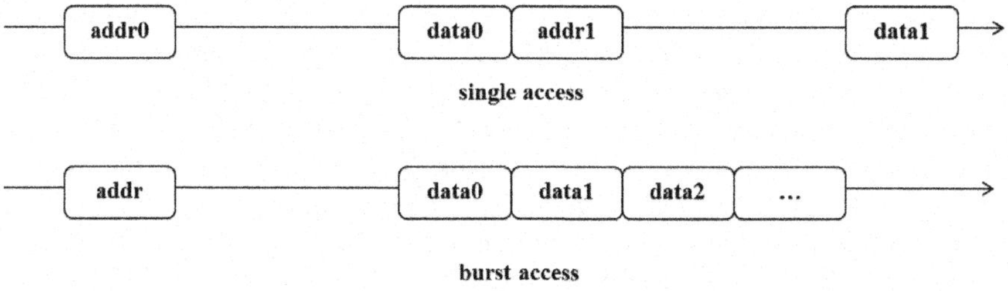

Fig. 8.18 Simplified off-chip memory access model

Fig. 8.18. There is a latency required between these two transactions, which is the major overhead in off-chip memory performance. Depending on the platform, the latency can be as high as several tens of cycles. A special data transfer mode (burst mode) allows sharing one address transaction among multiple data transactions, but its limitation is that the addresses of the burst data must be continuous. To improve the off-chip memory access efficiency in HLS code, the address of the array access needs to be continuous and the burst length needs to be long enough. The Merlin Compiler detects the possible continuous off-chip memory accesses and generates the local buffers for the burst

Memory Coalescing is another technique for improving the utilization of the off-chip memory bandwidth. As the kernel clock frequency is typically lower than the DDR interface frequency, a wide memory word can be accessed in each kernel clock cycle. Memory coalescing combines general data types (such as 32-bit integer and 64-bit double precision floating-point) into a large memory word (such as 256-bit or 512-bit), and accesses the off-chip memory using the large data bitwidth. Multiple data with general data type can be accessed in one cycle. To enable the automated memory coalescing, the address of the accessed data is required to be continuous as well.

Data Reuse is a widely used technique—not just in software program optimization and hardware design for FPGA. On-chip memory (or cache) is used to store the data temporarily to avoid the repeated accesses from off-chip memory. In HLS, there is a larger design space of data reuse optimization because FPGAs have thousands of on-chip memories which can be used to generate the data reuse buffers in a customized way. More details about data reuse automation can be found in [PZSC13, CZZ12].

Data Prefetching is also a common technique which overlaps the data transfer with the computation in the kernel. When the computation part is running at a certain data set, the communication part is fetching the data for the next data set. A double-buffer is automatically inserted to enable the parallelism between computation and communication. In source-to-source optimizations for HLS, data prefetching is performed automatically when the coarse-grained pipeline is applied. The data prefetching process is just an additional pipeline stage in the coarse-grained pipeline generation.

Loop and Data Transformations Loop and data transformations are crucial techniques in compiler optimizations for CPU and GPUs. Most of these techniques are directly applicable to HLS for FPGA as well, such as parallelism exploiting and data locality improvement. Due to the characteristics of FPGAs, there are also several additional techniques dedicatedly required in the source-to-source optimization for HLS. Loop and data transformation is triggered in the Merlin Compiler using *auto* pragmas. The Merlin Compiler incorporates various loop and data transformation techniques, and this section demonstrates some basic and widely applicable ones with examples.

Loop Interchange (or loop permutation) exchanges the order of loops in a loop nest, e.g., moving an innermost for-loop to an outer position in the loop nest. The result of the loop interchange is that the execution order of the instances of the loop body computation is changed. In general, loop interchange is beneficial, not only in enhancing fine-grained parallelism as in this case, but also in many different aspects, such as improving coarse-grained parallelism, enabling burst memory access, exploiting data reuse, reducing streaming buffer size, etc.

Loop Fusion merges two consecutive loops or loop nests into one. Loop fusion and its inverse, transformation distribution, are very useful transformations—not only for CPU program optimization but also for HLS. By changing the loop structure in the imperfectly nested loops, these transformations can achieve various purposes in the optimization, such as parallelism exploiting and data locality. In many cases, they are also used as a preprocessing transformation to form the regular full nested loops for other useful loop transformations such as loop interchange or loop tiling.

Loop Tiling is a very useful loop transformation which partitions the loop iteration space into small chunks or blocks. By performing the iterations chunk by chunk, the data locality is achieved because the data accessed in a chunk are typically neighboring data in different array dimensions. Loop tiling is very helpful in exploiting data reuse with a small reuse buffer. The tiling sizes are the configurable parameters in loop tiling which need to be selected carefully for a good trade-off between the performance improvement and resource utilization.

Data Layout Transformation changes the addresses (indices) of the data elements stored in the array. Data layout transformation can also achieve data locality by putting the data accessed by adjacent loop iterations in the adjacent locations in the array. The transformation is used in CPU program optimization to improve the cache performance. And in HLS, it can also help to reduce the on-chip buffer size and increase memory access efficiency.

8.3.5 Design Space Exploration

Design space exploration will be triggered at the code scope where the *auto* pragma is inserted. In the Merlin Compiler, the design space includes (1) the loop and data transformation options, and (2) the pipeline and parallel pragmas and their arguments. Exploring different design alternatives helps programmers find the best design point that achieves the highest performance at the lowest cost. However, manually exploring all these alternatives is very time-consuming. Thus, it is desirable to perform such a design space exploration by source-to-source automation at compile time.

The design space of module section and replication can be explored automatically using the techniques in [CHL+12] and [CHZ14]. Based on the research, each module has several design options with different trade-offs between performance

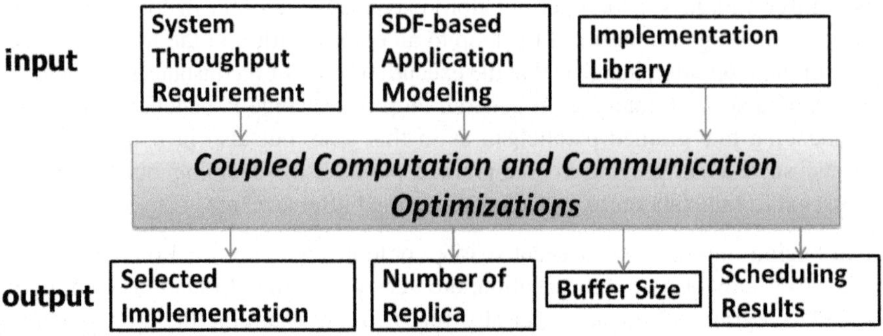

Fig. 8.19 Framework for automated module selection and replication

and area utilization. And these modules can be further replicated for better performance. The system design space exploration problem is to find a proper design option and number of replications for each module so that the system performance can be maximized with the given resource (Fig. 8.19).

In the Merlin Compiler, module duplication can be used to model the fine-grained and coarse-grained parallelization, and module selection can be used to model other design parameters such as loop transformation, and whether to apply pipelining. The module selection and duplication algorithm is applied in different loops in a bottom-up hierarchical way. At a certain level of loop nest, the Pareto-optimal results of current loop implementations are used as module selection options for the exploration at the upper loop levels.

The module selection and replication problem is formulated into a mathematical form. When the design space is huge, the integer linear programming (ILP) usually takes a very long time to solve the problem. Therefore, an iterative algorithm is applied to solve the problem: in each refinement iteration, the compiler first searches for the current system bottleneck, then, among all the possible ways to improve the system, the one that can improve the system with minimal cost will be inserted into the result set.

8.4 Case Study: Logistic Regression

The Merlin Compiler has been used successfully to accelerate many real-world applications. In this section we show how the compiler can significantly accelerate logistic regression using FPGAs.

Logistic regression [Bis06] is a method of choice for building predictive models for many complex pattern-matching and classification problems. It is used widely in such diverse areas as medical and social sciences. It is also one of the most popular machine learning techniques. In its binomial form, logistic regression seeks to find a model, represented as a vector w of weights, such that for a given input data x, where

x is a vector of features, the following linear predictor can be used to determine if x belongs to a class or not ($p = 0$ means x is in the class):

$$p = \begin{cases} 1 & \text{if } wx \geq 0 \\ 0 & \text{if } wx < 0 \end{cases}$$

What we are given is a (training) set $(x_0, y_0), (x_1, y_1), \ldots, (x_{n-1}, y_{n-1})$, where y_t is the binary label for input data x_t indicating whether it belongs to the class or not. Logistic regression tries to find a w of good prediction accuracy by minimizing the following loss/error function:

$$-\sum_t \left(y_t \log(h(wx_t)) + (1 - y_t) \log(1 - h(wx_t))\right), \text{ where } h(a) = \frac{1}{1 + e^{-a}}$$

The optimization problem is solved using gradient descent over the training set, starting with an initial weight, iteratively adding the gradient to the current weight until the stopping criteria are met. For a given weight w, the gradient is calculated using the following formula:

$$\sum_i (h(wx_t)) - y_i)\, x_i$$

Figure 8.20 shows the pseudocode to find the gradient (vector) for the multi-class logistic regression problem, where the input data are mapped to one of several classes:

```
for (c = 0; c < NUM_CHUNKS; c++)
{
    // Linear combination of weights and features
    for (i = 0; i < FEATURE_SIZE; i++)
        for (j = 0; j < CHUNK_SIZE; j++)
            for (k = 0; k < LABEL_SIZE; k++)
                result1[j][k] += w[k][i]*x[j][i];
    // Nonlinearization using the logistic function h()
    for (j = 0; j < CHUNK_SIZE; j++)
        for (k = 0; k < LABEL_SIZE; k++)
            result2[j][k] = h(result1[j][k]);
    // Differentiation with labels for residuals
    for (j = 0; j < CHUNK_SIZE; j++)
        for (k = 0; k < LABEL_SIZE; k++)
            result3[j][k] = result2[j][k] - y[j][k];
    // Linear combination of residuals and features
    for (i = 0; i < FEATURE_SIZE; i++)
        for (j = 0; j < CHUNK_SIZE; j++)
            for (k = 0; k < LABEL_SIZE; k++)
                gradient[j][k] += result3[j][k]*x[j][i];
}
```

Fig. 8.20 Pseudocode for gradient calculation

The main loop of gradient calculation consists of four loop nests, corresponding to the four operations in the gradient formula. The four loop nests are dependent as the output of one loop feeds to the next. The first step of the compiler is to break the dependences so that the loop nests can be run concurrently on hardware. This is accomplished with the coarse-grained pipelining feature that tries to isolate the loop nests by properly inserting buffers so that different loop nests compute for different chunks of training data.

Within each loop nest, the compiler applies multiple optimizations to improve its latency. Taking the first loop as an example, the compiler will exploit data parallelism by calculating `result1` concurrently for all labels, namely unrolling the k loop. In order to overcome the port limit (at most two ports for each on-chip memory on FPGAs), the compiler will also automatically partition array `result1` along the k dimension. Next, the compiler tries to pipeline the remaining i and j loops by overlapping the successive iterations of the parallelized loop body. The final hardware architecture generated by the compiler is shown in Fig. 8.21.

There are a few other automatic optimizations done by the compiler for this design, such as off-chip data access coalescing and burst inference. The optimized design, when implemented on a mid-range Kintex7 device from Xilinx, can achieve 14× speedup over a single threaded CPU implementation for a decimal digit recognition problem.

In another experiment, we built a classification model with 784 features and 50 labels using 60k training samples. We tried different levels of source-to-source optimizations and observed different speedups as shown in Table 8.1. "solution1" with 1.5× speedup, is generated with minimal optimizations; "solution2" with 37× speedup is generated with fine-grained pipelining and parallelization; "solution3"

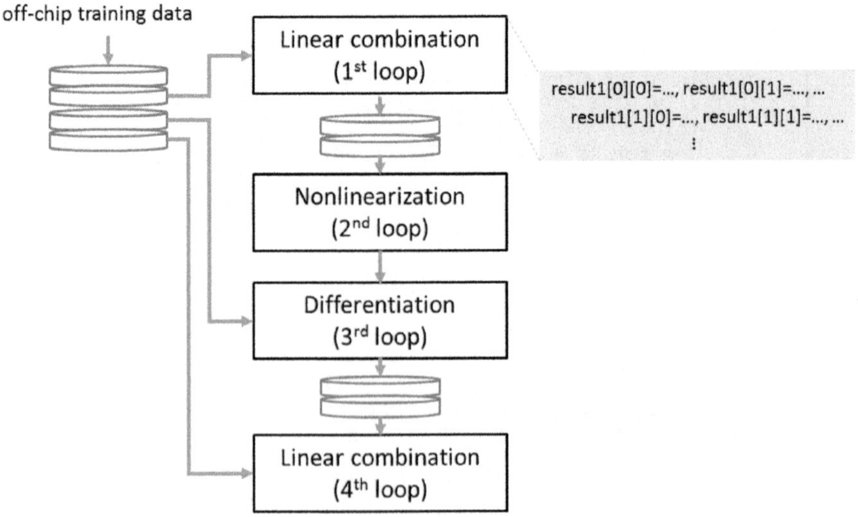

Fig. 8.21 Hardware architecture of gradient calculation

Table 8.1 Performance impacts of different optimizations

	FPGA		
CPU	Solution1	Solution2	Solution3
1.0	1.5	37.0	108.0

with 108× speedup is generated by enabling coarse-grained pipelining in addition to fine-grained pipelining and parallelization.

8.5 Related Work

This section discusses work that is relevant to our source-to-source transformation tool described in this chapter. We divide that work into six categories. The first one discusses the source-code level transformation with HLS, while the remaining five categories cover the source code transformation techniques on top of HLS.

Code Transformation with High-Level Synthesis (HLS) Code transformation is used extensively as part of high-level synthesis to improve the quality of results. An example of early work is the SPARK system [GGDN04b], which provides a number of source-level pre-synthesis transformations that include common sub-expression elimination (CSE), copy propagation, dead code elimination, and loop-invariant code motion, along with more coarse-level code restructuring transformations such as loop unrolling, together with their HLS engine. The AutoPilot HLS tool [ZFJ+08], which was the precursor of the widely used Vivado HLS tool from Xilinx [CLN+11], performs a number of code transformations. These include not only the traditional optimizations, such as constant propagation, dead code elimination, and common subexpression elimination that avoid functional redundancy, but also strength reductions that replace expensive operations with simpler low-cost operations, transformations such as if-conversion and tree height reduction that explicitly expose fine-grain operator-level parallelism, and coarse-grain code restructuring by loop transformations such as loop unrolling, loop flattening, loop fusion, etc. It also provides a number of analysis capabilities, such as bitwidth analysis, alias analysis, and dependence analysis that help to reduce the data widths and analyze the data and control dependences. These transformations are either performed locally within a function body or applied inter-procedurally across the function call hierarchy as LLVM [LLV15] passes prior to the invoking of their HLS synthesis engine. These optimizations in Vivado HLS are discussed in details in Sect. 7.2.2.

Loop Transformations Loop transformations have been studied over several decades for compiler optimization on general-purpose processors (e.g., [WL91, Fea92]). Recent work applies loop transformations for exposing more parallelisms and/or enabling more efficient on-chip memory reuse in HLS (e.g., [CZZ12, PZSC13]). For example, the work in [CZZ12] analytically models the on-chip buffer size and off-chip bandwidth after affine loop transformation, loop fusion/distribution and code motion, and then uses a branch-and-bound based exploration together with a knapsack-based reuse optimization to search for optimal solutions for on-chip memory allocation of the imperfectly nested loops. The work in [PZSC13] leverages the power and expressiveness of the polyhedral compilation model to develop a multi-objective optimization system to guide loop transformations to consider the trade-off of on-chip buffer usage versus off-chip communications. Both works take the user C code as inputs and generate the transformed C code with intended optimization for the Vivado HLS tool for synthesis to Xilinx FPGAs. Some other loop transformation techniques are addressed in Sect. 2.4.

Memory Partitions Memory partitioning for HLS was formulated and investigated systematically in [CJLZ11, CJLZ09] with the goal of enabling the maximum throughput for pipelining. It was originally developed for one-dimensional arrays, and later on extended to circular reuse buffers [WZCC12] and multidimensional arrays [WLZ$^+$13]. A unified theory was proposed in [WLC14] which covers all these cases efficiently based on the polyhedral formulation. Recent work on memory partitioning includes the use of lattice-based partitioning [CG15] and separation-logic based analysis [WBC14, WFY$^+$15]. The latter can support some dynamic memory allocation, such as heaps.

Source Code Generation As the polyhedral model is more often used in the source-level transformation for HLS, a rather unique work focused on automated code generation developed from a polyhedral framework for the resource usage optimization [ZLC$^+$13]. In particular, it focuses on CLooG, a widely used generic and scalable code generator [Bas04] for the polyhedral framework. It extensively studied the impact of alternative loop-bound computation techniques, and ended with significant resource savings when compared to the direct use of CLooG.

Inter-Module Optimization Although modern HLS tools can perform module-level synthesis efficiently, considerable effort is needed to compose these modules efficiently to achieve system-level optimization. A heuristic method was used in [HWBR09] for maximal replication of all the stateless actors (modules) for throughput maximization; it then iteratively fuses the actors that do not affect the throughput. A more recent work combines module selection/replication and buffer size optimization and minimizes the area cost for both logic and memory under the required throughput constraint [CHZ14]. In another direction, the open-source LEAP (Latency-insensitive Environment for Application Programming) framework [YFAE14] provides a flexible way to compose different modules in multiple FPGAs asynchronously via coherent scratchpads. Inter-module optimization is also applied in the other HLS tools, and the related information can be found in Sects. 7.2.2 and 12.4.

Design Space Exploration (DSE) Another dimension of source- level transformation and optimization is generating various configurations to feed into a HLS system for design space exploration. The PICO system provides a "spacewalker" to traverse different design point generation of efficient compute kernels for the underlying VLIW architecture [AK08]. Another example is the work in [SW10, SW12], which reduces the complexity of the DSE problem by clustering, i.e., grouping the components (array, loop, and functions) of original source code into smaller clusters and then running the DSE for each cluster. A more recent work explores the structure of the design space by analyzing the dependencies between loops and arrays, and uses such knowledge to reduce the dimensions of the design space [PSKK15].

This list of related works is not meant to be comprehensive. The intention is to highlight some of the most relevant studies as a basis for further reading.

8.6 Conclusion

The current status of high-level synthesis demonstrates the great need and opportunities for source-to-source transformation for automated compilation to FPGA. As a pioneer contributor in this challenging area, the Merlin Compiler provides an end-to-end compilation flow with a friendly programming environment for software engineers. Comprehensive design automation and optimizations are performed automatically in a source-to-source framework, and significant acceleration on FPGA can be achieved without dealing with the hardware design details. We expect that the Merlin Compiler will greatly facilitate the widespread deployment of customized computing—especially in data centers with programmers who do not have a detailed knowledge of FPGAs.

Acknowledgements This work is partially supported by the National Science Foundation Small Business Innovation Research (SBIR) Grant No. 1520449 for project entitled "Customized Computing for Big Data Applications".

Chapter 9
Bluespec SystemVerilog

Oriol Arcas-Abella and Nehir Sonmez

Bluespec SystemVerilog (BSV) is a rule-based language, where hardware is described as object-oriented modules [Nik04]. Other high-level synthesis approaches try to hide the complexity of hardware (clock cycles, data movement, concurrency, etc.) under the appearance of a sequential and centralized execution. Instead, BSV exposes it to the user as an intuitive high-level metaphor. However, what it tries to hide are some of the tedious details of traditional hardware design: synchronization signaling (ready/enable signals), poor typing systems, etc.

It is a challenging language and radically different than other hardware description paradigms, but once learned it can substantially improve the productivity, reduce the errors through correctness by design, and achieve identical results to hand-coded Verilog designs [ANS+14]. This language is a good candidate for expert hardware designers with a background on Register-Transfer Level (RTL) languages, such as Verilog or VHDL, for designers that have to develop critical hardware components, or for keeping a very tight control over the performance and the resources used.

The framework is available under a commercial license by Bluespec Inc., a Massachusetts-based spin-off from the MIT and the work by Professor Arvind.

O. Arcas-Abella (✉)
Barcelona Supercomputing Center, Barcelona, Spain

Universitat Politecnica de Catalunya, Barcelona, Spain
e-mail: oriol.arcas@bsc.es

N. Sonmez
Barcelona Supercomputing Center, Barcelona, Spain

D. Koch et al. (eds.), *FPGAs for Software Programmers*,
DOI 10.1007/978-3-319-26408-0_9

9.1 Guarded Atomic Rules

The computation model of BSV is based on Term-Rewriting Systems (TRS) [HA00]. This formalism has been successfully used to describe concurrent systems. A TRS consists in a set of terms and a set of rewriting rules.

$$pat_{lhs} \text{ if } p \rightarrow exp_{rhs}$$

The rewriting rules have a pattern (pat_{lhs}) to match terms, a condition (p) and an expression (exp_{rhs}) to replace them with new values. Each iteration of the system, all the rules with valid conditions that match one or more terms are executed.

A simple example of a TRS is the Euclidean Greatest Common Divisor (GCD) algorithm:

$$GCD\,(a,b) \quad \text{if} \quad a \geq b \wedge b \neq 0 \quad \rightarrow \quad GCD\,(a-b,b) \qquad \text{(GCD Sub Rule)}$$

$$GCD\,(a,b) \quad \text{if} \quad a < b \quad \rightarrow \quad GCD\,(b,a) \qquad \text{(GCD Swap Rule)}$$

In this example there are two terms (a and b) and two rules (the subtraction rule and the swapping rule). Both rules match the $GCD\,(a,b)$ terms, the first one replacing them with the new expression $GCD\,(a-b,b)$ if $a \geq b \wedge b \neq 0$. The second rule swaps the terms if $a < b$.

In BSV, these rules are called *atomic guarded rules*, and the terms are the state of the system, ie. hardware registers and memories. Each clock cycle, the rules whose conditions are valid are executed or *fired*, updating the state of the hardware system. The previous TRS example would be described as follows in BSV syntax:

Listing 9.1 BSV rules for GCD

```
1 rule gcd_subtract ( a >= b && b != 0 );
2   a <= a - b;
3 endrule
4
5 rule gcd_swap ( a < b );
6   a <= b;
7   b <= a;
8 endrule
```

Similar elements to the TRS example are present in the BSV code: two rules, with their conditions or *guards*, modifying the terms of the system. The syntax is similar to Verilog, with opening and closing keywords for code blocks, and the right arrow operator for register assignment.

The statements in the rules are applied sequentially. Software programmers will note that the swapping rule is assigning the registers a and b simultaneously, which is invalid in software routines. In BSV this is possible because the rules are *atomic*: all the changes to the state are performed *atomically, at the end of the rule*. This explains why the register a, which is updated in line 6, keeps its old value until the end of the rule, and its old value can be read in line 7.

9.2 Modules and Interfaces

The rules are enclosed in modules. Each module contains state elements, rules and methods. The methods allow modules to input and output data to other modules. Interfaces define the methods, allowing for different *modules* to have the same *interface*, or set of methods.

This structure is based on the object oriented programming paradigm: modules are objects, and their interfaces define the public methods that other objects can use. The state elements are the internal state of the objects, and the rules are routines that are fired under certain conditions. Popular software languages, such as C++ and Java, have similar metaphors, and even hardware description languages like Verilog and VHDL have similar structures.

The updated GCD example, within a module with methods, state elements and an interface, would look like the following:

Listing 9.2 GCD example with interface and module definitions

```
interface GCD;
  method Action start(int newa, int newb);
  method int result;
endinterface

module mkGCD (GCD);
  Reg#(int) a <- mkRegU;
  Reg#(int) b <- mkReg(0);

  rule gcd_subtract ( a >= b && b != 0 );
    a <= a - b;
  endrule

  rule gcd_swap ( a < b );
    a <= b;
    b <= a;
  endrule

  method Action start(int newa, int newb) if ( b == 0 );
    a <= newa;
    b <= newb;
  endmethod

  method int result if ( b == 0 );
    return a;
  endmethod
endmodule
```

The interface `GCD` describes two methods, one with two parameters (two integers named *newa* and *newb*) with no returning value, and another method with no parameters but returning an integer. The module `mkGCD` implements this interface. By convention, BSV module names start with the prefix "mk", meaning *make* because they *implement* the interfaces.

The module contains two registers, *a* and *b*. These elements have an interface or type (`Reg#(int)`), and their functionality is implemented by a given module (in this case, `mkRegU`, a register with undefined initial value, and `mkReg(0)`, a register with the initial value 0). The interface and module in this example, `GCD` and `mkGCD`, will become available types and constructors for other modules exactly like `Reg` and `mkRegU`:

Listing 9.3 Example instantiation and use of the GCD module

```
GCD gcd <- mkGCD;

rule calc_gcd;
  gcd.start(35, 7);
endrule
```

The methods, defined by the interface, are implemented by the module. Each method has parameters and/or a returning value. Methods that update the state of the module are called *actions*, or *action values* if they also return a value. Methods that do not update the state are simple *methods*. In this example, the `start` *action method* updates the registers *a* and *b*. The `result` *method*, which does not update any state elements, returns the value of the register *a*. Like rules, methods can have a condition or *guard*. In the example it is defined right after the header of the methods, and in both cases it is the condition $b = 0$ (the algorithm has finished).

9.3 Rule Scheduling and Implicit Guard Lifting

In the BSV programming paradigm, rules should be considered internally atomic and externally sequential. The total set of rules of the hardware model should be thought as sequentially ordered, as if every cycle only one rule was fired. That would be extremely inefficient, losing all the parallel performance that characterizes hardware. In practice, the compiler tries to fire all the possible rules every cycle. This is called *rule composition*, and it is performed automatically.

There are three types of conditions to fire a rule: the explicit guard, the conflicts with other rules, and the implicit guards. The explicit guard is defined by the user. In case of potential conflict between rules, the concurrent execution of rules follows the *one rule at a time* law: two rules are *composable* (can be fired in the same cycle) if their conditions are valid and their parallel behavior is equivalent to their sequential behavior. One side effect of this law is that reads are always scheduled *before* writes.

The third condition, the implicit rules, is caused by methods triggered inside rules. For instance, BSV provides a rich set of First-In-First-Out (FIFO) buffers, or queues, to communicate data between rules. Each FIFO has different capacities and scheduling semantics, but three main methods: enqueue, read and dequeue. These methods are *guarded*: for safety, it is not possible to read or dequeue an empty FIFO. Likewise, it is not possible to enqueue to a full FIFO. If a module implements some FIFO elements, a rule trying to enqueue or dequeue them will be affected by the explicit guards of these methods. This is called *implicit guard lifting*: the guards of the inner operations of a rule are "lifted", or appended, to its explicit rule.

Listing 9.4 Implicit condition handling example with FIFOs

```
Reg#(int) input_value <- mkReg(0);
FIFO#(int) queue <- mkFIFO;
Reg#(int) output_value <- mkRegU;

```

```
rule producer ( input_value > 0 );
  queue.enq(input_value);
endrule

rule consumer;
  output_value <= queue.first();
  queue.deq();
endrule
```

The rule `producer` has an explicit condition, `input_value` > 0 and the rule `producer` apparently can be always fired.

$$EG_{producer} = \texttt{input\_value} > 0$$
$$EG_{consumer} = \textit{true}$$

However, the implicit conditions of the `enq`, `first` and `deq` methods are *lifted* by the compiler when generating the final guard of the rules:

$$G_{producer} = \texttt{input\_value} > 0 \ \wedge \ \neg\texttt{queue.full()}$$
$$G_{consumer} = \neg\texttt{queue.empty()}$$

This automatic management of implicit conditions allows the programmer to abstract from tedious details, such as not trying to read data from an empty FIFO or not reading the result of the GCD operation until the algorithm has finished.

9.4 Elaboration and Laziness

In addition to rules and state elements, modules can contain conditional, iterative and recursive constructs. Inside rules, `if`, `case`, `for` and `while` blocks allow for conditional and iterative descriptions. These operations are unfold during compilation and converted into hardware. For instance, the two branches of an `if` are unfolded into actual hardware circuits, but only one of them is active depending on the condition. Likewise, all the iterations of a `for` loop are unrolled and executed simultaneously.

The same constructs can be used outside the context of the rules. In this case, their behavior defines not how the hardware works, but how it is constructed. Similar to macros in C, which change the shape of the code, the BSV evaluates the module-context constructs in a compilation stage called *elaboration*. For instance, a `for` loop can replicate the effect of a rule.

Another characteristic of BSV is the *laziness*. This concept is inherited from functional languages like Haskell, and allows to perform some operations even if not all the type information is explicit, in a "lazy" manner. Java and C++ have limited versions of lazy evaluation: polymorphic classes can operate with arbitrary data types. BSV interfaces and modules can be polymorphic, but it is much more extended and many constructs can be defined in a lazy manner. The hash notation, as in `Bit#(32)` or `Vector#(32, Reg#(int))`, denotes a polymorphic type and the inner types.

Listing 9.5 Elaboration with functional and lazy constructs

```
Vector#(32, Reg#(int)) vectorRegs <- replicateM(mkRegU);

function Bit#(n) incmsb(Bit(n) x) = x + x[valueOf(n)-1];

for ( Integer i = 0; i < 32; i = i + 1 ) begin
  rule inc_vector;
    let a = incmsb(vectorRegs[i]);
    vectorRegs[i] <= a;
  endrule
end
```

In this code there are examples of iterative elaboration and lazy evaluation. The loop generates 32 rules that increase a vector of registers simultaneously. In addition to loops, elaboration can be performed with conditional constructs (i.e. hardware is generated selectively) or recursion (i.e. a module that instantiates itself). The registers are increased using a polymorphic function, `incmsb`, which takes an arbitrary-length bitstream and adds the value of its most significant bit.

9.5 Types and Typeclasses

Data types in BSV are based on primary types and data structures. Primary types include scalar data types like booleans, integers and bit streams. As in software languages, primary data types can be composed to form new data types. The C-like `typedef`, `struct`, `union` and `enum` constructs can be used for that purpose. More advanced data structures allow for field selection using pattern matching, like *tagged unions*. In the following example, the primary Bool type is defined as an enumeration with two values, and a standard IEEE double precision floating point is defined as a structure with 3 fields:

Listing 9.6 Data type construction in BSV

```
typedef enum {True, False} Bool deriving(Bits, Eq);

typedef struct {
  Bit#(1) sign;
  Bit#(11) exp;
  Bit#(52) mantissa;
} IEEE754_FP deriving(Bits, Eq);
```

The `deriving` keyword specifies what *typeclasses* this type belongs to. *Typeclasses* specify properties that their associated types must comply. For instance, all the types that belong to the `Bits` *typeclass* can be converted back and forth to a bit stream, and the `Eq` *typeclass* allows a data type to be used with the equal and not equal operators (`==`, `!=`). In the previous example, the Bool type can be converted to a 1-bit (`Bit#(1)`) value. The `deriving` command asks the compiler to automatically generate this translation, but the user could manually implement the characteristics required by the *typeclasses* (the conversion to bits, or the equal operator).

There are 15 standard *typeclasses*, among them `Bits`, `Eq`, `Arith` (the type must be compatible with the typical arithmetic operations), `Ord` (comparison operators like $<$ or $\geq$), `Bitwise` (typical bitwise operations like and, or, or shift) or `BitExtend` (the data type can be truncated or extended to match a given bit stream). But the user can define its own *typeclasses*. This functionality allows the programmer to impose characteristics to certain types. Such characteristic is useful when working with unknown types. That is the case of polymorphic interfaces and modules, like FIFOs, which can force the type argument to comply with certain desired properties, such as being convertible to bits or supporting arithmetic operations.

The required *typeclasses* in polymorphic modules are specified with the *provisos* construct. This mechanism allows not only to specify *typeclasses*, but also to perform some other advanced properties over the types.

Listing 9.7 Polymorphic interface and module with type provisos

```
interface GCD#(type t);
  method Action start(t newa, t newb);
  method t result;
endinterface

module mkGCD (GCD#(t))
  provisos(
    Bits#(t, t_sz),
    Arith#(t),
    Add#(t_sz, unused, 32));
  ...
endmodule
```

In Verilog and VHDL, this operations are defined as generics and parameters. In the previous example, the `mkGCD` module imposes several typeclasses over the generic type `t`: it should be convertible to bits (resulting in a bit stream of size `t_sz`), it should support arithmetic operations, and its size in bits should not be greater than 32: using the special `Add` provisos, we imposed the property $t\_sz + x = 32$, where x is a non-negative integer deduced by the compiler.

The language comes with a set of standard types, called the Standard Prelude package. This library has dozens of types, from basic scalar types to FIFOs, vectors, a finite state machine sublanguage, clock management functions or pseudo-random number generators.

9.6 Compiling, Debugging and the Verilog Interface

The BSV compiler can generate C++ and SystemC code for simulation, and Verilog code for hardware synthesis. This compilation process can be guided with *attributes*, similar to C *pragmas*. These attributes are specially useful to give hints to the rule scheduler, for instance to give priority to a rule over another. BSV also supports some debugging statements, inherited from SystemVerilog, like the `$display` printf-like command.

The Verilog interface is specially useful when interfacing already existing hardware. In contrast to other languages, which can interface Verilog *by default* (if the Verilog modules are adapted to the computation model), BSV supports importing external Verilog modules through a well-defined interface. The Verilog external ports are groped into methods. The enable and ready ports are hidden as implicit conditions, which are taken into account when scheduling the rules.

9.7 Summary

Bluespec SystemVerilog is a rule-based language specifically targeted to hardware design. The *guarded atomic rules* are intuitive metaphors for hardware, and the language hides most of the tedious details such as the implicit conditions and safety signaling. The functional laziness of BSV allows to fold the code in succinct descriptions, and the advanced typing system (along with an exhaustive standard library) raises the abstraction level and increases the productivity. The static scheduling performed by the compiler and the strong type system guarantees correctness by design, reducing the verification time.

It is a complex and singular language, and it may be challenging for new users. But it provides a good balance between productivity and a very tight control over the resulting hardware: BSV models can generate identical results, in terms of performance and resource usage, to hand-coded Verilog designs, while keeping a high level of abstraction [ANS$^+$14].

Part II
Academic HLS Solutions

Chapter 10
LegUp High-Level Synthesis

Andrew Canis, Jongsok Choi, Blair Fort, Bain Syrowik, Ruo Long Lian, Yu Ting Chen, Hsuan Hsiao, Jeffrey Goeders, Stephen Brown, and Jason Anderson

LegUp is a high-level synthesis (HLS) tool under active development at the University of Toronto since 2011. The tool is on its fourth public release, is open source and freely downloadable. LegUp has been the subject of over 15 publications and has been downloaded by over 1500 groups from around the world. In this section, we overview LegUp, its programming model, unique aspects of the tool versus other HLS offerings, and conclude with a case study.

The input to LegUp is a `C`-language program, as well as constraints specified in a `Tcl` file. LegUp is implemented as back-end passes of the open-source LLVM compiler framework [LA04]—the same framework used by Altera and Xilinx for their OpenCL software development kit (SDK) and Vivado HLS products, respectively. By default, `-O3` optimizations are executed prior to HLS, including dead-code elimination, constant propagation, loop unrolling, and others [HLC$^+$15]. LegUp HLS then operates in one of two modes: (1) pure hardware or (2) hybrid. In the former, the *entire* input program is synthesized to a hardware circuit, in Verilog register-transfer level (RTL). In the latter, the program is synthesized to a hybrid system comprising a processor and one or more hardware accelerators. In the hybrid mode, a portion of the program runs in software on the processor and a portion is executed in hardware. We elaborate on the hybrid flow further below.

The pure hardware flow of LegUp produces generic Verilog that is FPGA-vendor agnostic. However, the performance models included in the LegUp distribution are specifically for Altera FPGA families, and thus, better performance is to be expected when the target platform is Altera. The hybrid flow, on the other hand, uses Altera-

---

A. Canis • J. Choi • B. Fort • B. Syrowik • R.L. Lian • Y.T. Chen
H. Hsiao • S. Brown • J. Anderson (✉)
Department of Electrical and Computer Engineering, University of Toronto, Toronto, ON, Canada

J. Goeders
Department of Electrical and Computer Engineering, University of British Columbia, Vancouver, BC, Canada
e-mail: janders@ece.toronto.edu

D. Koch et al. (eds.), *FPGAs for Software Programmers*,
DOI 10.1007/978-3-319-26408-0_10

specific bus primitives to connect the processor with accelerators and thus, at present it can *only* be used for Altera FPGAs.

With respect to `C` language support for HLS, LegUp supports pointers, structs, loops, arrays, integer and floating point computations. Synthesis of recursive functions is not supported, however, in the hybrid flow, such functions may run on the processor. Likewise, LegUp requires that memory be statically allocated; there is no automated synthesis of `malloc/free`.

10.1 Hybrid Flow

A unique feature of LegUp is the ability to generate hybrid systems incorporating a general-purpose processor and one or more custom hardware accelerators. Hybrid system generation requires no source code modification, only that the user specify the functions to accelerate in the constraints `Tcl` file. LegUp can target both MIPS I and ARM Cortex-A9 processor architectures, and provides application profiling for both of these architecture to help the programmer choose which function to accelerate.

10.1.1 Hybrid Flow Overview

The LegUp hybrid design flow is illustrated in Fig. 10.1. The designer starts with a software implementation of the application in standard `C` code. At step `1`, the `C` program is compiled to a binary executable targeting a processor. At step `2`,

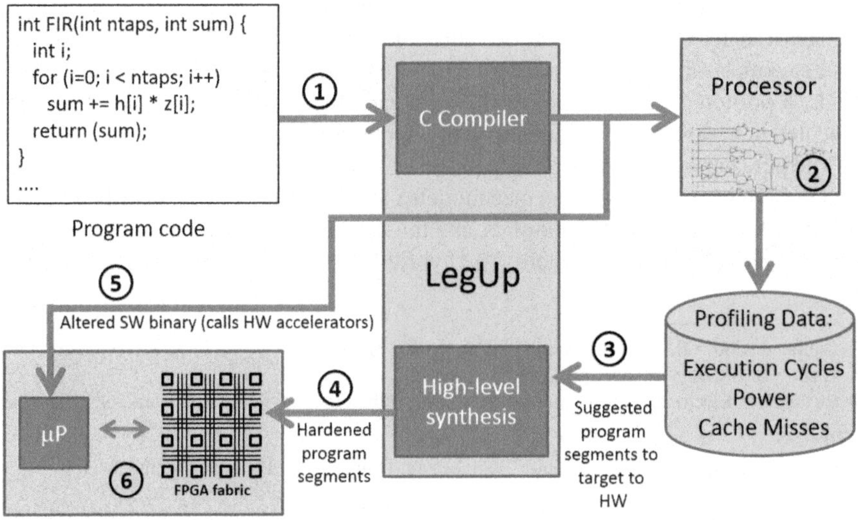

Fig. 10.1 Hybrid design flow with LegUp [CCA+13]

the application runs on the processor and profiling information is collected. The profiling data can be used to identify the critical sections of the program that can benefit most from hardware acceleration. At step `3`, the user designates which functions should be synthesized into hardware accelerators. At step `4`, LegUp's high-level synthesis engine is invoked to synthesize these functions into hardware accelerators described in Verilog RTL. Next, in step `5`, the `C` source is re-compiled with the accelerated functions replaced by wrapper functions, which are used to invoke the hardware accelerators. Lastly, the hybrid processor/accelerator system executes on the FPGA in step `6`.

10.1.2 MIPS Hybrid System Architecture

The MIPS I hybrid system is composed of the soft processor, an instruction cache, a data cache, and one or more hardware accelerators. The accelerators are connected directly to the data cache, ensuring data coherency with the processor. The system also contains profiling hardware [AABC11] that collects data about the application running on the processor. For each function in the program, the profiler collects the number of instructions executed, the number of cycles due to instruction and data stalls, and the total cycle count for the function. The user is presented with a table showing this information, and can then decide which functions to accelerate.

10.1.3 ARM Hybrid System Architecture

Altera and Xilinx provide soft processors called Nios II and Microblaze, respectively, which are implemented in the FPGA fabric. The performance of these processors is limited and highly dependent on the FPGA family being used. Recently, the FPGA vendors have incorporated hard processors tightly coupled with the FPGA fabric. The current generation of low-cost SoC FPGA devices include a hard dual-core ARM Cortex-A9 processor, while high-performance SoC FPGA devices include a quad-core 64-bit ARM Cortex-A53 processor.

The ARM-based hard processor system (HPS) on Altera's SoC devices contains the dual-core processor, a memory controller, a memory management unit, and peripherals. Each processor has private L1 caches, and access to a shared L2 cache. The hard processor system also allows FPGA peripherals to make cache-coherent memory accesses.

The hard processor system also includes hardware performance monitoring units for each processor. During compilation in step `1` of Fig. 10.1, the `C` program is automatically instrumented with calls to profiling helper functions that maintain event and cycle counts for each function. The profiling information is collected while the application runs on the processor, and is displayed to the user once the application has finished.

10.1.4 Hybrid Flow Summary

The hybrid flow allows partitioning of a software program into both hardware accelerators and the remaining software. This is particularly beneficial for programs that contain constructs such as recursion, system calls, and dynamic memory allocation that cannot be synthesized into hardware; code containing these constructs can be run on the processor.

10.2 Programming Model

LegUp, as with many HLS tools, exhibits *syntactic variance* in that the quality of results produced depends on the style of the `C` input to the tool. A few rules of thumb with respect to coding style are helpful to achieving a higher quality implementation.

Eliminating control flow: Loop pipelining, as described below, exploits parallelism by allowing loop iterations to overlap with one another during hardware execution. LegUp HLS is only able to pipeline loops wherein the loop body is a single basic block (i.e. it has no control flow). Avoiding `if-else` in the loop body, and replacing such constructs with the `C` ternary operator (`<condition> ? <expr1> : <expr2>`) is encouraged to allow loop pipelining to succeed. Broadly speaking, eliminating control flow is useful as it simplifies the finite state machine controller synthesized by LegUp.

Memory parallelism: The underlying memories in Altera and Xilinx FPGAs are dual-ported, implying that at most two memory operators are permitted per cycle. For memory-intensive code, this limitation may result in longer schedules and worse performance. To mitigate this, LegUp generally locates each array in a different memory, each of which can be accessed in parallel.[1] Hence, partitioning arrays into sub-arrays that are independently accessed in different parts of the code may improve performance.

Strength reduction: Where possible, costly arithmetic instructions should be replaced by lightweight, reduced precision, or logical operations. For example, using fixed point instead of floating point will bring significant area and performance benefits. Similarly, implementation of unsigned division with shift (where possible) will offer better performance and area.

Unlike many HLS tools, LegUp uses no specialized datatypes, nor does it make use of pragmas inserted in the code. The `C` program is unaugmented ANSI `C`. Constraints are provided to LegUp through a `Tcl` file. For example, the following `Tcl` command is used to constrain the number of dividers in the hardware to one:

```
set_resource_constraint divide 1
```

[1]An exception to this is for cases wherein it cannot be statically determined which array is pointed to.

A constraints guide is available on the LegUp website, describing the dozens of different `Tcl` constraints supported by LegUp HLS.

10.3 LegUp Differentiators

Aside from the capability to automatically generate a hybrid processor/accelerator system in a push-button manner, LegUp HLS has several features not commonly found in HLS tools.

10.3.1 Pthreads and OpenMP Support

LegUp provides HLS support for *Pthreads* and *OpenMP* [CBA13], standard parallel programming methodologies that software engineers are already likely familiar with. Pthreads can be used to execute the *same* or *different* functions in parallel by using `pthread_create` and `pthread_join`. With OpenMP, the user is able to parallelize a section of code by simply using a pragma, `omp parallel`. In LegUp, parallelism specified in software is automatically synthesized into parallel hardware accelerators. Each software thread is automatically mapped into a concurrent hardware module.

In multi-threaded programming, synchronization can be used to ensure that the program executes correctly. To this end, LegUp provides support for two key thread synchronization constructs, *mutexes* and *barriers*. Mutex and barrier variables in software are replaced with hardware mutex and barrier modules, which connect to parallel hardware modules to ensure atomicity and provide synchronization among concurrent modules.

The use of Pthreads and OpenMP are supported both in the pure hardware flow, as well as in the hybrid flow. In the pure hardware flow, the entire program, both the sequential and the parallel segments of software, are synthesized to hardware. The sequential module invokes multiple parallel modules, and retrieves their return values (if any) when they have completed execution. In the hybrid flow, which includes an embedded processor (*soft* MIPS or *hard* ARM) in the system, parallel code segments are synthesized to concurrent hardware accelerators. The remaining (sequential) portions of the program are executed in software on the processor. The processor calls parallel accelerators and retrieves their return values by using *wrapper functions*, which are automatically generated to replace the original parallel functions in software. From software/hardware partitioning, to compilation of software threads to hardware, to generation of the complete system including interconnect, on-chip cache, and off-chip interface, all steps are completely automatic in LegUp. This allows the user to easily create parallel hardware and exploit the spatial parallelism available on an FPGA. It also enables

Table 10.1 Pthreads/OpenMP support in LegUp

Pthreads functions	Description
pthread_create(..)	Invoke thread
pthread_join(..)	Wait for thread to finish
pthread_exit(..)	Exit from thread, can be used to return data
pthread_mutex_lock(..)	Lock mutex
pthread_mutex_unlock(..)	Unlock mutex
pthread_barrier_init(..)	Initialize barrier
pthread_barrier_wait(..)	Synchronize on barrier object

OpenMP pragmas	Description
omp parallel	Parallelize a section of code
omp parallel for	Parallelize a for loop
omp master	Parallel section executed by master thread only
omp critical	Specify a critical section
omp atomic	Specify an atomic section
reduction(operation: var)	Reduce a var with operation

OpenMP functions	Description
omp_get_num_threads()	Get number of threads
omp_get_thread_num()	Get thread ID

design space exploration, where the user can simply vary the number of threads in software, to increase/decrease parallelism in hardware with different area trade-offs.

Table 10.1 shows a list of Pthreads and OpenMP library functions which are supported in LegUp. An input software program using any of the functions/pragmas shown in the table can be used *as is*, requiring no manual code changes by the user.

10.3.2 Multi-Cycling

Multi-cycling is a well-known optimization technique in sequential circuit design, wherein selected combinational paths are permitted to have delay larger than a one clock period, for the purpose of raising the overall clock frequency of the circuit. The optimization can be applied in cases when the result being computed by a combinational sub-circuit (within a larger circuit) is not required in the clock cycle immediately following the cycle in which the sub-circuit's computation commenced. Consider a circuit containing a combinational sub-circuit A with delay 15 ns, and other sub-circuits having delays less than or equal to 10 ns. If sub-circuit A can be multi-cycled, the overall circuit can operate with a 10 ns clock period; otherwise, a 15 ns period must be used.

From the HLS perspective, multi-cycling opportunities can be discovered in the scheduling results, when an operation scheduled in cycle i produces an output that is not consumed until cycle j, where $j - i > 1$. LegUp automatically discovers such opportunities, avoids registering i on the next clock edge, and prints out multi-cycle constraints for the Altera downstream tools, making them aware of the multi-cycle paths (for proper timing analysis, and also for timing optimization).

Besides automatically identifying multi-cycle paths, LegUp incorporates the capability to optimize schedules to create more multi-cycling opportunities, based on profiling an application in software [HCS+15]. In this flow, an application is first executed in software and the number of executions of each basic block is ascertained. For infrequently executed basic blocks, their schedules are then *stretched* to produce multi-cycled paths, ensuring the logic for such paths will not reside on the overall critical path of the synthesized circuit.

10.3.3 Beta HLS Debugging

LegUp includes a debugger tool which allows a user to debug a circuit, implemented on an FPGA, in the context of the original source code. When in-system debug is enabled, LegUp will automatically insert debugging circuitry into the Verilog RTL. This circuitry allows a debugger application, running on a workstation, to connect to, control, and observe the HLS circuit. The debugger application, shown in Fig. 10.2, is designed to look and behave like a software debugger, and provides features such as single-stepping, breakpoints, and inspecting variable

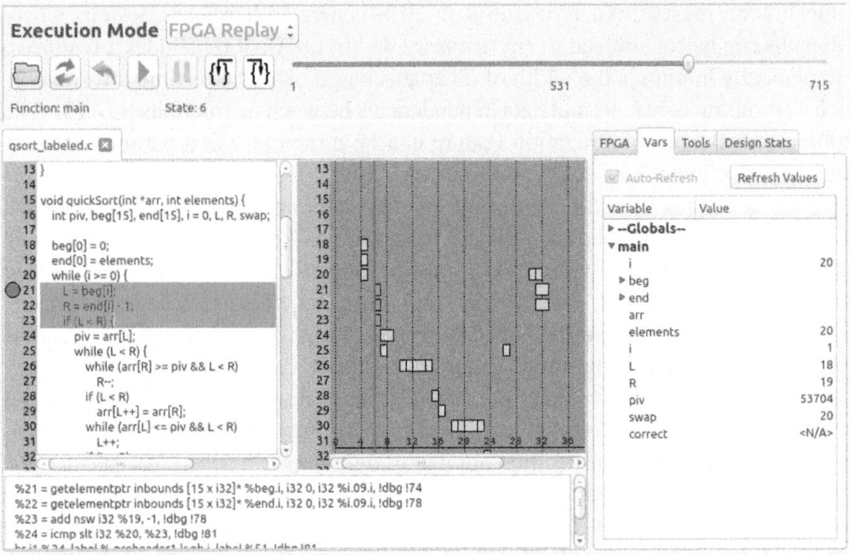

Fig. 10.2 Screenshot of the source-level debugger application

values. The left pane provides the source-code, with currently executing instructions highlighted, and the right pane contains the value of source-code variables. A Gantt chart is included to illustrate the scheduling of the underlying LLVM IR instructions, and the bottom pane lists the currently executing IR instructions. Presently, the debugger tool is in "beta" and only works with the pure hardware flow of LegUp.

A key feature of the debugger is the *Record and Replay* mode of operation. This allows the circuit to be run at-speed, while important signals are recorded into memories within the FPGA. Once a breakpoint is hit, the debugger tool connects to the FPGA, retrieves the execution trace from memory, and allows the user to perform debugging using the recorded data. The user can step forward and back through the recording, and inspect variables. The memories within the FPGA can only store a portion of the entire ciruit execution; however, significant effort has been spent on efficiently storing and compressing the execution trace to increase the length of execution that can be captured, thus making it easier to locate the root cause of a bug [GW14, GW15].

The debugger also features a `C`/RTL verification tool, which executes the `C` code using *gdb* and simulates the Verilog RTL code using *ModelSim*, and reports any discrepancies to the user [CBA14a].

10.3.4 Automated Bitwidth Minimization

Software programs utilize standard datatypes of predefined widths to represent variables, regardless of whether or not all bits in the variable will be used. This is acceptable for programs that are meant to be run on the processor where fixed-width datapaths are present, but is wasteful in HLS-generated hardware designs where datapaths can be customized to any arbitrary width. LegUp 4.0 includes a feature to automatically minimize the width of datapaths based on compile-time information, such as program constants and data dependencies between instructions [GA13]. The automated bitwidth minimization feature can be turned on via a parameter in the constraints `Tcl` file.

In the automatic bitwidth minimization pass, a bitmask is associated with each variable, denoting whether each bit in the variable is known (logic-`0` or logic-`1`), unknown, or a sign bit. The full width of the datapath can be shrunk to *number of unknown bits + 1 sign bit*, and all other bits can be statically assigned to the correct value. The bitmask of each variable is propagated through the control data-flow graph (CDFG) of the program to update the bitmask of the result variable in any instruction that uses the propagated variable as an operand. A series of forward and backward propagations through the CDFG is performed until there are no changes in any bitmasks, or until a specified number of iterations is reached. As an example, assume there exists two variables of unknown 16-bit values, *A* and *B*, and consider the `C`-language statement: `Z = A & (B << 2)`. In this case, the two rightmost bits of the intermediate result of `(B << 2)` are guaranteed to be logic-`0`, the two rightmost bits of *Z* are guaranteed to be logic-`0`, and the two rightmost bits of *A* can safely be ignored. In addition to minimizing datapath size, the size of functional units can also be minimized (i.e. if Z feeds into a multiplier, the two rightmost bits of the product are guaranteed to be logic-0).

10.3.5 Loop Pipelining

Loop pipelining is a high-level synthesis scheduling technique that overlaps the execution of loop iterations to achieve higher performance. We use this schedule to generate a pipelined datapath in hardware for operations within the loop, increasing parallelism and hardware utilization.

In many C applications, the majority of run time is spent executing critical loops. Consequently, loop pipelining is crucial for generating a hardware architecture with comparable performance to hand-designed RTL. Furthermore, complex loops usually have resource constraints, typically caused by limited memory ports, in combination with constraints imposed by cross-iteration dependencies. The interaction between multiple constraints can pose a challenge for loop pipelining scheduling algorithms, which, if not handled properly can lead to a loop pipeline schedule that fails to achieve the best performance.

LegUp uses a novel loop pipelining scheduling algorithm with backtracking to handle complex loops with competing resource and dependency constraints [CBA14b]. This scheduler is based on the SDC scheduling formulation [CZ06] allowing for a flexible range of user constraints. During scheduling, LegUp will abandon any infeasible partial schedules and then backtrack by attempting other possible scheduling combinations. Backtracking can lead to better schedules than the prior greedy approach [ZL13] in cases where the priority ordering prevents the discovery of a valid schedule in a single pass. LegUp also applies algebraic transformations to the loop's data dependency graph using operator associativity to reduce the length of recurrences.

10.3.6 Multi-Pumping of DSP Units

Modern FPGAs contained hardened Digital Signal Processor (DSP) units that can be used to realize multiply/accumulate functionality with higher speed and lower power relative to implementing the same functionality with soft logic (look-up-tables and flip-flops). The DSP units can typically operate very fast—in the 500+ MHz range—which is 2× the clock speed of an average FPGA design. In LegUp, we exploit this property by allowing the DSP units to be "multi-pumped" at 2× the system clock frequency [CAB13]. In this flow, a single DSP unit can be used to perform 2 multiply operations in one system clock cycle, thereby mimicking the presence of extra DSP units. This provides two benefits: (1) higher performance can be achieved for a constrained number of DSP units, as the HLS scheduler is able to schedule more multiply operations across fewer cycles, and (2) fewer DSP units can be used for a given level of performance, saving area.

10.4 Current Results

Across the releases of LegUp HLS, the quality of results produced has consistently improved. To illustrate this, this section compares the quality of results produced by the current version of LegUp against the first LegUp release. The comparison uses 13 benchmarks, which include all 12 benchmarks from the CHStone high-level synthesis benchmark suite [HTHT09] and Dhrystone [Wei84], a popular synthetic benchmark. The target device is an Altera Cyclone II FPGA, as it was the only device supported by LegUp 1.0. The study uses the pure hardware flow, and the timing constraint is set to an aggressive target to achieve the highest possible clock frequency.

Table 10.2 shows the geometric mean results for LegUp 1.0 compared to the current 4.0 version. The area results are given in terms of number of Cyclone II LEs, number of memory bits, and 9×9 multipliers, followed by performance metrics in execution cycles, Fmax (MHz), and wall-clock time (μs). The last row presents the ratio of the current LegUp geometric mean vs. LegUp 1.0.

The quality-of-results on all metrics are significantly improved since the first LegUp release. On average, wall-clock time is improved by 48 %, cycle count by 38 %, FMax by 19 %, LEs by 41 %, memory bit usage by 15 %, and multipliers by 8 %.

The majority of these improvements in circuit quality can be traced back to the following changes: Firstly, scheduling of `phi`[2] and `branch` instructions, were handled differently such that they could be chained to other operations to decrease the number of clock cycles. The circuit clock frequency was also improved by removing the combinational loops that could occur in the binding step of LegUp. Dual-port memories are now used instead of single-port memories, allowing for greater instruction-level parallelism. In 2010, the `jpeg` benchmark from the CHStone benchmark suite was updated to contain an approximately 50 % smaller

Table 10.2 LegUp 1.0 vs. current LegUp 4.0 geomean results (hardware-only implementation)

	LEs	Mem bits	Multipliers	Cycles	Freq (MHz)	Time (us)
1.0	14,328	31,236	9.6	19,738	73	272
Current	8492	26,676	8.9	12,200	86	142
Ratio	0.59	0.85	0.92	0.62	1.19	0.52

[2]An instruction in the LLVM intermediate representation which selects incoming values based on the control flow (i.e. predecessor basic block).

image, which sped up the benchmark. Experimentation with different clock period constraints, which affects the scheduling of instructions, also led to greater geomean performance across the CHStone benchmarks. Finally, data accessed by a single function is put into localized memories inside the corresponding hardware module as opposed to global memory. This improves performance due to increased memory bandwidth. Also, small memories with the same data width are combined into grouped memories to save area.

10.5 Case Study: Sobel Filter

In this section, we attempt to answer the questions: how close is LegUp-generated hardware to hand-designed hardware? Or perhaps more importantly, can our proposed methodology generate circuits that can meet realistic FPGA design constraints?

To investigate these questions, we present a case study of a Sobel image filter. This filter is typically used in edge detection, which is important for computer vision applications. We will first describe the Sobel algorithm and present a straight-forward `C` implementation. We then describe the hand-written hardware implementation of the filter. Next, we provide an implementation using LegUp, showing the transformations we made on the `C` code in order to match the performance of the RTL implementation. For every implementation, we assumed a 512 by 512 pixel image. We show that LegUp can produce a filter with a wall-clock time within 2 % of the custom implementation, but with about 65 % more circuit area. This case study illustrates the types of code transformations we must currently apply for high-level synthesis to create a circuit of peak performance. Some of these transformations are non-obvious to a software engineer, which we hope will motivate future work.

10.5.1 Sobel Filter: `C` Code vs. Hardware

The Sobel filter is performed by convolution, using a three pixel by three pixel stencil window. This window shifts one pixel at a time from left to right across the input image, and then shifts one pixel down and continues from the far left of the image as shown in Fig. 10.3. At every position of the stencil, we calculate the edge value of the middle pixel `e`, using the adjacent pixels labeled from `a` to `i`. Two three by three gradient masks, Gx and Gy, are used to approximate the gradient in both x and y directions at each pixel of the image using the eight neighbouring pixels. The `C` source code for the Sobel filter is provided in Fig. 10.4. The outer two loops ensure that we visit every pixel in the image, while ignoring image borders (line 3). The stencil gradient calculation is performed on lines 4–11. The edge

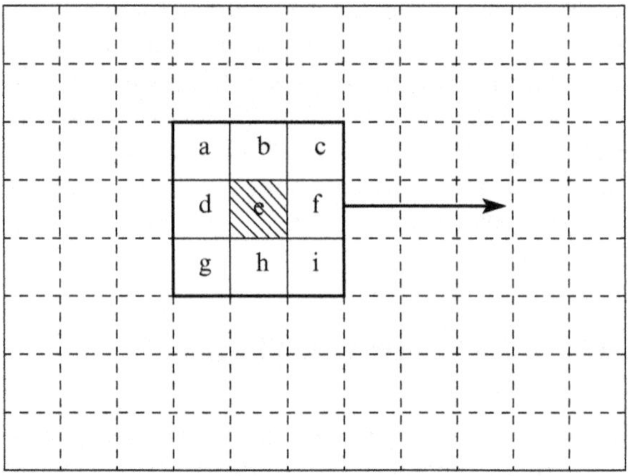

Fig. 10.3 Sobel stencil sliding over input image

```
1 for (y = 0; y < HEIGHT; y++) {
2   for (x = 0; x < WIDTH; x++) {
3     if (not_in_bounds(x, y)) continue;
4     x_dir = 0; y_dir = 0;
5     for(xOffset = -1; xOffset <= 1; xOffset++) {
6       for(yOffset = -1; yOffset <= 1; yOffset++) {
7         pixel = input_image[y+yOffset][x+xOffset];
8         x_dir += pixel   Gx[1+xOffset][1+yOffset];
9         y_dir += pixel   Gy[1+xOffset][1+yOffset];
10      }
11    }
12    output_image[y][x] = 255 - (bound(x_dir) + bound(y_dir));
13  }
14 }
```

Fig. 10.4 C code for Sobel filter

weight is calculated on line 12, where we bound the x and y directions to be from 0 to 255. Finally, we store the edge value in the output image.

A hand-coded RTL implementation of the Sobel Filter was provided by an experienced hardware designer. The hardware design assumes that a "stream" of image pixels are being fed into the hardware module at the rate of one pixel every cycle. The hardware implementation stores the previous two rows of the image in two shift registers, or line buffers. These line buffers can be efficiently implemented in FPGA block RAMs. Using the two 512-pixel wide line buffers, the hardware can retain the necessary neighbouring pixels for the stencil to operate, and update this window of pixels as each new input pixel arrives every cycle.

The hardware has some additional control to wait until the line buffers are full before the output edge data is marked as valid, and additional checks that set the output to zero if we are on the border of the image. This hardware implementation does not need an explicit FSM. To summarize, in steady state, this hardware pipeline receives a new pixel every cycle and outputs an edge value every cycle, along with

```
// line buffer shift registers
unsigned char prev_row[WIDTH] = {0};
int prev_row_index = 0;
unsigned char prev_prev_row[WIDTH] = {0};
int prev_prev_row_index = 0;

// stencil buffer:
unsigned char stencil[3][3] = {0};

inline void receive_new_pixel(unsigned char pixel) {

    // shift existing stencil to the left by one
    stencil[0][0] = stencil[0][1]; stencil[0][1] = stencil[0][2];
    stencil[1][0] = stencil[1][1]; stencil[1][1] = stencil[1][2];
    stencil[2][0] = stencil[2][1]; stencil[2][1] = stencil[2][2];

    int prev_row_elem = prev_row[prev_row_index];

    // grab next column (the rightmost column of the sliding stencil)
    stencil[0][2] = prev_prev_row[prev_prev_row_index];
    stencil[1][2] = prev_row_elem;
    stencil[2][2] = pixel;

    // shift in new pixel
    prev_prev_row[prev_prev_row_index] = prev_row_elem;
    prev_row[prev_row_index] = pixel;

    // adjust shift register indices
    prev_row_index++;
    prev_prev_row_index++;

    prev_row_index = (prev_row_index==WIDTH) ? 0 : prev_row_index;
    prev_prev_row_index = (prev_prev_row_index==WIDTH) ? 0 : prev_prev_row_index;
}
```

Fig. 10.5 C code for the stencil buffer and line buffers synthesized with LegUp

an output valid bit. The first valid output pixel is after 521 clock cycles, after which point an edge will be output on every cycle for the next 262,144 cycles (512×512). The first 513 valid edge output values will be suppressed to zero because we are still on the border of the image. Therefore, the custom RTL circuit has a total cycle count of 262,665, which is 0.2 % worse than optimal, where optimal is finishing right after the last pixel (after 262,144 cycles). The latency of 521 cycles is due to time spent filling up the line buffers in the hardware before we can begin computation.

We now describe the LegUp synthesized circuit, starting from the original C code in Fig. 10.4. By default, compiler optimizations built into LLVM will automatically unroll the innermost 3×3 loop (lines 5–11) and constant propagate the gradient values (lines 8–9). During constant propagation, the LLVM optimizations can detect the zero in the middle of each gradient mask allowing us to ignore the middle pixel during the iteration. Consequently, there are eight loads from the input image required during each outer loop iteration (lines 1–14), one for each pixel adjacent to the current pixel (line 7). The outer loop will iterate 262,144 (512×512) times. We have nine total memory operations in the loop, eight loads (line 7) and one store (line 12). We found that LegUp schedules the unmodified code into nine clock cycles per iteration, mainly due to the shared memory having only two ports and a latency of two cycles. This circuit takes 2,866,207 cycles to execute.

```
int sobel_opt(
    unsigned char input_image[HEIGHT][WIDTH],
    unsigned volatile char output_image[HEIGHT][WIDTH])
{
    int i, errors = 0, x_offset =-1, y_offset =-1, start = 0;
    unsigned char pixel, edge_weight;
    unsigned char   input_image_ptr = (unsigned char  ) input_image;
    unsigned char   output_image_ptr = (unsigned char  ) output_image;

    loop: for (i = 0; i < (HEIGHT) (WIDTH); i++) {
        pixel =  input_image_ptr++;

        receive_new_pixel(pixel);

        int x_dir = 0, y_dir = 0, xOffset, yOffset;
        for(yOffset =-1; yOffset <= 1; yOffset++) {
            for(xOffset =-1; xOffset <= 1; xOffset++) {
                x_dir += stencil[1+yOffset][1+xOffset]   Gx[1+xOffset][1+yOffset];
                y_dir += stencil[1+yOffset][1+xOffset]   Gy[1+xOffset][1+yOffset];
            }
        }
        edge_weight = 255- (bound(x_dir) + bound(y_dir));

        // we only want to start calculating the value when
        // the shift registers are full and the window is valid
        int check = (i == 512 2+2);
        x_offset = (check) ? 1: x_offset;
        y_offset = (check) ? 1: y_offset;
        start = (!start) ? check : start;
        int border = not_in_bounds(x_offset, y_offset) + !start;

        output_image[y_offset][x_offset] = (border) ? 0 : edge_weight;

        x_offset++;
        y_offset = (x_offset == WIDTH - 1) ? (y_offset + 1) : y_offset;
        x_offset = (x_offset == WIDTH -1) ? -1 : x_offset;
    }

    return errors;
}
```

Fig. 10.6 Optimized C code for synthesized Sobel Filter with LegUp

10.5.2 Sobel Filter: Enhanced C Code

We can perform various transformations on the C code to improve the generated hardware and bring its performance in line with the hand-coded hardware. The first transformation we can make is to use a stencil and two line buffers holding the previous two rows. The C code for this is given in Fig. 10.5, with the stencil stored in a nine element two-dimensional array on line 8. We shift the stencil after each new pixel arrives on lines 13–15, and shift new data into the stencil on lines 20–22. The two line buffers are implemented on lines 25–33 using arrays: `prev_prev_row` and `prev_row`. We have to manually keep track of an index to indicate where to shift data into and out of the arrays, with the index rolling over to zero when reaching the end of the array (lines 32–33). We can now calculate an edge value using only the stencil buffer, without reading from memory eight times every loop iteration. We also enable local memories, so that we are not constrained by the global memory controller ports.

Next, we need to enable loop pipelining, to overlap iterations of the outermost loop of the algorithm. But first we must merge the two outer loops into one loop and add a label "loop". We also change the array accesses to use pointer dereferencing to avoid unnecessary index calculations. Second, we must manually remove any control flow in the loop body to allow loop pipelining because automatic if-conversion is Beta functionality in LegUp. We do this by replacing any if statements with the ternary operator, (`<cond> ? <expr1> : <expr2>`). We show the new `C` code in Fig. 10.6, where the incoming pixel is read on line 11, the stencil and line buffers are shifted on line 13, and the edge weight is calculated on lines 15–22. We have added some new control variables, such as a check for when the stencil has been filled (line 26), a calculation of the output image x and y indices (lines 27–28 and lines 34–36), whether the output is now valid (line 29), and an additional check for whether we are on the image border (line 30). If we are on the border, we output a zero, otherwise we output the edge weight (line 32). There is only one load (line 11) and one store (line 25) in the loop body that go to the dual-ported shared global memory controller. Therefore, we can pipeline the transformed loop with an initiation interval of one. The circuit now finishes after 262,156 cycles, only 12 cycles worse than optimal. Although we do assume that the output image is already initialized to zero for the very last row of the image (the bottom border). We also set the LegUp clock period scheduling constraint as low as possible, to ensure a better final circuit FMax.

Some of the `C` transformations we have just described would be unintuitive to software developers, particularly using the line buffers and stencil to reduce memory operations in the loop. The user would have to be familiar with the concept of a pipeline initiation interval and the strategy of reducing memory contention in the loop body. Also, the user would have to rewrite all control flow in the loop to use the ternary operator.

We measured the results of the LegUp vs. hand RTL case study by targeting the Stratix IV [Alt10b] FPGA (EP4SGX530KH40C2) on Altera's DE4 board [Alt10a] using Quartus II 13.1SP2 to obtain area and FMax metrics. Quartus timing constraints were configured to optimize for the highest achievable clock frequency. The results are summarized in Table 10.3, with the custom hardware implementation shown in the first column side-by-side with the LegUp-synthesized results in the second column. In the third column, we compute the ratio of the two results: LegUp/Hand-RTL.

We found that after performing manual code transformations, LegUp produced a circuit with a wall-clock time within 2 % of the hand-written hardware implementation. However, the synthesized circuit area was larger, consuming 64 % more ALUTs and 66 % more registers.

We observed a few reasons for this increase in area. First, we are using many unnecessary additional registers due to the low LegUp clock period constraint, which causes scheduling to not chain any operations. We needed the clock period constraint to achieve an acceptable FMax but this indicates that our timing analysis and estimation needs improvement. Also, the pipeline produced by LegUp is needlessly complex and includes additional array indexing that did not exist in

Table 10.3 Experimental results

Metric	Hand-RTL	LegUp	LegUp/Hand-RTL
FMax (MHz)	191.46	187.13	0.98
Cycles	262,665	262,156	1.00
Time (ms)	1.37	1.40	1.02
ALUTs	495	813	1.64
Registers	382	635	1.66
Memory (bits)	6,299,616	6,299,616	1.00

the custom implementation. LegUp also generates a standard FSM when this application does not need one. Generally, the custom hardware implementation is very minimalistic and fits in a few short pages of Verilog (296 lines without comments). In contrast, LegUp's Verilog is 2238 lines and includes many unnecessary operations from the LLVM intermediate representation, such as sign extensions and memory indexing.

We expect that LegUp will perform well for hardware modules that are control-heavy and fairly sequential. For hardware modules with pipelining, we will also perform well, as long as the user can express the pipelining in a single C loop. These could include video, media, and networking applications. LegUp, and all HLS tools, will struggle with highly optimized hardware designs with a known structure, such as a fast Fourier transform butterfly architecture. Also, LegUp cannot generate circuits that have exact cycle-accurate timing behaviour such as a bus controller.

10.6 Summary and Future Work

LegUp is a high-level synthesis tool that is under active development and currently on its fourth public release. It can synthesize a C language program to a hardware circuit, or alternately, to a hybrid system comprised of a processor and one or more custom hardware accelerators. The unique aspects of LegUp relative to other HLS offerings include support for the synthesis of software threads, bitwidth optimizations, multi-cycling support, multi-pumping of DSP units, debugging, and advanced loop pipelining.

Future work for the project includes support for streaming applications, which are composed of feed-forward pipelined kernels interconnected by FIFO buffers. We also are exploring synthesis of memory architectures to permit greater memory-access parallelism in the Pthreads/OpenMP flow. Another research thrust is the high-level synthesis of approximate circuits, honoring precision requirements specified by the user. More about the LegUp project can be found on the project website: http://legup.eecg.toronto.edu

Chapter 11
ROCCC 2.0

Walid A. Najjar, Jason Villarreal, and Robert J. Halstead

Riverside optimizing compiler for configurable computing (ROCCC) was started as a project at the University of California, Riverside in 2002. To put in a historical context: FPGAs were much smaller, and slower, then they are today (2015); Graphics Processing Units (GPUs) were used exclusively for *graphics*; *reconfigurable computing* was taking shape as a research area but not yet within the main stream of academic research, let alone in industrial production. However, multiple research projects had already demonstrated, many times over, the clear advantages and potentials of this nascent paradigm as an alternative that combines the re-programmability advantages of fixed data path devices (Central Processing Units (CPUs), DSPs and GPUs) with the high speed of custom hardware (ASICs). Within that time frame, the nearly exclusive focus of reconfigurable computing was on signal and image processing because of their streaming nature. Video processing was considered a future possibility to be realized when the size (area) and bandwidth capabilities of FPGAs got larger.

In this section we provide an overview of the ROCCC toolset, its approach to the compilation of FPGA-based accelerators from HLL and examples. It is beyond the scope of this section to cover every aspect of this 10 years long project.[1]

[1]The source code and documentation (including User and Developers Manuals), on ROCCC can be found on https://github.com/nxt4hll/roccc-2.0 and http://roccc.cs.ucr.edu.

W.A. Najjar (✉) • J. Villarreal • R.J. Halstead
University of California, Riverside, CA, USA
e-mail: najjar@cs.ucr.edu

D. Koch et al. (eds.), *FPGAs for Software Programmers*,
DOI 10.1007/978-3-319-26408-0_11

11.1 The ROCCC Approach

From its inception, ROCCC was designed as a C to HDL compilation tool explicitly intended for generating *code accelerators* rather than a general purpose High-Level Synthesis (HLS) tool. Why this distinction? Typically, accelerators have their semantic root in a *loop nests* while a general HLS tool can target any arbitrary C code. Hence, the focus of the ROCCC compiler transformations has been on loop nests.

The objectives of the code transformations and optimizations implemented in ROCCC are three folds:

1. *Parallelism.* Code transformation techniques, such as loop unrolling and others, can potentially extract massive parallelism from loop nests at both the iteration and operation (instruction) levels. A major impediment to the extraction of parallelism is the presence of memory aliases which is why ROCCC does not support arbitrary pointer operations within loops. ROCCC supports an expensive set if loop transformations including: unrolling, fusion, temporal sub-expression elimination, window narrowing, systolic array generation.

   The two limitations on parallelism on an FPGA are *area* and *bandwidth.* Loop transformations can have a major and non-intuitive impact on the area utilized by the generated circuit [BCVN10]. Loop fusion, in some instances can actually shrink the occupied area below that of the larger of the two loops [NBD+03]. The area limitation is strictly associated with the target device (i.e. the FPGA itself) while the bandwidth limitations is most often determined by the architecture of the target platform.[2]
2. *Data Reuse.* Exploiting locality in streaming data requires on-chip storage mechanism customized for each specific loop nest. A sliding window over a 1-, 2- or 3-d array is typical in signal, image and video processing. Locally storing fetched data for future reuse can same significant bandwidth at the cost of higher area utilizations and possibly a lower clock cycle. However, it can boost parallelism at practically no cost: the data for the next iteration is already on chip.

   The *smart buffer* [GBN04, GNB08] is a compiler generated mechanism for re-using fetched data for sliding windows loop nests. The structure of the buffer is determined by the window size, array size and dimensions and the stride of the reuse in each dimension.
3. *Clock Cycle Time.* Exploiting parallelism through loop transformations and locality through data reuse has negative impacts on the clock cycle time because of the implied longer wires. Hence the importance of pipelining and other low-level circuit optimizations, such as tree balancing etc.

[2] We use the term target platform to indicate the architecture of the board around the FPGA device(s) including memory banks, I/O ports etc.

In the early 2000s it was clear that no first- or even second-generation HLS tool will be able to completely abstract away the complexities of programming FPGAs and make the process as streamlined as traditional software development. However, in designing ROCCC we sought to make that process easier, improve it productivity and, most of all, provide the user with easy to use tool to undertake a *design space exploration* of target application on a given FPGA device and platform. From experience, we knew how tedious and time consuming the design space exploration process is when done manually at the HDL level. But we also had extensive experience with the non-intuitive impact of loop transformations on the circuit's area and throughput as shown in [NBD$^+$03] and [BCVN10]:

ROCCC was therefore designed with the following principles:

- *User in control.* The ROCCC toolset is designed for a user who has some knowledge of the digital design process and the FPGA programming tool chain but most of all has an intimate knowledge of the application at hand as well as the target platform. The toolset gives the user control over which transformation to apply to each individually labeled loop within a loop nest as well as the order of application of these transformations. Under some conditions, it would have been possible to let the compiler automatically detect which transformations are feasible and select an acceptable option, however the scope and implications of unintended consequences was too large. Early on, we had realized that the impact of loop transformations was not always intuitive to most users [BCVN10].
- *Design space exploration.* As mentioned above, the ultimate objective in the design of an accelerator is to obtain an optimal tradeoff between area, bandwidth and an achievable clock frequency. The ROCCC user interface relies on a GUI, rather than in code pragmas, to indicate which optimizations should be applied to which loop. The objective being to allow the user to generate different projects for different sets of optimizations applied to the same unmodified source code. The GUI allows the user to specify what high-level optimizations are to be applied (e.g. loop transformations) and modify the default settings on low-level transformations (e.g. pipelining).
- *Code reuse.* The Eclipse-based ROCCC GUI supports the saving and reuse of code structures, such as modules and systems (see Sect. 11.2), in a MySQL database. A kernel or a whole application can be saved either as a C source, as VHDL code or as IP core netlist and reused, by drag-and-drop, as is in other programs. Obviously, a C source would benefit from the compiler transformations while the VHDL or IP core forms would not.This feature gives the user the flexibility of fully integrating the imported code into an application or using a entity with know properties (e.g. area and clock cycle time).
- *Platform independent.* The platform architecture surrounding the FPGA device(s) plays a crucial role in determining the achievable performance. This architecture includes (1) one or more memory banks and the memory interconnect, (2) the I/O ports to/from the board, and (3) the interface to the host software (typically via a PCIe bus). The code generated by the ROCCC toolset

is platform independent: all the inputs and outputs to that code are buffered (in BRAM when available). It is up to the user to determine the compiler transformations that best fit a given platform. The ROCCC generated code can be interfaced to the physical components on a platform either manually or by using a wrapper customized for that platform. The ROCCC distribution includes wrappers for the Pico Computing M501 board and the Convey Computers HC-1 and HC-2ex. The Developer Manual includes instructions on how to build a new wrapper.

11.2 Structure of ROCCC Code and Examples

ROCCC supports two styles of C programs, which we refer to as modules and systems. Modules represent concrete hardware implementations of purely computational functions. Modules can be constructed using instantiations of other modules in order to create larger components that describe a specific architecture. System code performs repeated computation on streams of data. It consists of loops that iterate over arrays. System code may or may not instantiate modules. System code represents the topmost perspective and generates hardware that interfaces to memory systems.

ROCCC is not designed to compile entire applications into hardware and has certain general restrictions on both module and system code. Constructs that ROCCC 2.0 does not support include: generic pointers, non-component functions, including C-library calls, non-for loops, stream accesses other than those based on a constant offset from loop induction variables

Module code represents a hardware building block to be used in larger applications. Modules are computational data-paths and are written as computational functions. All inputs to modules are passed in by value and all outputs are passed by reference. Inputs must only be read from and output ports can only be written to inside the function. ROCCC does not support writing to an output port multiple times inside the function. Modules can only process scalar values and cannot have arrays as input or output variables. Internal variables may be created but are not visible outside of the module.

System code performs computation on streams of data and produces streams of data. Scalars may also be read as input and generated as output, but as opposed to modules, input scalars are read once at the beginning of computation and output scalars are only generated once at the end of computation.

Similar to module code, system code is written as a void function that takes input and output parameters. Input scalars are passed by value, output scalars are passed by reference, and both input and output streams are passed as pointers. The function definition must declare inputs before outputs. Although passed as pointers, the internal use of streams must be through array accesses.

Figure 11.1 show an example an FIR module source code (a) and the corresponding data-flow graph (b) with constant weights. The system code invoking the FIR

a

```
// Example module code
//  Input parameters must
//  come before output
//  parameters
void FIR(int A0, int A1,
         int A2, int A3,
         int A4,
         int& result)
{
  const int T[5] =
    {3,5,7,9,11} ;
  result = A0 * T[0] +
           A1 * T[1] +
           A2 * T[2] +
           A3 * T[3] +
           A4 * T[4] ;
}
```

b

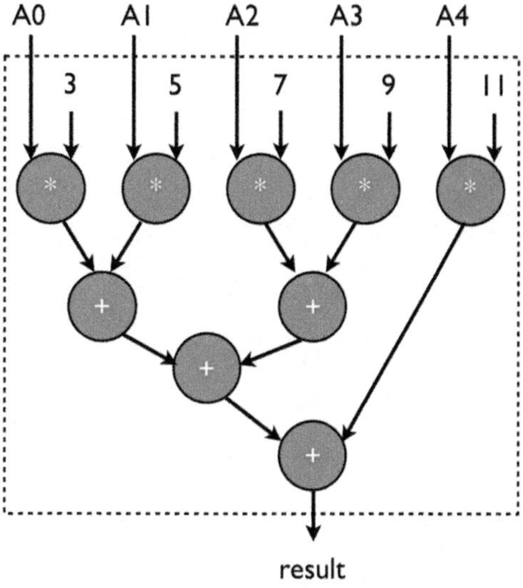

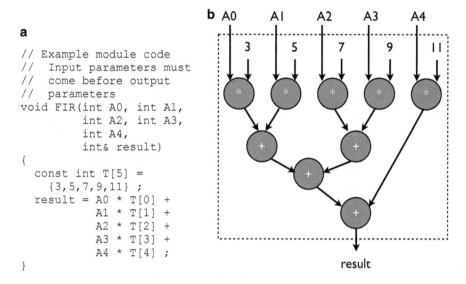

Fig. 11.1 Example of FIR module, C code (**a**) and data-flow graph (**b**)

module is shown in Fig. 11.2 as source C code (a), its data-flow graph (b) and the data-flow graph after inlining and full unrolling of the loop (c). Note that instead of a black box the top level design has all of the individual operations exposed and may perform additional optimizations on this code. The schematic of the FIR system generated in hardware is shown in Fig. 11.3.

The high-level code transformations supported by ROCCC are summarized in Fig. 11.4.

Systolic Array Example the ROCCC optimizations can be used for the generation of a systolic array when compiling wavefront-style codes such as dynamic programming. An example is shown in Fig. 11.5. The source code (a) would be executed on a two dimensional systolic array (b). After the optimizations, it is compiled into a linear systolic array (c). The occupancy of the elements of the linear array is shown in (d). The schematic of the overall complete systolic system is shown in Fig. 11.6.

11.3 Compiled Hardware Accelerated Threads (CHAT)

The Compiled Hardware Accelerated Threads (CHAT) tool is designed to assist developers with implementing irregular applications on FPGAs [HVN14]. CHAT is built using the ROCCC toolset. Like ROCCC it is platform and memory independent allowing for easy porting to emerging FPGA architectures. However, while the focus of ROCCC is on generating highly optimized kernels for streaming applications, the focus of CHAT is on kernels for irregular applications (Fig. 11.7).

a

```
void FIRSystem(int* A, int* B)
{
  int i ;
  int tmp ;
  for (i = 0 ; i < 10; ++i)
  {
    // Module instantiation
    FIR(A[i], A[i+1], A[i+2],
        A[i+3], A[i+4], tmp) ;
    B[i] = tmp ;
  }
}
```

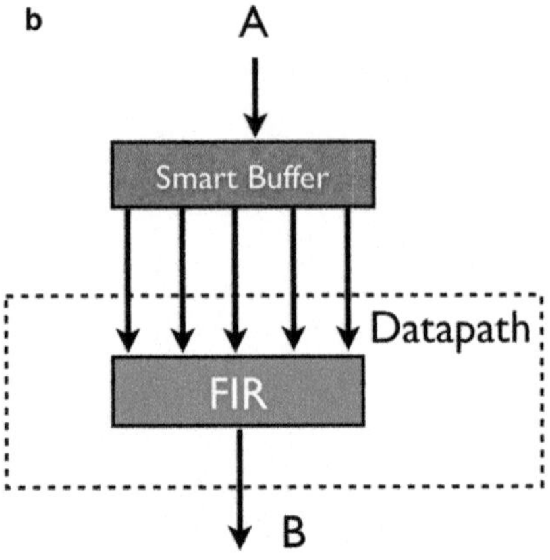

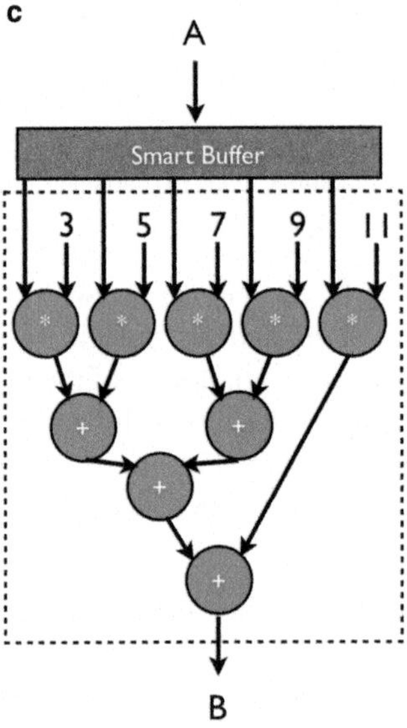

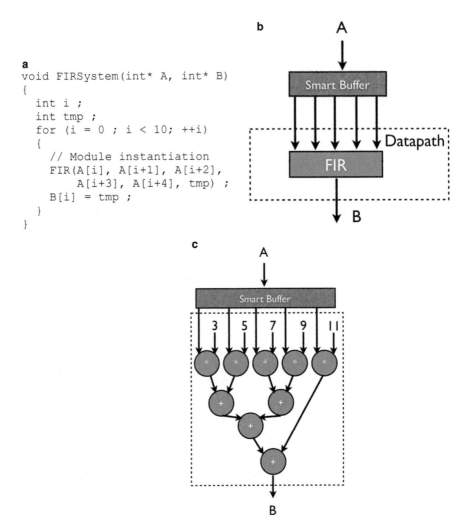

Fig. 11.2 FIR system code and data-flow graph. (**a**) C code. (**b**) Generated hardware. (**c**) After inlining

The CHAT Execution Model. By definition irregular applications have poor spatial and temporal localities, their performance is limited by memory latency. MT-FPGA (Multithreading on FPGAs) [HVN14] is an execution model that combines the memory masking ability of multithreaded execution with a customized data path. The execution model is depicted in Fig. 11.5: a ready thread executes on the customized data-path until it performs a memory read at which point it relinquishes execution and is moved to the waiting queue. The data-path being pipelined allows a new ready thread, if available, to start executing every cycle. Buffering the returned data in the order it was requested allows the decoupling of the data fetching from

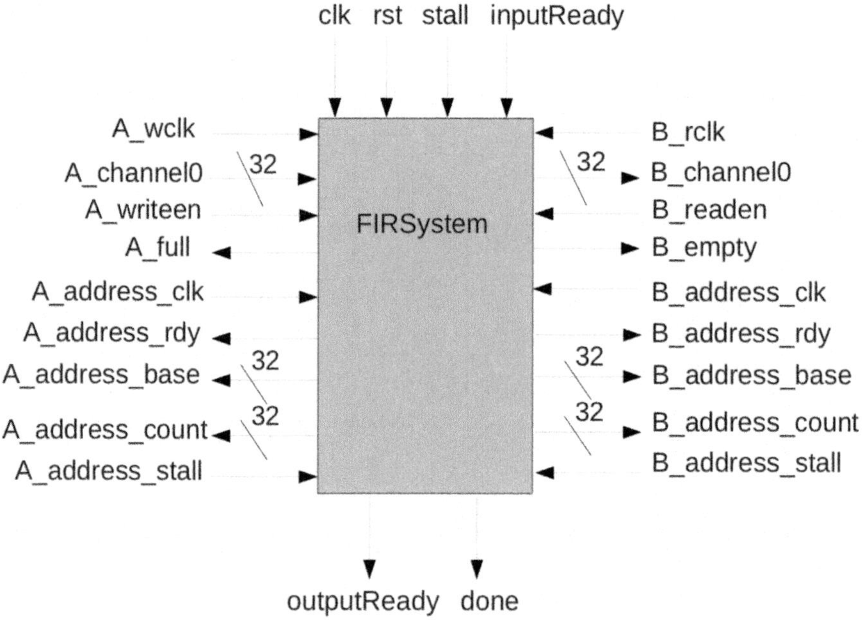

Fig. 11.3 Schematic of the hardware generated code for FIR

Loop	Procedure	Array
• *Normalization*	• *Code hoisting*	• *Scalar replacement*
• *Invariant code motion*	• *Code sinking*	• *Array RAW/WAW elimination*
• *Peeling*	• *Constant propagation*	• *Array renaming*
• *Unrolling*	• *Algebraic identities simplification*	• *Constant array value propagation*
• *Fusion*	• *Constant folding*	• *Feedback reference elimination*
• *Tiling (blocking)*	• *Copy propagation*	
• *Strip mining*	• *Dead code elimination*	
• *Interchange*	• *Unreachable code elimination*	
• *Un-switching*	• *Scalar renaming*	
• *Skewing*	• *Reduction parallelization*	
• *Induction variable substitution*	• *Division/multiplication by constant approximation*	
• *Forward substitution*	• *If conversion*	
• *Temporal common sub-expression elimination*		

Fig. 11.4 ROCCC supported high-level code transformations

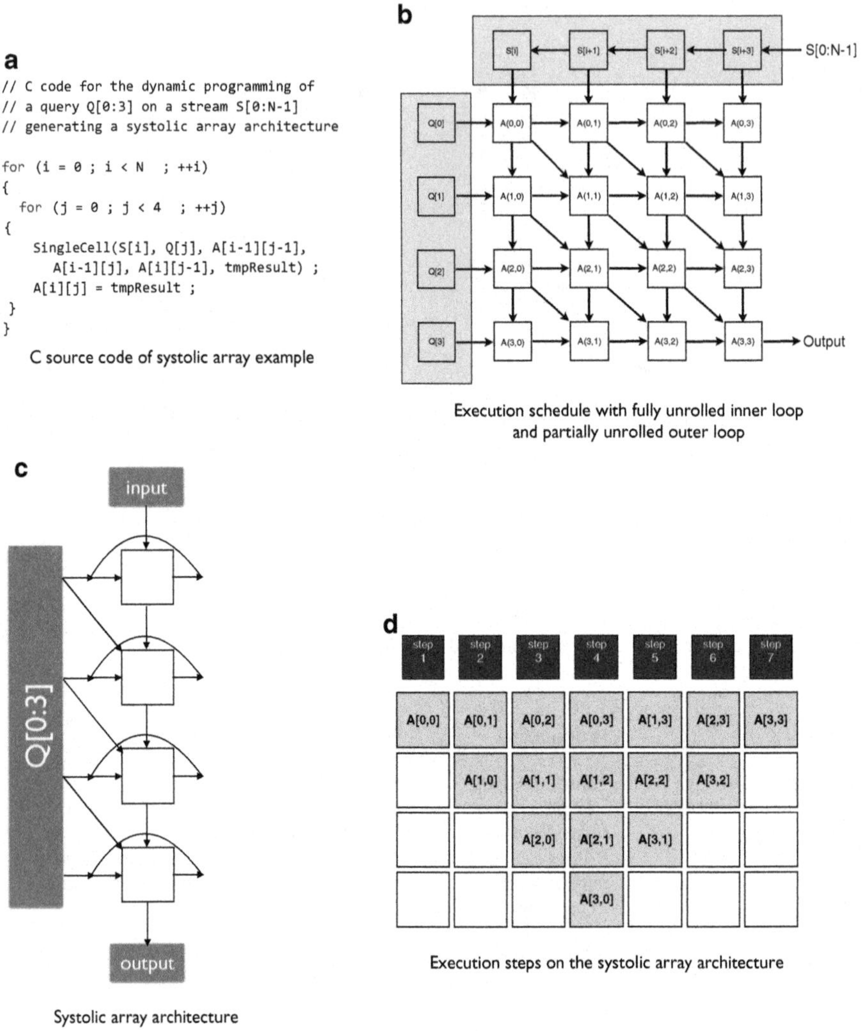

Fig. 11.5 Systolic figure 1 and 2

the actual computation. For example, in the MT-FPGA implementation of SpMV [HVN14], the row value from the sparse matrix and the column value from the vector are read separately from memory but in the same order, when both values are returned they are multiplied and added to the partial sum for that row. The process is repeated for all the non-zero values in that row at which point the thread terminates.

The advantages of the MT-FPGA execution model, and its CHAT implementation, are: (1) Massive parallelism: the parallelism is equal to the number of outstanding memory requests. (2) Full utilization of a customized data path: after an initialization period, both the multiplier and adder are fully utilized by a ready

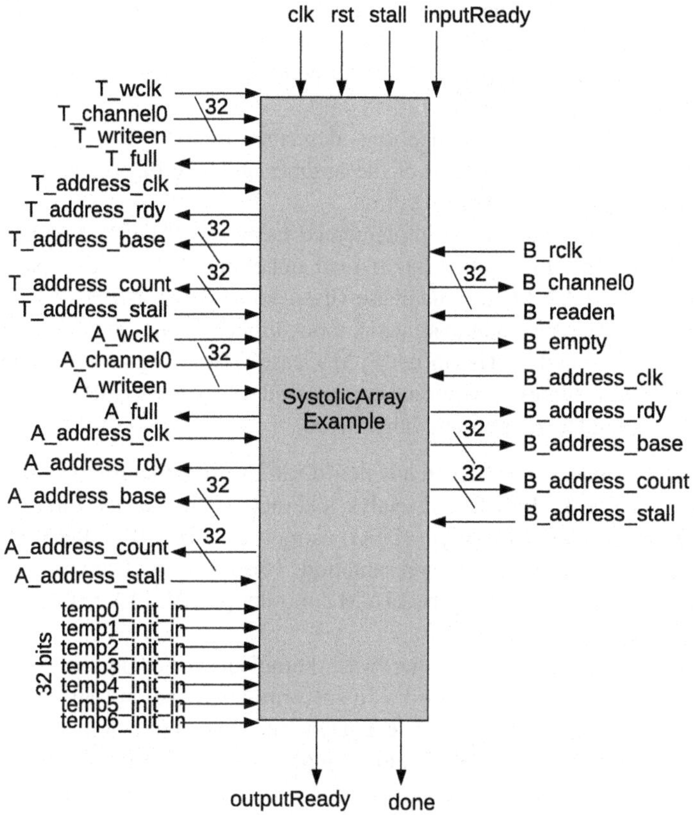

Fig. 11.6 Schematic of the hardware generated code for the systolic array example

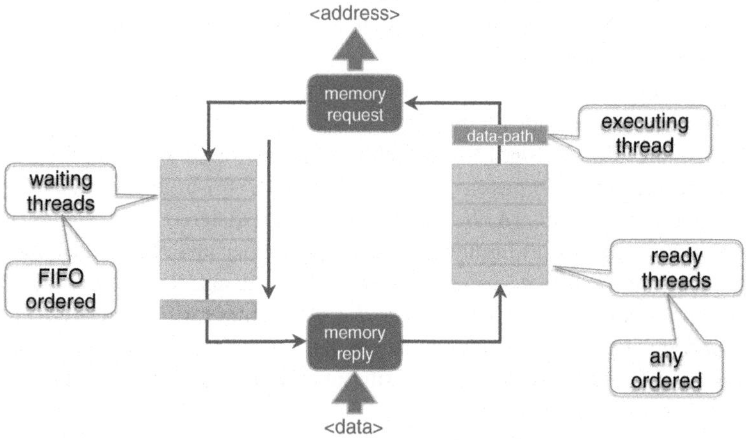

Fig. 11.7 CHAT execution model

thread. (3) Minimal state: the state of a thread is extremely small (for example in SpMV it consists of the partial sum, 64 bits, and the number of products still to be computed for that row); hence the number of pending threads, stored on a FIFO, can be very large (several thousands).

A taxonomy of irregular applications, described in [HVN14], shows the relationship between the determinism of the number and size of each thread and the approach to implementing their execution.

Three applications have been implemented using the MT-FPGA model on the Convey HC-2ex: bioinformatics (short read matching) [FVLN15], sparse linear algebra (SpMV) [HVN14] and database (hash join operation) [HANT15]. In all three implementations we demonstrated a much higher throughput than state of the art high-end CPUs and GPUs (in the SpMV case). The SpMV was implemented using CHAT and assumes a traditional compressed sparse row (CSR) storage model but can be modified to support any other model.

The CHAT Compiler CHAT extends the ROCCC infrastructure to support irregular applications. Like ROCCC, it analyzes kernels at two levels. First high-level analysis builds a Data Flow Graph (DFG) using the SUIF 2.0 toolset [WFW+94] and generates an intermediate representation. Next low-level analysis creates a Control flow graph (CFG) using the LLVM compiler [LA04, LLV15] and generates the VHDL design.

High-level analysis reads the hardware kernel described in the C language. However, because C was developed as a software language, not all its constructs make sense in a hardware context; e.g. Dynamic Allocation, Recursion, pointer arithmetic, etc. Therefore, CHAT only supports a subset of the language. For example arrays are treated as streams of data. They occupy contiguous blocks of memory, but they can be randomly accessed. CHAT does not currently support pointer chasing. Rows for multi-dimensional arrays are assumed to be stored sequentially. Constant values are stored in hardware registers. Variable values are temporarily stored in registers, and the values are moved from queues to the data-path and back. Branches in execution will generate multiple data-paths that are filtered through a multiplexer.

The goal of high-level analysis is to identify the hardware components, and create a DFG. New passes are added to the SUIF 2.0 [WFW+94] compiler to achieve this goal. We use the code in Fig. 11.8 to highlight the major steps in CHAT, but each step may require multiple passes in SUIF. First CHAT's high-level analysis will identify three data streams (*A*, *B*, and *C*), and one registered value (*length*). Two streams will be identified as input streams (*A*, and *B*) because they read values, and one stream will be identified as an output stream (*C*) because it is written too. CHAT does not support streams that both read, and write. Temporary registers are created, called *suifTmps*, to direct the data-path. Finally, the CIRRF [GBC+08] is output. An example of the C source code for a simple passthrough with its corresponding Hi-CIRRF is shown in Fig. 11.8.

User inputs are used to further customize the Hi-CIRRF pass. Parallelism can be increased by unrolling the *for loops* which will duplicate the data-paths in the DFG.

```
void passthrough (int  A, int  B, int  C, int i;
{
for (i = 0; i < length; ++ i) C[i]=B[A[i] ];
}
```

```
...
CHAT init inputscalar ( length ) ; CHATInputStreams(A,B) ;
...
for (i = 0 ; (i < length) ; i = i + 1)
{
CHATInputFifo1 0 (A, i , suifTmp2 , 0) ; CHATDataFifo1 2 (B, suifTmp2 ,
suifTmp3 , 0 ) ; suifTmp4 = CHATIntToInt(suifTmp3 , 32) ;
...
CHATOutputFifo1 2(C, i , suifTmp4 , 0) ; CHAT ? output ? C ? scalar () ;
CHATOutputStreams (C) ;
}
```

Fig. 11.8 C code and Hi-CIRRF examples for a simple irregular application

Doing so could cause redundant hardware. CHAT will analyze the unrolled design, and merge components working on the same exact data.

CHAT's low-level analysis reads the DFG from the Hi-CIRRF pass. It creates a control flow graph (CFG), and then generates the synthesizable VHDL. Thread management is a key consideration during this phase. Each thread must maintain its state locally on the FPGA, but because of the FPGAs parallelism multiple threads can be changing states in the same clock cycle. All thread data is stored on-chip in BRAMs, which are configured as FIFOs. This can be done because CHAT assumes all memory requests are returned in-order. However, the compiler could be extended to support out-of-order memory requests. In this case a design would implement CAMs instead of FIFOs.

The Lo-CIRRF compilation is implemented by a number of different passes in the LLVM [LA04] compiler. Here we give a high level overview of what happens, but just as in the previous section each step may require many different passes. First the compiler assigns each CIRRF statement into their own basic block, which allows parallel scheduling for non-sequential operations. Elements of A must block until they have valid data, but other elements are free to execute. Buffers are placed throughout the data-path to limit stalling, and alleviate back-pressure. Paths requiring memory requests will also be padded with large buffers to allow multiple outstanding requests.

Portability is another goal for the CHAT tool: its design assume nothing about the FPGA platform it will be implemented on. However, the compiler does create hooks for developers to leverage for performance. They are important for complex operations (i.e. division, or floating-point operations) where custom DSPs blocks are often available. The compiler will generate simple FIFOs for small buffers, but for larger buffers the compiler provides hooks to the developer. Custom IP cores can therefore be used to improve timing, and area utilization.

Simple Irregular Applications In this section we use the CHAT tool to build two simple irregular application examples and their implementation on a Convey HC-

```
void summation(int  A, int  B, int  C, int m, int p) {
   int i, j;
   for(j = 0; j < m; ++j)
      for(i = 0; i < p; ++i)
         C[j] += A[j][B[i]];
}
```

Fig. 11.9 Summation kernel with a 1-dimensional index stream

2ex heterogeneous machine[Con15]. They use a 1-dimensional and 2-dimensional streams as the index respectively.

One Dimensional Indexing We implement the basic irregular application expressed in Eq. (11.1). The actual CHAT code for this equation is shown in Fig. 11.9: array B has a regular access pattern, which iterates from 0 to m, array A has an irregular access pattern because it uses values from B in its index.

$$C[m] = \sum_{i=1}^{p} A[m, B[i]] \tag{11.1}$$

The CHAT tool will generate two input controllers (A, and B), and a single output controller (C). It will also generate two counter components for variables i and j. Finally the compiler will create a summation data-path. Data from the i counter will be used for the B input controller to issue memory requests. As the results return they will be routed to the A input controller where they are combined with the j counter for a new memory request. The memory requests for A are routed into the summation data-path, and once p elements have been accumulated the result is sent to the C output controller. The output controller writes the final result to memory.

The design requires only three memory channels, and should be replicated to utilized the Convey's 16 memory channels. Replicating the design is done through the compiler by unrolling a for loop. Unrolling the inner for loop increases thread level parallelism. Each cycle a single thread issues multiple request, which go into a summation tree. Unrolling the outer for loop increases application level parallelism. Each cycle multiple threads are executing in parallel. However, this also means the same B values can be used by all A controllers. The compiler identifies this, and generate a single B input controller, and routes the value to all A input controllers.

Two Dimensional Indexing We also implement an irregular application that is indexed by a 2-dimensional stream as shown in Fig. 11.10. CHAT will optimize this kernel very differently than the 1-dimensional kernel. The B input controller will provide different values for each row in A, and therefore it cannot be merged into a single shared memory channel. Multiple B input controllers must be created for each A input controller.

Convey HC-2ex Implementation We implement both kernels on a Convey HC-2ex, and explain our design considerations.

The CHAT tool does no limit how many times a loop can be unrolled, but in practice architecture limitations must be considered. The Convey HC-2ex has 16

```
void summation(int   A, int   B, int  C, int m, int p) {
   int i, j;
   for(j = 0; j < m; ++j)
      for(i = 0; i < p; ++i)
         C[j] += A[j][B[j][i]];
}
```

Fig. 11.10 Summation kernel with a 2-dimensional index stream

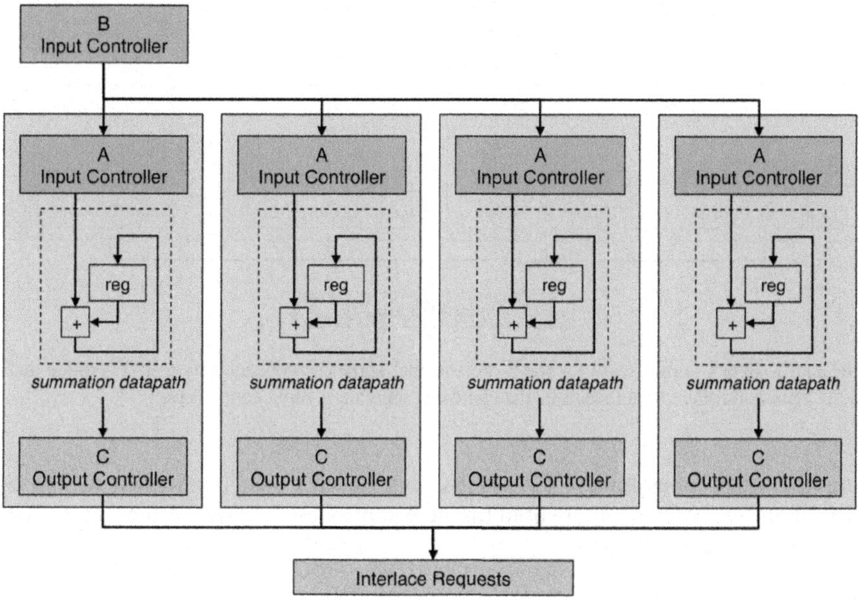

Fig. 11.11 FPGA components for the 1-dimensional kernel as generated by CHAT. Notice that the B input controller is optimized to shared its data between all A input controllers

memory channels per FPGA. Since the summation circuit is very space efficient, our design is memory channel limited, but larger designs may be area limited. To best utilize the memory channels we must consider how CHAT will optimize the 1-dimensional kernel. The layout of components is shown in Fig. 11.11. Each input memory controller (*A* and *B*) continually requests data from memory (every cycle unless stalled by a full buffer), and therefore need their own dedicated channel. The design has a single shared *B* controller. Since the output is a single value per row, the *C* controllers can be merged onto a single channel. Considering the HC-2ex's restrictions we can unroll the design 14 times. One memory channel is used by the *B* controller, another channel is used for all 14 *C* controllers, and 14 channels are used for the *A* controllers.

In the 2-dimensional kernel it is not possible to share a single *B* input controller. The layout of components is shown in Fig. 11.12. All controllers for *A* and *B* still continually request data, but now there are multiple *B* controllers. The *C* controllers can still be merged into one memory channel. Therefore, the design can only be

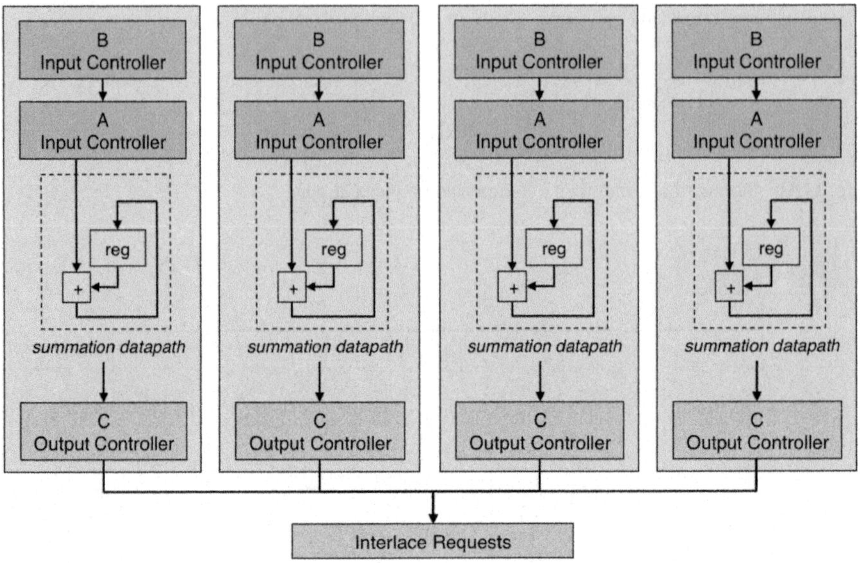

Fig. 11.12 FPGA components for the 2-dimensional kernel as generated by CHAT. Notice that the B input controller cannot share its data between separate A input controllers

unrolled seven times for the Convey HC-2ex: seven channels are used for the *B* controllers, seven channels are used for the *A* controllers, and two channels are used by the merged *C* controllers.

Acknowledgements The development of ROCCC at UC Riverside was supported, in part, by NSF Grants CCF-0745490, CCF-0811416, CCF-0905509, IIS-1144158, IIS-1161997, CCF-1219180. The development of ROCCC 2.0 from the initial ROCCC toolset was carried out at Jacquard Computing Inc. supported by the US Air Force Research Lab contract FA945309C0173.

Chapter 12
HIPA^{cc}

Moritz Schmid, Oliver Reiche, Frank Hannig, and Jürgen Teich

As today's computer architectures are becoming more and more heterogeneous, a plethora of options including CPUs, GPUs, DSPs, reconfigurable logic (FPGAs), and other application-specific processors come into consideration for close-to-sensor processing. Especially, in the domain of image processing on mobile devices, among numerous design challenges, a very stringent energy budget is of utmost importance, making embedded GPUs and FPGAs ideal targets for implementation. Although there has been tremendous progress in making the respective programming models more approachable, a deep understanding of the algorithmic details and the hardware architecture are necessary to achieve good results. Over the past decades, C-based HLS focusing on FPGAs has become very sophisticated, producing designs that can rival hand-coded RTL. A drawback is that these frameworks must be very flexible and although being able to create an efficient hardware design from a C-based language can significantly shorten the development time, architectural knowledge and specific coding techniques are still a must.

To ease the burden on developers, DSLs aim at combining architecture- and domain-specific knowledge, thereby delivering performance, productivity, and portability. HIPA^{cc} is a publicly available framework for the automatic code generation of image processing algorithms on GPU accelerators. Starting from a C++ embedded DSL, HIPA^{cc} delivers tailored code variants for different target architectures, significantly improving the programmer's productivity. In this chapter, we use HIPA^{cc} as foundation and extend it to be able to generate C++ code specific to the C-based HLS framework Vivado HLS from Xilinx. The proposed design flow is depicted in Fig. 12.1. The contributions presented in this chapter are an extension of the work presented in [RSH^{+}14] and can be summarized as follows.

M. Schmid (✉) • O. Reiche • F. Hannig • J. Teich
Friedrich-Alexander-Universität Erlangen-Nürnberg (FAU), Erlangen, Germany
e-mail: moritz.schmid@fau.de; frank.hannig@fau.de

D. Koch et al. (eds.), *FPGAs for Software Programmers*,
DOI 10.1007/978-3-319-26408-0_12

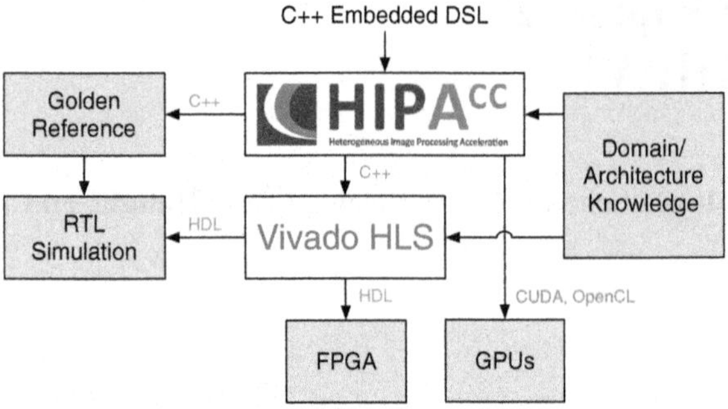

Fig. 12.1 Design flow of the combination of HIPAcc and Vivado HLS

- We introduce FPGAs as a novel target to HIPAcc in order to generate code for C-based HLS from a DSL.
- In detail, we discuss how the framework must be modified to cope with the design challenges of the FPGA target.
- We evaluate our approach by assessing the performance and energy requirements of the generated FPGA designs in contrast to other hardware targets, supported by HIPAcc.

As new contributions in this work, we present a more detailed treatise of HIPAcc, including specific transformations to generate streaming pipelines for FPGA accelerators, and evaluate the Nvidia Tegra K1 as an additional accelerator target.

The generated target code is derived from a high-level description for image processing algorithms. Therefore, this work uses the high-level description presented in [MRH+16].

12.1 Background

Due to their high computational power in combination with excellent energy efficiency, FPGAs have not only recently been a target for the implementation of signal processing systems [TB01]. In order to keep up with the high processing rates of software-based accelerators, it is vital to exploit the massive parallelism of the platform. An often desired, but so far unfortunately unreached goal for HLS frameworks is to be able to generate high throughput hardware implementations from code that was developed by software developers for sequential execution on a microprocessor. Whereas the typically sequential software execution model greatly benefits from the high clock frequencies of general purpose CPUs, hardware execution must exploit parallelism contained in the program to allow lower clock frequencies and therefore a significant reduction in power requirement. Low and medium level image processing algorithms are often composed of loop programs, where the same instructions will be executed over and over again for each pixel of the image. A simple example are edge and feature detectors in image processing pipelines. One of the well-known edge detectors is the Laplacian operator, where multiple neighboring pixels are accessed within a given window. We denote such an algorithm as a *local operator*. In the same way as caching can significantly reduce the effects of the memory bottleneck on CPU-based architectures, data streaming using FIFO interfaces is a means on FPGAs to considerably reduce the amount of cycles to fetch new pixels in comparison to reading from and writing to DPRAM. Pipelining and overlapping of the execution of these instructions is usually the next step to generate high performance implementations. Here, *throughput* (how much data per time unit) and *local latency* (how many clock cycles to complete one iteration) are contrasting design goals that also affect design speed as well as required area and energy. For streaming pipelines in image processing, a high throughput is often considered more important, since the local latency often becomes negligible in comparison to the amount of pixels to process. Furthermore, local operators, such as the Laplacian operator, often need to access a pixel more than once. Thus, handling streaming data on an FPGA requires a memory architecture to retain data for multiple accesses. Memory resources on modern FPGAs can be broadly categorized into DPRAMs and FFs. An efficient memory architecture for streaming data uses a combination of *line buffers* for storage of complete image lines (DPRAMs) and *memory windows* for the actual processing of local neighborhoods (implemented using FFs), as illustrated in Fig. 12.2.

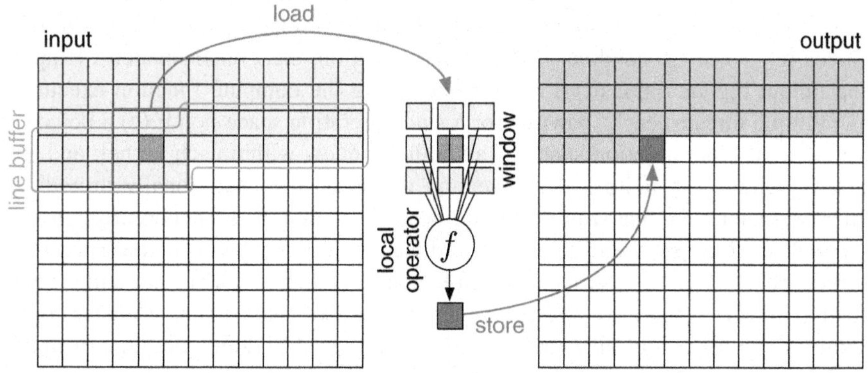

Fig. 12.2 A combination of line buffers and memory windows is typically used to process local operators on streaming data

12.2 The HIPA^{cc} Framework

The Heterogeneous Image Processing Acceleration (HIPA^{cc}) framework [MHT$^+$12, MRH$^+$16] consists of a DSL that is embedded into C++ and a source-to-source compiler. Exploiting the compiler, image filter descriptions written in DSL code can be translated into multiple target languages such as Compute Unified Device Architecture (CUDA), Open Computing Language (OpenCL), or Renderscript as used in Android [MRHT14]. In this work, our approach is to generate code for C-based HLS from a high-level description.

In the following sections, we will use the Laplacian operator as a simple example for describing image filters and briefly describe which source code transformations are applied and how code generation is accomplished.

12.2.1 Domain-Specific Language

Embedded DSL code is written by using C++ template classes provided by the HIPA^{cc} framework. Therefore, DSL code is written in a similar manner as using a framework API. In fact, those compiler-known classes are fully functional and can be compiled by a normal C++ compiler, serving as a reference CPU implementation. However, with the HIPA^{cc} source-to-source compiler, code generation is involved that targets mainly GPU accelerators.

The most essential C++ template classes for writing 2D image processing DSL codes are: (a) an `Image`, which represents the data storage for pixel values; (b) an `IterationSpace` defining the Region Of Interest (ROI) for operating on the output image; (c) an `Accessor` defining the ROI of the input image and enabling filtering modes (e.g., nearest neighbor, bilinear interpolation, etc.) on mismatch of input and output region sizes; (d) a `Kernel` specifying the compute function executed by multiple threads, each spawned for a single iteration space point; (e) a `Domain`, which defines the iteration space of a sliding window within each kernel; and (f) a `Mask`, which is a more specific version of the `Domain`, additionally providing filter coefficients for that window. Image accesses within the kernel description are accomplished by providing relative coordinates. To avoid out-of-bound accesses, kernels can further be instructed to implement a certain boundary handling (e.g., clamp, mirror, repeat) by specifying an instance of class `BoundaryCondition`. Template parameters of all mentioned classes are used to define the type of pixel data that is processed (e.g., represented as integer or floating point), enforcing kernels to be consistent and sound in terms of type conversion within the host language's type system.

To describe a Laplacian operator, we need to define a `Mask` and load the appropriate filter coefficients, defined as constants, see Listing 12.1 (lines 6–10). It is further necessary to create an input and output image for storing pixel data and load initial image data into the input image (lines 11–15). The input image is

Listing 12.1 Example code for the Laplacian operator

```
1 // input image
2 const int width = 512, height = 512;
3 uchar4 *image = (uchar4*)read_image(width, height, "input.pgm");
4
5 // coefficients for Laplacian operator
6 const int coef[3][3] = { {  0,  1,  0 },
7                          {  1, -4,  1 },
8                          {  0,  1,  0 } };
9
10 Mask<int> mask(coef);
11 Image<uchar4> in(width, height);
12 Image<uchar4> out(width, height);
13
14 // load image data
15 in = image;
16
17 // reading from in with mirroring as boundary condition
18 BoundaryCondition<uchar4> bound(in, mask, BOUNDARY_MIRROR);
19 Accessor<uchar4> acc(bound);
20
21 // output image
22 IterationSpace<uchar4> iter(out);
23
24 // define kernel
25 Laplacian filter(iter, acc, mask);
26
27 // execute kernel
28 filter.execute();
```

bound to an `Accessor` with enabled boundary handling mode *mirroring* (lines 18–19). After defining the iteration space, the kernel can be instantiated (line 25) and executed (line 28).

Kernels are implemented by deriving from the framework's provided `Kernel` base class, inheriting a constructor for binding the iteration space and a `kernel()` method. Within that method the actual kernel code is provided, see Listing 12.2 (lines 14–20). Mask and input image are accessed using relative coordinates (line 17). Thereby, it is ensured that out-of-bound accesses are caught and handled appropriately according to the specified boundary handling mode. To write the output pixel value to the iteration space, the convolution result must be assigned to the `output()` method (line 20).

Because the Laplacian operator is a local operator performing a standard convolution, it is also possible to describe the kernel using the `convolve()` method shown in Listing 12.3. This method takes three arguments: (a) the mask defining window size and filter coefficients; (b) the reduction mode used to accumulate the results of multiple iterations; and (c) a C++ lambda function describing the computational steps that should be applied in each iteration.

Additional HIPA^cc language constructs for local operators also provide more general operations for iteration and reduction steps over local window regions. Using a `Domain`, these regions do not have to be rectangular. Besides local operators, global operators are also supported for describing reductions (*min*, *max*, or *sum*) of images or an image region.

Listing 12.2 Kernel for the Laplacian operator

```
class Laplacian : public Kernel<uchar4> {
  private:
    Accessor<uchar4> &input;
    Mask<int> &mask;

  public:
    Laplacian(IterationSpace<uchar4> &iter,
              Accessor<uchar4> &input, Mask<int> &mask)
        : Kernel(iter), input(input), mask(mask) {
      addAccessor(&input);
    }

    void kernel() {
      int4 sum = { 0, 0, 0, 0 };
      for (int y = -1; y <= 1; ++y)
        for (int x = -1; x <= 1; ++x)
          sum += mask(x, y) * convert_int4(input(x, y));
      sum = max(sum, 0);
      sum = min(sum, 255);
      output() = convert_uchar4(sum);
    }
};
```

Listing 12.3 Alternative convolution kernel

```
int4 sum = convolve(mask, HipaccSUM, [&] () -> int4 {
              return mask() * convert_int4(input(mask));
            });
sum = max(sum, 0);
sum = min(sum, 255);
output() = convert_uchar4(sum);
```

12.2.2 Code Generation

The HIPA^{cc} compiler is based on the Clang/LLVM 3.5 compiler infrastructure.[1] Utilizing the Clang front end, HIPA^{cc} parses C/C++ code and generates an internal Abstract Syntax Tree (AST) representation. Operating on this representation, HIPA^{cc} will generate two kinds of code: Host code for managing kernel launches and memory transfers, and device code containing the actual kernel description in the specified target language (e.g., CUDA, OpenCL, Rendescript).

12.2.3 Generating Code for Vivado HLS

Considering Vivado HLS as a target for code generation involves numerous challenges to overcome. Convolution masks provided in DSL code must be translated in a more suitable version for FPGAs and hardware accelerators. The same applies

[1] http://clang.llvm.org.

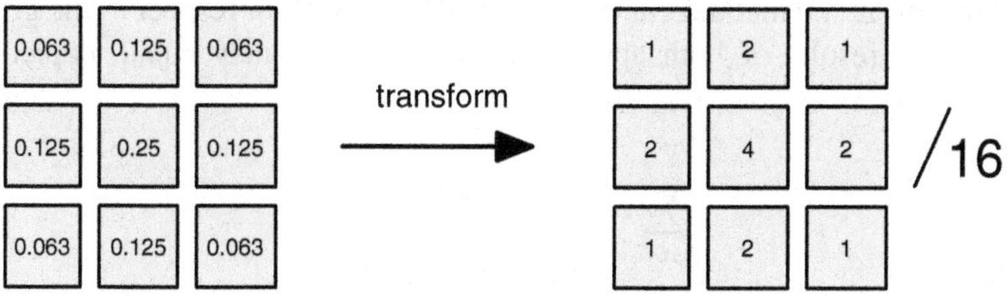

Fig. 12.3 Transformation of a Gaussian convolution mask from floating point to integer coefficients

to DSL vector types that need to be wrapped into integer streaming buffers for pipelining. Such a pipelined structural description has to be inferred from the linear execution order of kernels. Hereafter, kernel implementations need appropriate placement of Vivado HLS pragmas depending on the desired target optimization.

12.3 Transforming Masks

Mask coefficients of a convolution f, where its result remains in the same range as the input, can be described in two ways: Either by providing floating point coefficients, which in total yield exactly 1, or by multiplying with integer values followed by a division step for normalizing to the original range. On GPUs, the former is the preferred choice, as issuing integers or floating points instructions makes no difference in terms of latency. It is even more beneficial to use floating point coefficients, since avoiding the normalization step at the end requires less instructions. Therefore, the DSL framework follows this approach.

On FPGAs however, the multiplication with integer coefficients and the additional normalization step reduces the impact on resource requirements. Ideally the whole convolution can be covered by simple shift operations.

As the HIPAcc framework has been designed for GPUs and therefore follows the former approach, it is favorable to transform given floating point coefficients to integer values. An example for transforming a simple Gaussian mask is provided in Fig. 12.3. For the transformation, certain constraints need to be met. The mask size must be constant and the coefficients must be known beforehand at compile time. Mask transformation can only be applied if the condition in Eq. (12.1) holds, which states that every coefficient c_i scaled by the normalization factor N results in a natural number (with respect to a small error ε, which can be defined as a compile time parameter),

$$\forall c_i \in C, \exists n \in \mathbb{N}_0 : N \cdot c_i \pm \varepsilon = n \tag{12.1}$$

where N is defined as the inverse of the smallest coefficient $N = {}^1/_{\min_{c_i \in C}\{c_i\}}$.

The transformation of every floating point coefficient c_i to its integer representative x_i is done by scaling each coefficient by the normalization factor and correct rounding, defined as $x_i = \lfloor N \cdot c_i + 0.5 \rfloor$. It can be shown, given the constraint defined in Eq. (12.1), that the convolution is still valid. With respect to the granted small error ε, results of both approaches are approximately equal, depicted in Eq. (12.2).

$$\sum_{c_i \in C} c_i \cdot d_i \approx \frac{\sum_{c_i \in C} x_i \cdot d_i}{N} \tag{12.2}$$

As those results are not exact, they might not be suitable for every algorithm or use case. Therefore, this feature can be controlled by a compiler switch.

12.4 Streaming Pipeline

High-level programs given in HIPA^{cc} DSL code process image filters buffer-wise. Each kernel reads from and writes to buffers sequentially, running one after another with buffers serving as synchronization points (so called host barriers). Buffers can be read and written, copied, reused, or allocated only for the purpose of storing intermediate data.

Throughout this section, we will demonstrate our method using the Harris corner detector [HS88], depicted in Fig. 12.4. This filter consists of a pipeline of kernels, implemented as point operators and local operators, which are described in detail in Sect. 12.6.2. Although, every kernel writes to its own buffer, in total only four buffers are necessary by applying smart reuse of existing buffers containing out-dated intermediate data.

This buffered concept is fundamentally different from streaming data through kernels, processing a computational step as soon as all input dependencies are available. Kernels are therefore interconnected with each other using stream objects implementing FIFO semantics. Such a streaming concept requires a structural description, resolving direct data dependencies unconstrained from the exact sequential ordering of kernel executions.

We can transform the buffer-wise execution model into a structural description suitable for streamed pipelining by analyzing the DSL host code, replacing memory transfers by stream objects, and generating appropriate kernel code. Vivado HLS can then be instructed to run all kernels in parallel, as shown in Fig. 12.5, which can

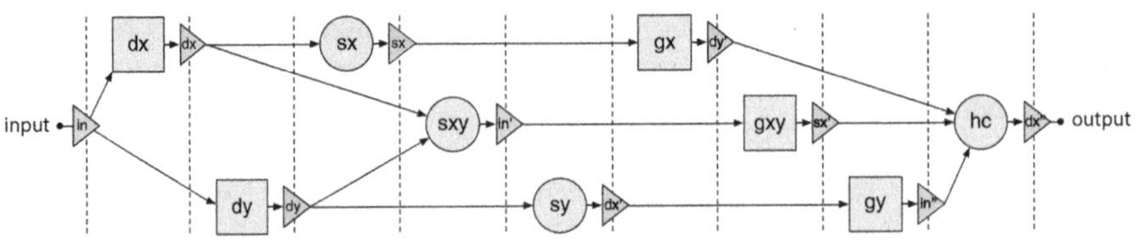

Fig. 12.4 Sequential execution of HIPA^{cc} kernels for the Harris corner detector implemented as a pipeline of point operators (*circles*) and local operators (*squares*). *Triangles* mark buffers and *dashed lines* represent host barriers between kernel executions

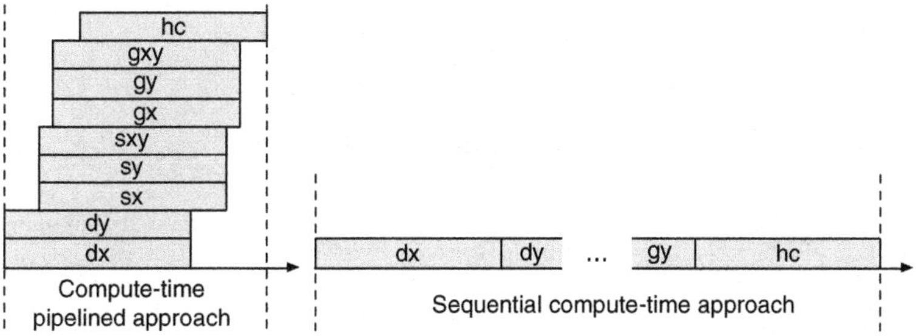

Fig. 12.5 Pipelined and sequential execution

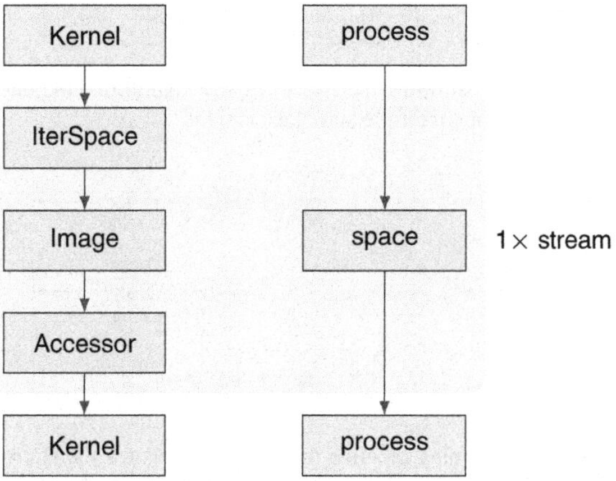

Fig. 12.6 Transforming the HIPA^{cc} representation into the internal representation by applying a one-to-one mapping of kernels and images to processes and spaces

deliver a significantly shorter processing time. We first introduce how the structural description is transformed from a given host code and then give insight into device code generation.

12.4.1 Host Code Analysis

The host code is translated into an AST representation that is traversed by HIPA^{cc}. During this traversal process, we track the use of buffer allocations, memory transfers and kernel executions by detecting compiler-known classes. For each kernel, the direct buffer dependencies are analyzed and fed into a dependency graph.

Given this graph, we can build up our internal representation, a simplified AST-like structure based on a bipartite graph consisting of two vertex types: *Space* representing images and *process* marking kernel executions. Kernel executions are traversed in the sequential order, in which they are specified.

In the simplest case, each kernel maps to exactly one *process* and each image to exactly one *space*. Wrapper objects for accessing image data, such as `IterationSpace` and `Accessor` are directly translated to edges in the internal representation, as shown in Fig. 12.6. Writes to images are transferred to the internal representation in Static Single Assignment (SSA) manner.

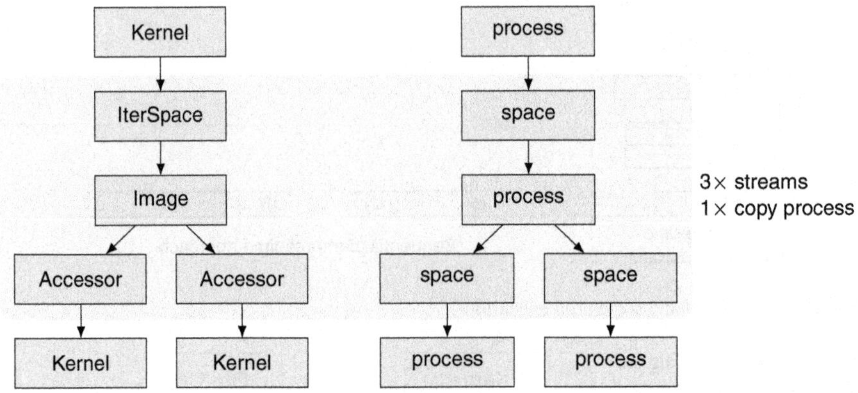

Fig. 12.7 Transforming the HIPA^{cc} representation into the internal representation for images that are shared among multiple kernel executions

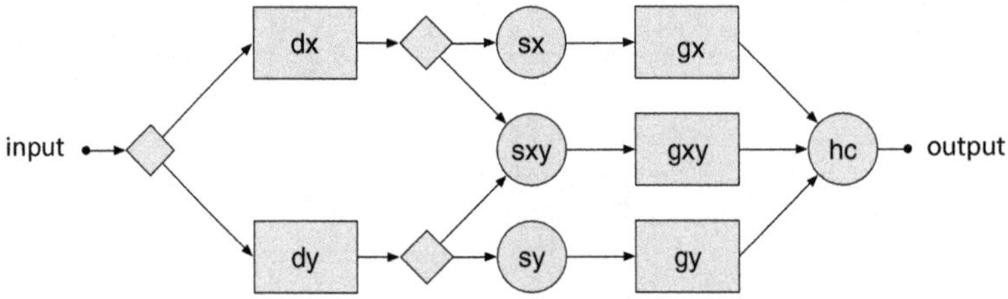

Fig. 12.8 Streaming pipeline of Vivado kernels for the Harris corner detector. *Diamond shapes* represent kernels for splitting data of a single stream object into multiple stream objects

Hereby, reusing the same image multiple times will form multiple new *space* vertices in the graph.

In the more complex case, where the inputs of multiple kernels depend on the same image and the same temporal instance of intermediate data, the corresponding image will be translated to a *copy process* surrounded by an additional *space* vertex for each wrapper object used for reading or writing the image. An example for two kernels reading from one image is illustrated in Fig. 12.7. This way, it is guaranteed that streaming data later on will be copied before handing it over to the next computation steps.

Considering the running example of the Harris corner detector, the reuse of images in Fig. 12.4 is indicated by colors. Images of the same color are translated into separate streaming objects for pipeline creation. Furthermore, intermediate data distributed among multiple succeeding kernel executions (e.g., dx) can also be found. Here, the insertion of a *copy process* will be applied for creating the internal representation.

Once our internal representation has been created, we can infer the structural description for the streaming pipeline shown in Fig. 12.8. The graph is linearized for code generation by traversing backwards through the graph in Depth-First Search (DFS), originating from the output *spaces*. Herby, parts of the graph that are not contributing to the output will be pruned. For code generation, every *process* vertex is translated to a kernel execution and every *space* vertex marks the insertion of a unique Vivado HLS stream object. The resulting code embodies the structural description of the filter, which is written to a file serving as *entry function*.

Device Code Generation

For generating device code, an AST is extracted and extended from the DSL's kernel description. Additional nodes are added to fulfill the requirements of the target language such as applying index calculations and ROI index shifts. Depending on optimizations specified by compiler options, vast portions of the extracted AST are modified and replicated. For example when enabling the use of texture or local on-chip memory on GPUs, the appropriate language constructs must be inserted. Further optimizations like thread-coarsening (similar to partial loop unrolling), where multiple iterations of the same kernel are merged into a single one, are applied by cloning the extracted AST and inserting it multiple times with slightly adjusted memory access indices. The resulting AST is transformed back to source code by utilizing Clang's *pretty printer* and written to separate files for each kernel. These files will be included (CUDA, Vivado HLS) or loaded (OpenCL) by the rewritten host code file.

For Vivado HLS, we needed to introduce some additional constraints. As described earlier, masks must be constant, not only for mask transformation, but also for constant propagation,[2] which we enforce for FPGA targets. Further constraints are that image dimensions must be known beforehand at compile time, in order to process it by Vivado HLS synthesis right away. The resulting C++ code is based on a highly optimized library for image processing, as proposed in [SAHT14], and is still human-readable. Globally affecting parameters (such as image dimensions) are defined at a single central spot, so that those parameters can be conveniently altered for further syntheses without rerunning the HIPAcc compiler. Similar to HIPAcc's other targets, separate files are created for each kernel. These files will be included by the *entry function*, which already embeds all executions in a structural description, described in the previous section. Kernel optimizations applied for device code are explained in the next section.

12.4.2 Parallelization and Design Optimization

Although the possibility to use a C-based language for design entry lowers the hurdle for acceptance of a HLS framework, algorithms stated in such a language are inherently sequential and must be parallelized and optimized in order to efficiently use the FPGAs resources. As opposed to the world of HPC, where the fastest

[2]Replacing memory loads of constant values by inserting the numerical literals.

processing speed is paramount, developing hardware accelerators may be subject to several contrasting design goals. Of course, high throughput and short clock periods are important achievements, but often it is also necessary to comply with a certain resource budget. A central element of Vivado HLS for this task are the synthesis directives, which allow to specify how the input design is to be parallelized and optimized.

Placing Vivado Synthesis Directives

Synthesis directives in Vivado HLS can either be inserted in the code directly as *pragmas*, or collected in a script file which is applied during synthesis. Here, we use both approaches. Directives, which transform the sequential specification into a typical parallel hardware structure, such as *unrolling* a for-loop to model a shift-register, or which optimize hierarchical structures, for example *inlining* calls to small functions, are rarely changed and are thus placed directly in the code as pragmas. Other directives, we frequently use for parallelization and optimization, such as *pipelining* and specifying the size of FIFOs to interconnect a streaming pipeline can be used for design space exploration and to enforce a certain optimization strategy. Searching and altering the code to adapt these directives may become inconvenient for large designs. Therefore, we provide these directives in a script file, so that they can be readily changed.

Optimizing Loop Counter Variables

Software developers often tend to use the convenient `int` data type to specify loop counter variables. For image processing, the image dimensions seldom require the full range of a 32-bit two's complement. Moreover, using more bits for a variable than required causes undesired excessive use of resources and may degrade the achievable maximum clock frequency, especially if the variable must often be compared to another. In Vivado HLS, appropriate bit-widths can be explicitly specified if the range of the variable is known during the code generation. Another possibility is to let the synthesis tool automatically infer the required bit-width by inserting assertions on the range of the variable, similar to the range definitions for the `integer` data type in VHDL. We apply assertions to loop counter variables in our approach, so that bit widths do not need to be explicitly adapted if image dimensions change between code generation and synthesis. In this way, we ensure that the design always uses an optimal binary representation for loop variables, reduce the amount of required resources and achieve shorter critical paths for logical operations on such variables.

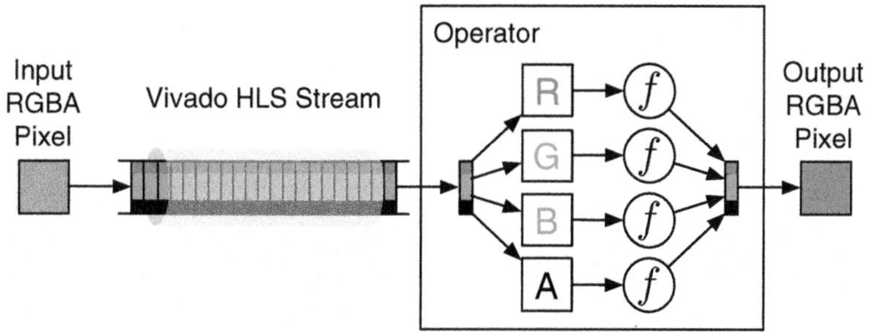

Fig. 12.9 Packing of vector types into Vivado streams. This example shows an RGBA pixel, the channels are packed together into a single stream. Although the channels must be processed individually, a common stream and line buffer can be used

Mapping Vector Types

Since HIPAcc was developed for GPUs, it supports numerous vector types, which are crucial for performance on specific GPU architectures. Besides performance concerns tackled by explicit vectorization, these types are in particular well-suited for expressing computations on common formats of image processing in a natural way.

To support vector types for the Vivado HLS target, basically two approaches are available: (a) treat all vector elements separately, resulting in multiple line buffers, multiple windows, and multiple streams for each local operator; or (b) pack all vector elements into a single integer of the same bit-width as the whole vector, resulting in only a single stream for each operator, depicted in Fig. 12.9. The former approach is realized by the OpenCV implementation provided with Vivado HLS 2014.1. The latter is followed by our code generation, because this might reduce the overall consumption of memory resources.

Existing vectorized DSL code does not need to be modified. We have implemented vector types for Vivado HLS as C structures and computations are realized by operator overloading. Reads and writes to DSL `Images`, which will be mapped to stream objects, are replaced by conversion functions, either extracting the vector types from packed stream elements or packing vector types into stream elements. Inter-kernel computations, i.e., operations on local variables, are described as vector operations in DSL code already and are therefore covered by overloaded operators.

Delaying Point Operators

As explained earlier, obeying causality when applying local operators to streaming data causes an increased group delay. If point operators are used in streaming pipelines before a local operator, the production rate of the point operator is higher than the consumption rate of the local operator. One way to address this issue is to insert an appropriately sized stream buffer between the two operators. Alternatively, a delay can be enforced on the point operator to equalize consumption and production rates. The generated code uses the latter approach in order to save memory resources.

12.5 Test Bench Generation

When developing architectures and FPGA layouts, it is of great importance to cover sources of errors by testing, at best starting in early design phases. A solution to achieve effortless testing is the automatic test bench generation from a high abstraction level. By applying such techniques, testing is more flexible, can be accomplished much more efficiently than on RTL, and requires less coding effort than, for example, in VHDL. Our approach allows to write DSL code from that HIPA^{cc} can derive a test bench that can be used by Vivado HLS.

In order to accomplish testing, HIPA^{cc} can utilize parts of the provided DSL host code, which is embedded into C++ and thereby already surrounded by functioning program code. The host code's AST will be traversed by HIPA^{cc} using Clang's *Rewriter* engine. For each node that is related to compiler-known classes, HIPA^{cc} rewrites the source code location by inserting appropriate runtime calls. On FPGAs for instance, assignments between pointers and `Images` will be replaced by runtime calls for initially filling array data into Vivado stream objects. Similar replacements are applied to all other occurrences of compiler-known classes within the input code.

The resulting rewritten host code is still very similar[3] to the original input and the surrounding program code is left untouched. Therefore, if the designer has specified testing routines in the C++ program that verifies the results computed by DSL code, the rewritten program can serve as a test bench without further modification. The same applies to GPU targets, which further emphasizes the consistency of our approach to cover fundamentally different targets from one and the same code base.

12.6 Experimental Results

We evaluate our results on several different hardware target platforms. All hardware targets are compared in terms of performance and energy efficiency. Our implementations are generated by HIPA^{cc} for each target, stemming from the *exact same code base*.

[3]For example, `image` names are maintained, only given a prefix when converting to streams.

12.6.1 Evaluation Environment

First, we briefly describe our environment by providing detailed information about the architectures and libraries we take into account for evaluation.

Nvidia Tegra K1 is a Multi-Processor System-on-Chip (MPSoC) hosting an ARM Cortex-A15 quad core CPU and an embedded General Purpose GPU (GPGPU) based on Nvidia Kepler architecture. The GPU is featuring 192 scalar compute cores, clocked at 960 MHz. The theoretical peak performance limits at 370 GFLOPs with doubled operation count, because of the Fused Multiply-Add (FMA) instruction, which is counted as two operations. Besides 2 GB of DDR3 RAM, which is shared among CPU and GPU, it also contains L1 and L2 caches and up to 48 kB of local on-chip memory. It further supports utilizing texture memory and texture caches.

Nvidia Tesla K20 is a discrete GPGPU solely for the purpose of computing. It contains 2496 scalar compute cores, distributed among 13 multi-processors, each running at 705 MHz. This results in approximately 3520 GFLOPs, again counting FMA as two instructions. Besides 5 GB of GDDR5 RAM, also includes L1 and L2 caches as well as up to 48 kB local on-chip memory.

Xilinx Zynq 7045 is a SoC, which tightly integrates an ARM Cortex-A9 dual core CPU and a Kintex FPGA, using an ARM AMBA interconnect. The included FPGA can be considered a mid-range device, offering 350K logic cells, comprised of 218,600 Lookup Tables (LUTs), 437,200 flip-flops, 2180 kB of on-chip memory, and 900 DSP slices. Furthermore, the device includes a Gen2 hardcore PCI Express block and up to 16 high-speed serial GTX transceivers, each capable of transmitting at 12.5 Gb/s.

12.6.2 Algorithms

For the evaluation, we consider three typical image processing algorithms: an edge detector based on the Laplacian operator, the Harris corner detector [HS88], and the computation of optical flow using the census transform [Ste04]. Those algorithms are of high relevance for richer imaging applications, such as augmented reality or driver assistance systems in the automotive sector. They all embody a high degree of parallelism and are equally well suited for every of our considered target architectures. Although these algorithms are well known, their implementation details may differ significantly, thus we briefly clarify the algorithm specifics used for our evaluation.

Laplace The Laplacian (LP) filter is based on a local operator, as already described in Sect. 12.1. Depending on the mask variant used, either horizontal and vertical edges or both including diagonal edges can be detected. In our results, we denote the first variant as LP3HV and the second as LP3D. Additionally, we evaluate a third variant, LP5, based on a 5×5 window, which can also detect diagonal edges.

Harris Corner (HC3) was first introduced by [HS88]. It consists of a complex image pipeline depicted in Fig. 12.4. After building up the horizontal and vertical derivatives (dx, dy), the results are squared and multiplied (sx, sy, sxy), and processed by Gaussian blurs (gx, gy, gxy). The last step (hc) computes the determinant and trace, which is used to detect threshold exceedances.

Optical Flow (OF) issues a Gaussian blur and computes signatures for two input images using the census transform. Therefore, for each image two kernels need to be processed. A third kernel performs a block compare of these signatures using a 15×15 window in order to extract vectors describing the optical flow. Regarding continuing streams of images (e.g., videos), for GPU targets the first image's signatures can be reused. Hereby, it is only necessary to process the second image and to reperform the block compare, resulting in the execution of 3 kernels per iteration. Whereas on FPGAs, the signatures for both images always have to be computed, resulting in the execution of 5 kernels per iteration. For fair comparison, we were considering this fact when evaluating our throughput results.

For the evaluation, we have used an 8-bit integer data types and images of size 1024×1024. The algorithms we synthesized in Vivado HLS with *high effort* settings for scheduling and binding. PPnR resource requirements were obtained by implementing the generated Verilog code in Vivado 2014.1. Power requirements were assessed by performing a timing simulation on the *post-route* simulation netlists and evaluation of the switching activity in Vivado (Table 12.1).

12.6.3 FPGA vs. Embedded GPU Implementation

One of the benefits of using a DSL for HLS is to have a target-independent high-level description at hand, which can generate code for several different hardware targets from exactly the same code base. In this section, we analyze the above presented algorithms for heterogeneous hardware platforms, including a Tegra K1 embedded GPU (eGPU), a Zynq 7045 FPGA, and a Tesla K20 server-grade GPU in terms of performance and energy consumption. The performance results are summarized in Table 12.2. An efficient configuration for the GPU implementations

Table 12.1 PPnR results for the Laplacian operator and Harris corner detector implementations

	II	LAT	SLICE	LUT	FF	BRAM	DSP	F (MHz)	P (mW)
LP3HV	1	1050638	141	288	521	2	0	349.9	232
LP3D	1	1050641	226	398	1034	2	0	341.1	232
LP5	1	1052768	3917	4521	23795	4	200	220.1	247
HC3	1	1063233	9349	23331	31102	8	254	239.4	498

Table 12.2 Comparison of execution times on Nvidia Tegra K1, Xilinx Zynq 7045, and Nvidia Tesla K20

	Tegra K1	Zynq 7045	Tesla K20
LP3HV	19.08	3.00	0.10
LP3D	22.73	3.08	0.16
LP5	39.53	4.78	0.38
HC3	29.85	4.38	0.91
OF	54.35	5.19	2.21

Time in ms for an image of size 1024 × 1024

Table 12.3 Comparison of throughput and energy consumption for the Nvidia Tegra K1, Xilinx Zynq 7045, and Nvidia Tesla K20

	Tegra K1		Zynq 7045		Tesla K20	
	TP (fps)	E (fpW)	TP (fps)	E (fpW)	TP (fps)	E (fpW)
LP3HV	52.4	11.8	333.3	1423.1	10,000.0	74.1
LP3D	44.0	9.9	324.7	1387.6	6250.0	46.3
LP5	25.3	5.7	209.2	846.9	2631.6	19.5
HC3	33.5	7.5	228.3	458.5	1098.9	8.1
OF	18.4	4.1	192.5	409.6	452.5	3.4

was found using HIPAcc's exploration feature. The Tesla K20 can exceed the performance of the embedded Tegra GPU by a factor between approximately 25 and 200, depending on the algorithm. Execution times for both GPUs increase with the window size and kernel complexity, clearly decelerated by the number of memory loads. For the FPGA, the impact of larger window sizes is by far less noticeable. We further provide a comparison of the throughput (refer to Fig. 12.10) and energy efficiency (refer to Fig. 12.11) of the platforms in and energy consumption in Table 12.3. The power consumption of the GPUs can be estimated by considering about 60 % of the reported peak power values (4.45 W for the Nvidia Tegra[4] and about 135 W for the Nvidia Tesla). The FPGA provides the most energy efficient solution.

Even though the achievable frame rate per watt decimates notably faster than the throughput, this effect is much more distinctive on GPUs, leaving FPGAs clearly as the architecture of choice for more complex algorithms.

We attribute the lower energy efficiency of the Tegra compared to the Tesla to the slower DDR3 memory. Tesla's dedicated GDDR5 memory introduces higher access latency, but is optimized for high throughput. Access latency can be hidden by heavily applying simultaneous multi-threading. In particular for the Laplacian filer, where the computational workload is rather small, the throughput bottleneck of the DDR3 memory becomes crucial. Therefore, power efficiency of the Tegra

[4]Jetson DC power analysis of the CUDA smoke particle demo, adjusted for fan and system power consumption: http://wccftech.com/nvidia-tegra-k1-performance-power-consumption-revealed-xiaomi-mipad-ship-32bit-64bit-denver-powered-chips/.

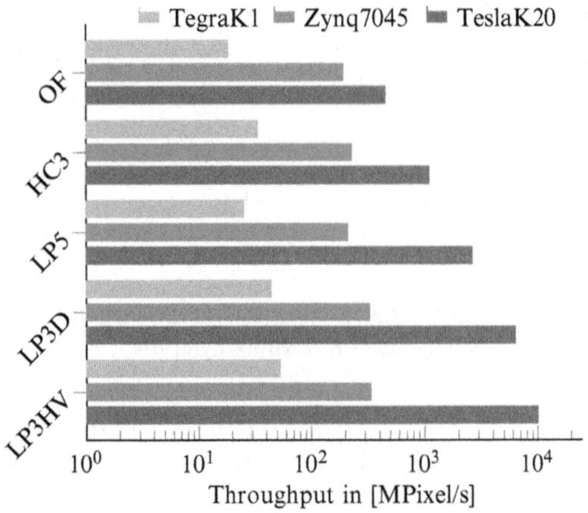

Fig. 12.10 Throughput comparison

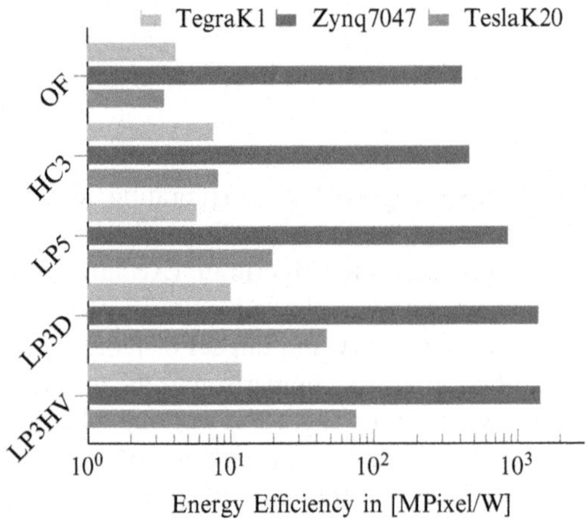

Fig. 12.11 Energy efficiency

GPU improves for more complex filters with higher computational demand, such as the Harris corner and the optical flow.

12.7 Conclusions

In this work, we have demonstrated how the DSL-based framework HIPA^{cc} can be used to automatically generate highly optimized code for the HLS of multiresolution applications for implementation on FPGAs. In this way, the specification of the design requires significantly less programming effort from the developer and thus also poses less chances for coding errors. The presented case studies from image processing demonstrate that the approach is applicable to a broad range of problem scenarios. As HIPA^{cc} also includes embedded GPGPUs as a hardware target [MRHT14], we have compared the proposed FPGA approach to a highly optimized GPU implementations, generated from the same code base. The assessment exposes the benefits of using a heterogeneous framework for algorithm development and can easily identify a suitable hardware target for efficient implementation.

Acknowledgements This work was partly supported by the German Research Foundation (DFG) as part of the Research Training Group 1773 "Heterogeneous Image Systems".

Part III
FPGA Runtime Systems and OS Services

Chapter 13
ReconOS

Andreas Agne, Marco Platzner, Christian Plessl, Markus Happe, and Enno Lübbers

In this chapter, we present an introduction to the ReconOS operating system for reconfigurable computing. ReconOS offers a unified multi-threaded programming model and operating system services for threads executing in software and threads mapped to reconfigurable hardware. By supporting standard POSIX operating system functions for both software and hardware threads, ReconOS particularly caters to developers with a software background, because developers can use well-known mechanisms such as semaphores, mutexes, condition variables, and message queues for developing hybrid applications with threads running on the CPU and FPGA concurrently. Through the semantic integration of hardware accelerators into a standard operating system environment, ReconOS allows for rapid design space exploration, supports a structured application development process and improves the portability of applications between different reconfigurable computing systems.

13.1 Introduction

A growing number of applications originating in and integrating diverse fields like pattern recognition, machine learning, large-scale data analytics, or communication systems feature a diverse mix of control-intensive and highly parallelizable regular data-centric operations. This heterogeneity of the composition of application com-

A. Agne • M. Platzner (✉) • C. Plessl
University of Paderborn, Paderborn, Germany
e-mail: platzner@upb.de

M. Happe
ETH Zurich, Zürich, Switzerland

E. Lübbers
Intel Labs Europe, Munich, Germany

D. Koch et al. (eds.), *FPGAs for Software Programmers*,
DOI 10.1007/978-3-319-26408-0_13

ponents increasingly prohibits the exclusive use of a single acceleration technology. Thus, in the case of fine-grained reconfigurable computing, applications are rarely mapped exclusively to an FPGA accelerator. Instead, individual application parts amenable to parallel execution, customization, and deep pipelining are often implemented as custom hardware to improve performance or energy efficiency. Other parts, especially code that is highly sequential or difficult to implement as custom hardware, are executed in software mapped to a CPU. This decomposition of applications into separate, communicating parts that require synchronization among them is also widely used in pure software systems for achieving a separation of concerns and concurrent or asynchronous processing.

In software systems the operating system standardizes these communication and synchronization mechanisms and provides abstractions for encapsulating the unit of execution (processes, threads), communication, and synchronization. Reconfigurable computing systems still lack an established operating system foundation that covers both software and hardware parts. Instead, communication and synchronization are usually handled in a highly system and application-specific way, which tends to be error prone, limit the productivity of the designer, and prevent portability of applications between different reconfigurable computing systems.

ReconOS offers an operating system, programming model and system architecture, which are closely geared to each other in order to provide unified operating system services for functions executing in software and hardware and a standardized interface for integrating custom hardware accelerators. To this end ReconOS leverages the multi-threading programming model which is widely used in software development and extends the host operating system with support for hardware threads. These extensions allow the hardware threads to interact with software threads using the same, standardized operating system mechanisms (system calls), for example, semaphores, mutexes, condition variables, and message queues. From the perspective of an application it is thus completely transparent whether a thread is executing in software or hardware.

The availability of an operating system layer providing symmetry between software and hardware threads yields the following benefits for reconfigurable computing systems: First, the application development process can be structured in a step-by-step fashion with an all-in-software implementation as a starting point. Performance-critical application parts can then be turned into hardware threads one-by-one to explore the hardware/software design space successively. Second, the portability of applications between different reconfigurable computing systems is improved by using defined operating system interfaces for communication and synchronization instead of low-level platform-specific interfaces. Finally, the unified appearance of hardware and software threads from the application's perspective allows for moving functions between software and hardware during runtime, which supports the design of adaptive computing systems that exploit partial reconfiguration.

In this chapter we extend our previous article [AHL+14b] to provide a general overview of ReconOS and to emphasize the aspects that make ReconOS particularly appealing for developers with a software background. We would also like to point out that ReconOS is part of a lineage of operating system approaches for FPGAs, which have been studied since the 1990s. In Sect. 13.9 we discuss the evolution of operating systems for reconfigurable computing and how ReconOS relates to this heritage.

13.2 Step-by-Step Application Development

ReconOS aims at simplifying the creation of applications for hybrid CPU/FPGA systems with a structured, step-by-step development process, which is more amenable to a software developer than a typical hardware-core centric design flow. This process is typically divided into three steps, which are outlined in Fig. 13.1:

- First, the developer prototypes the application's functionality in multi-threaded software using, for example, the Pthreads library on Linux. This first software-based implementation allows for functional testing. Communication and synchronization between threads is handled using the well-know operating system functions.
- Second, the multi-threaded software is ported to the embedded CPU on the targeted platform FPGA, e.g., a MicroBlaze running Linux. The developer can now use profiling to identify the application's potential for parallel execution, i.e., those threads that could benefit from the fine-grained parallelism of a hardware realization, and those code segments that are amenable to a coarser-grained parallel implementation with multiple threads.
- The third step includes creating the hardware threads and the ReconOS system architecture. At this point, ReconOS easily allows the developer to evaluate different mappings of threads to hardware and software and to quickly assess the overall performance on the target system.

Over the years, ReconOS has been used to implement numerous applications on hybrid CPU/FPGA systems. These experiences have confirmed that the hybrid multi-threading approach offered by ReconOS does indeed simplify the development process and is accessible also to developers with very little FPGA hardware design experience.

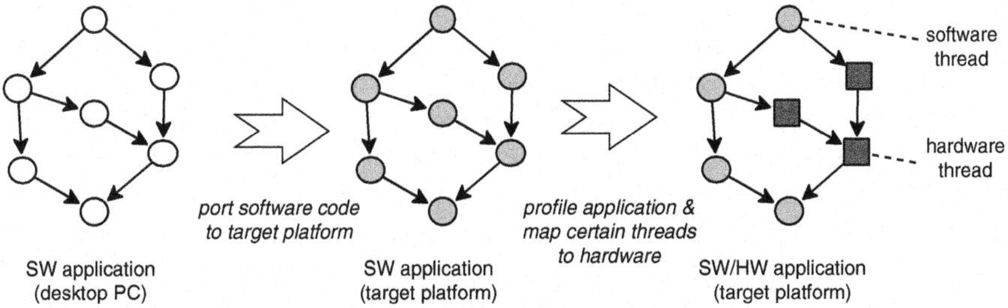

Fig. 13.1 The step-by-step development process: First, the software application is prototyped on a desktop machine. Then, it is ported to the target CPU platform. Finally, suitable threads are ported to hardware

13.3 Multi-Threading Programming Model

The key idea of ReconOS, which enables the previously mentioned benefits, is extending the multi-threading programming model and supporting operating system functions across the hardware/software interface. In multi-threaded programming applications are partitioned into several *threads* that implement the computational tasks and may be executed concurrently. In our case, threads can be either blocks of sequential software or parallel hardware modules. For orchestrating the concurrent execution of threads and communication between them, applications use dedicated operating system functions with well-defined semantics and interfaces. For example, threads can pass data using *message queues* or *mailboxes*, explicitly coordinate execution through *barriers* or *semaphores*, or implicitly synchronize access to shared resources by locking and unlocking mutually exclusive locks (*mutexes*). These objects and their interactions are widely used in well-established APIs for programming multi-threaded software applications, such as the POSIX API. One of the major advantages developers can draw from the ReconOS approach is that these abstractions can not only be used for software threads but also for optimized hardware implementations of data-parallel functions—the hardware threads—without sacrificing the expressiveness and portability of the application description.

An example for a multi-threaded software application is sketched in Listing 13.1. The application comprises a thread that receives packets streaming in via ingress mailbox `mbox_in`, processes them in a user-defined way, sends the processed packets to egress mailbox `mbox_out`, and updates a packet counter stored in a shared variable protected by lock `count_mutex`. Using standard APIs for message passing and synchronization, the software thread accesses operating system services in an expressive, straightforward, and portable way. As an added benefit, such a thread description manages to clearly separate thread-specific processing from operating system calls.

The same thread's ReconOS hardware implementation is shown in Fig. 13.2, partitioned into similar thread-specific logic and operating system interactions. The thread's data path is implemented in the thread-specific *user logic* and is only limited by available FPGA resources. All requests and interactions that the hardware thread needs to have with the operating system are captured in the *OS synchronization finite state machine* and channeled to the operating system kernel through the *OS interface* (OSIF). This mechanism enables seamless operating system calls from within hardware modules. The thread developer specifies the Operating system synchronization state machine (OSFSM) using a standard VHDL state machine description, as shown in Listing 13.2. For accessing operating system functions in this state machine ReconOS provides a VHDL library that wraps all operating system calls with VHDL procedures. A particular state-machine design pattern lets the OS-controlled signal `done` (line 46) temporarily, inhibit the transition of the OSFSM to the next state, effectively allowing blocking operating system calls—such as `mutex_lock()`—to suspend hardware thread operations.

Listing 13.1 Example of a stream processing software thread using OS services [AHL+14b]

```
extern mutex_t *count_mutex;              // mutex protecting packet counter
extern mqd_t mbox_in,                     // ingress packets
             mbox_out;                    // egress packets

void *thread_a_entry( void *count_ptr ) {
  data_t buf;                             // buffer for packet processing

  while ( true ) {
    buf = mbox_get( mbox_in );            // receive new packet
    process( buf );                       // process packet
    mbox_put( mbox_out, buf );            // send processed packet
    mutex_lock( count_mutex );            // acquire lock
    ( (count_t) *count_ptr )++;           // update counter
    mutex_unlock( count_mutex );          // release lock
  }
}
```

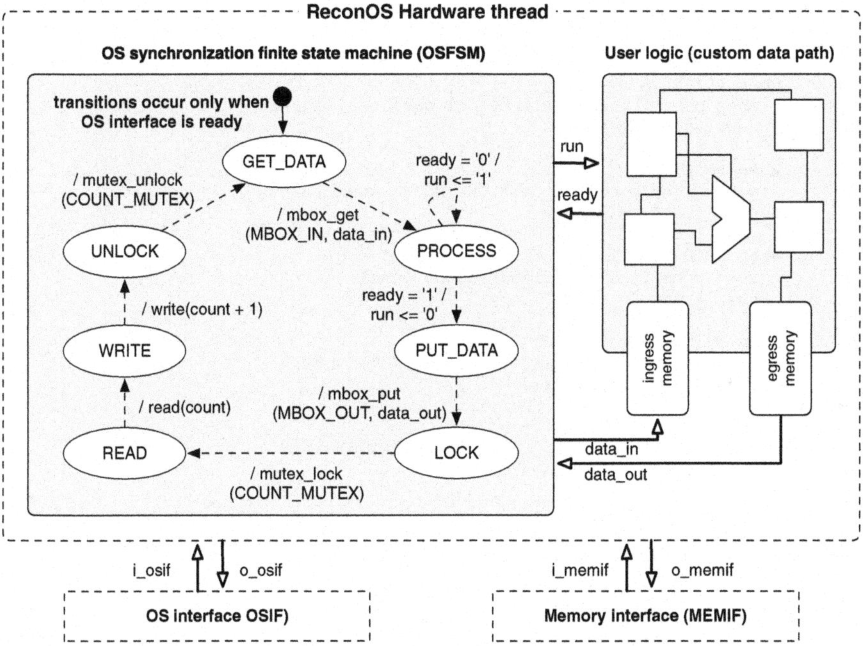

Fig. 13.2 A ReconOS hardware thread comprises the OS synchronization finite state machine and the user logic implementing the data path

As a consequence, the OSFSM in VHDL closely mimics the sequence of operating system calls within the equivalent software thread: it reads a packet from a mailbox, passes it to a separate module to be processed, writes the processed packet back to another mailbox, and increments a thread-safe counter. The description of the actual user logic, however, may well differ from the software realization, as

Listing 13.2 OS synchronization FSM for a stream processing hardware thread [AHL^{+}14b]

```
OSFSM: process(clk, reset)
begin

  if reset = '1' then
    state <= GET_DATA;
    run <= '0';
    osif_reset (o_osif , i_osif);
    memif_reset(o_memif, i_memif);
  elsif rising_edge(clk) then

  case state is

    when GET_DATA =>
      mbox_get(o_osif,i_osif,MB_IN,data_in,done);
      next_state <= COMPUTE;

    when COMPUTE =>
      run <= '1';
      if ready = '1' then
        run <= '0';
        next_state <= PUT_DATA;
      end if;

    when PUT_DATA =>
      mbox_put(o_osif,i_osif,MB_OUT,data_out,done);
      next_state <= LOCK;

    when LOCK =>
      mutex_lock(o_osif,i_osif,CNT_MUTEX,done);
      next_state <= READ;

    when READ =>
      read(o_memif,i_memif,addr,count,done);
      next_state <= WRITE

    when WRITE =>
      write(o_memif,i_memif,addr,count + 1,done);
      next_state <= UNLOCK;

    when UNLOCK =>
      mutex_unlock(o_osif,i_osif,CNT_MUTEX,done);
      next_state <= GET_DATA;

  end case;

  if done then state <= next_state; end if;

  end if;
end process;
```

this is the area where the fine-grained parallel execution of an FPGA-optimized implementation can realize its strengths—unhindered by the necessarily sequential execution of operating system calls.

13.4 System-on-Chip Architecture

The ReconOS run-time system-on-chip architecture provides the structural foundation to support the multi-threading programming model and its execution on CPU/FPGA platforms. Figure 13.3 shows a conceptual overview of a typical ReconOS system that is partitioned into three parts: (1) application software, (2) OS kernel and (3) hardware architecture. The application's software and the OS kernel are usually executed on a main CPU. The operating system provides APIs, libraries, programming model objects, device drivers, and further low-level functions such as memory management.

The ReconOS run-time environment consists of hardware components that provide interfaces, communication channels, and other functionality such as memory access and address translation to the hardware threads. Additionally, the run-time system comprises software components in the form of libraries and kernel modules that offer an interface to the hardware, the operating system, and the application's software threads.

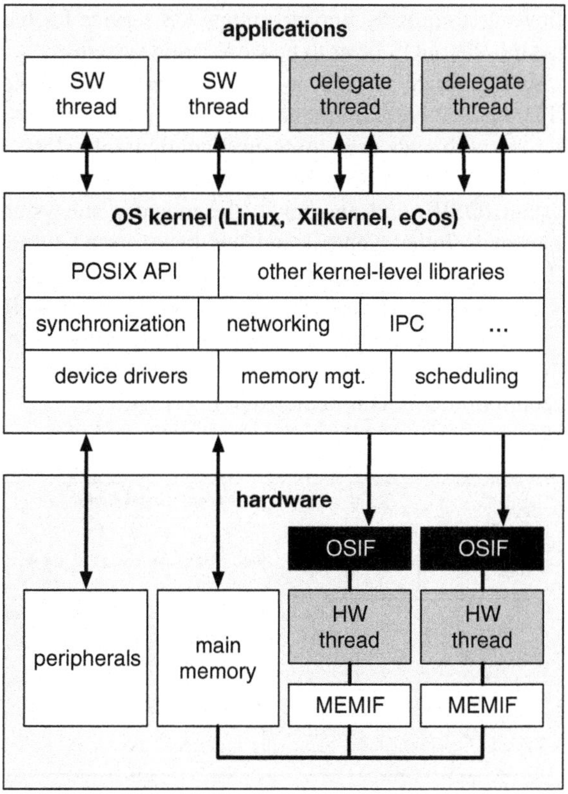

Fig. 13.3 Conceptual view of a typical ReconOS system-on-chip architecture. Hardware threads are connected to delegate threads over an OSIF, while software threads interact directly with the OS kernel

A key component for multi-threading across the hardware/software boundary is the *delegate thread*. It is a light-weight software thread that interfaces between the hardware thread and the operating system. When a hardware thread needs to call an operating system function, it forwards this request through the operating system interface (OSIF) to the associated delegate thread using platform-specific (but application-independent) communication interfaces. The delegate thread then calls the operating system functions on behalf of its corresponding hardware thread. From the OS kernel's point of view, hardware threads are completely hidden behind their respective delegate threads. From the application programmer's point of view, the ReconOS run-time environment completely hides the delegate threads and other ReconOS internals. Thus, an application programmer only needs to take care of the application's hardware and software threads.

This delegate mechanism together with the unified thread interfaces gives ReconOS exceptional transparency regarding the execution mode of a thread, i.e., whether it runs in software or hardware. While the delegate mechanism causes a certain overhead for executing OS calls, the resulting simplicity of switching thread implementations between software and hardware greatly facilitates system generation and design space exploration. Furthermore, the delegate mechanism allows us to quickly implement any OS service for hardware threads and to easily port the ReconOS layer to new operating systems.

A ReconOS system-on-chip architecture is typically composed of a main CPU, several reconfigurable slots, a memory subsystem, and various peripherals. Hardware threads are sitting in reconfigurable slots, which have been defined at design time. The reconfigurable slots have two static interfaces, one to the operating system (OSIF) and another to the memory subsystem (MEMIF). The ReconOS concept is rather general and has been ported to several FPGA families, main CPU architectures, and host operating systems (see Sect. 13.8). For the remainder of this chapter we describe the implementation of ReconOS v3.1, which is the most recent version of ReconOS targeting Xilinx Virtex-6 FPGAs and utilizing a MicroBlaze/Linux environment. An example architecture with one CPU and two reconfigurable slots is depicted in Fig. 13.4.

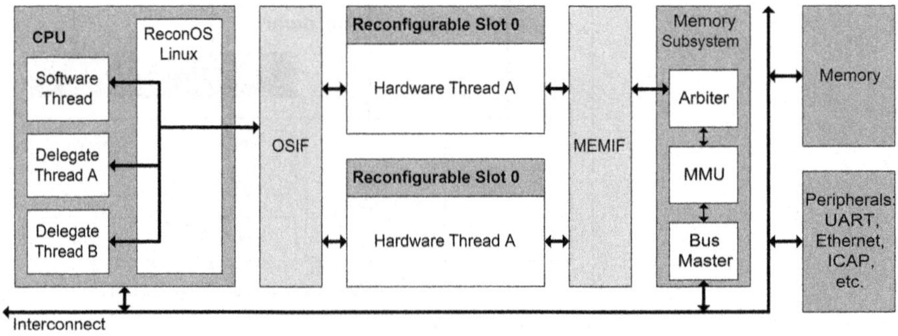

Fig. 13.4 A ReconOS hardware architecture with a CPU and two reconfigurable hardware slots

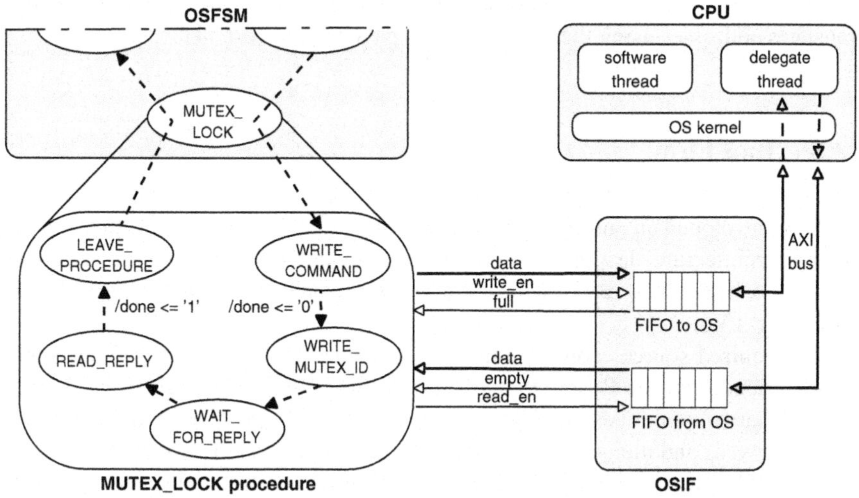

Fig. 13.5 A finite state machine nested within the OSFSM handles the communication between the hardware thread and the OS (via OSIF and delegate thread). The OSIF contains two FIFOs that connect the hardware thread with the CPU. The operating system forwards the hardware thread's requests to the respective delegate thread where the request is carried out

A POSIX-like library of VHDL procedures assists developers with creating the OSFSM for a hardware thread. We have already introduced four procedures in Listing 13.2, namely `mutex_lock/unlock()` and `mbox_get/put()`. Internally, these VHDL procedures contain nested finite state machines (FSMs) that communicate with the delegate threads over the OSIF. The relationship between the OSFSM, a nested FSM of the `mutex_lock` procedure and the OSIF is outlined in Fig. 13.5.

We use a dedicated communication protocol that encodes an OS request as a sequence of words that is written into the outgoing FIFO `o_osif`. The sequence contains the identifier of the OS call and a call-specific number of parameters. The delegate thread reads the request and calls the required OS function, i.e. `mutex_lock()`, on behalf of the hardware thread. When the mutex has been locked, the delegate sends an acknowledgement over the thread's incoming FIFO `i_osif`. Other functions such as `mbox_get` also send a return value back to the hardware thread. The nested state machine and the OSFSM are synchronized via the handshaking signal `done`.

Developers can conveniently access the main memory using dedicated VHDL library of read/write procedures. These procedures encode the requests and automatically transfer data between the hardware thread's local memory and the memory subsystem using two unidirectional FIFO queues. The memory subsystem arbitrates and aligns the hardware threads' memory requests. It supports single word as well as burst accesses of arbitrary lengths. For Linux systems, it implements a fully-featured memory management unit (MMU) that translates virtual to physical addresses.

The MMU includes a translation lookaside buffer (TLB), which autonomously translates addresses using the Linux kernel's page tables [APL11].

13.5 Tool Flow

We have developed an automated tool flow that generates, if necessary, the system-on-chip architecture design and compiles the application software for ReconOS systems. A high-level overview of the current ReconOS tool flow (version 3.1) is shown Fig. 13.6.

The required sources comprise the software threads, the hardware threads and the specification of the ReconOS hardware architecture. We code software threads in C and hardware threads in VHDL, using the ReconOS-provided VHDL libraries for OS services and memory access. An automatic synthesis of hardware threads is not part of the ReconOS project; developers are, however, free to use any hardware description language or high-level synthesis tool to create hardware threads.

ReconOS extends the process for building a reconfigurable system-on-chip using standard vendor tools. On the software side, the delegate threads and device drivers for transparent communication with hardware threads are linked into the application executable and kernel image, respectively. On the hardware side, components such as the OS and memory interfaces as well as support logic for hardware threads are integrated into the tool flow. The ReconOS System Builder scripts assemble the base system design and integrate the hardware threads into a specified reference design. Bus interfaces, interrupts, and I/Os are connected automatically. The build

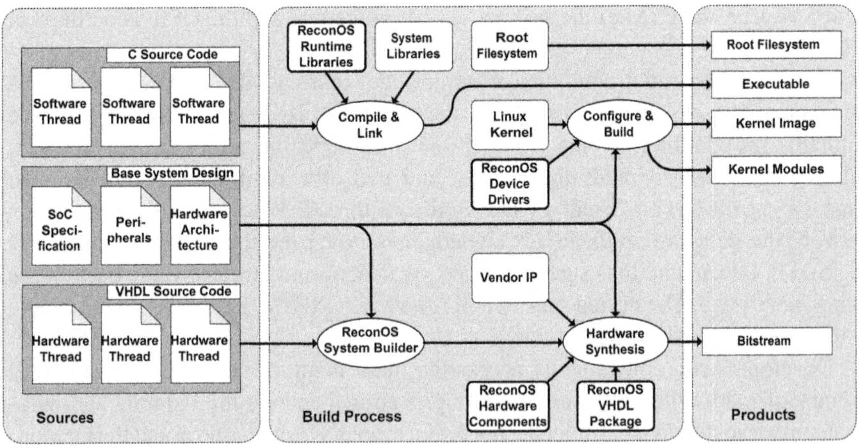

Fig. 13.6 Tool flow for assembling a ReconOS system on a Linux target. ReconOS-specific steps are outlined bold

process then creates an FPGA configuration bitstream for the reference design using conventional synthesis and implementation tools.

During design space exploration, the developer will usually create both hardware and software implementations for some of the threads. Switching between these implementations is a matter of replacing a single thread instantiation statement, e.g., using `rthread_create()` instead of `pthread_create()`. Such a decision for software or hardware can even be taken during runtime, see Sect. 13.7.

Recently, our efforts have focused on the development of a more abstract interface to the ReconOS build tools. We have created the ReconOS Development Kit (RDK) that further automates the design and implementation of ReconOS based systems. Once configured, the RDK is able to map a system consisting of hardware and software thread to a supported platform, defined by hardware architecture, vendor tool-flow, and host operating system. Figure 13.7 illustrates the vision behind the RDK. Currently, we support Virtex 6 and Zynq-based FPGA boards with Linux as the operating systems.

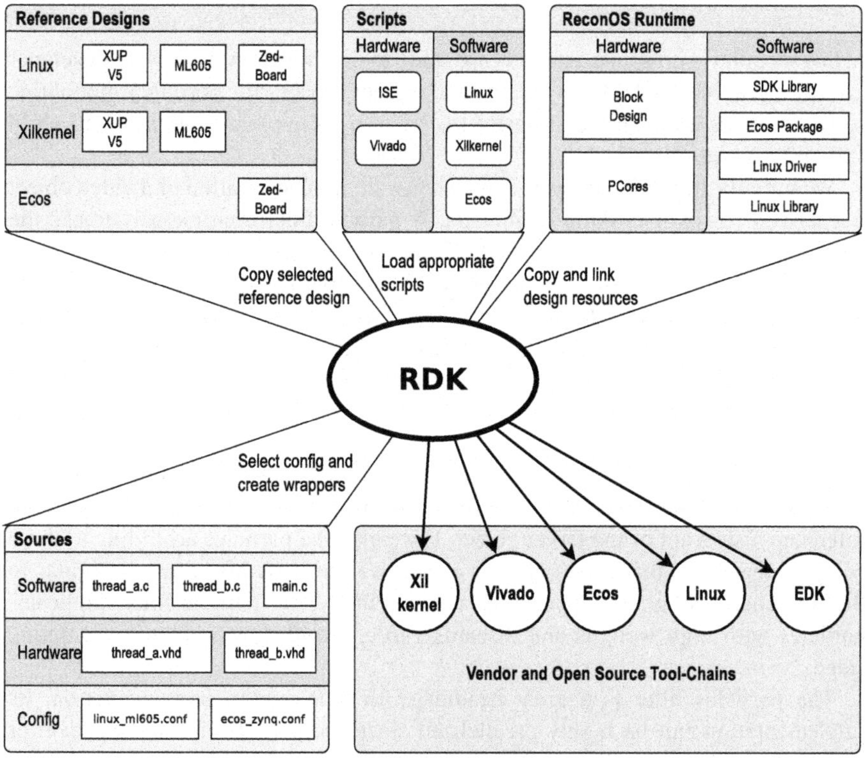

Fig. 13.7 The ReconOS development kit (RDK) maps source descriptions of hardware and software threads to back-end tool-chains supporting different target platforms and operating systems

While software threads can be debugged in-system with relative ease, for instance by setting breakpoints or cycling though the instructions step-by-step, this is generally not possible for hardware threads. Instead simulation is used to gain insight into the inner workings of a hardware module in order to identify bugs and incorrect behaviour. Simulation of hardware is usually performed in a so-called testbench environment that supplies the necessary inputs (stimulus) to the tested module. The RDK helps in that respect by mapping hardware threads to simulation targets, supplying a fully functional testbench environment supporting operating system interactions and memory accesses. This also allows for simulating a ReconOS system with software and hardware threads without requiring a hardware synthesis and place and route process. This capability does not only improve the ability to debug hardware threads, but also speeds up the design and verification process by omitting the hardware implementation flow.

13.6 Case Study: Video Object Tracker

In this section we present a detailed case study to illustrate how a software developer can use our structured, step-by-step development process for porting a monolithic software application—a video object tracker—to an implementation for a hybrid CPU/FPGA platform.

We initially start with an open-source software implementation of a video object tracker [HLP11] for desktop computers. A particle filter continuously tracks the current position and approximate outline of a selected object in a video sequence. It is a stochastic method that tracks several estimates (particles) of the current position and outline in parallel. The filter evaluates each particle by comparing its color histograms with a reference histogram of the target object. It processes the video frames sequentially and highlights the currently best particle in each frame, as can be seen in Fig. 13.8.

The particle filter is divided into three stages. First, it samples the particles according to a given system model (sampling stage). Second, it computes the color histograms of all particles for the current video frame and compares them with the reference histogram of the target object. It weights the particles according to these comparisons, such that a particle has a high weight if its histogram is similar to the reference histogram (importance stage). Finally, the particle filter duplicates particles with high weights and discards particles with low weights (resampling stage).

The particles filter is a great candidate for a ReconOS implementation. Its implementation can be highly parallelized since the particles are independent of each other and can therefore be processed in parallel. The first two filter stages, sampling and importance, can also be pipelined. In the following, we describe the transformation of a monolithic software implementation to a multi-threaded, hybrid ReconOS implementation step-by-step.

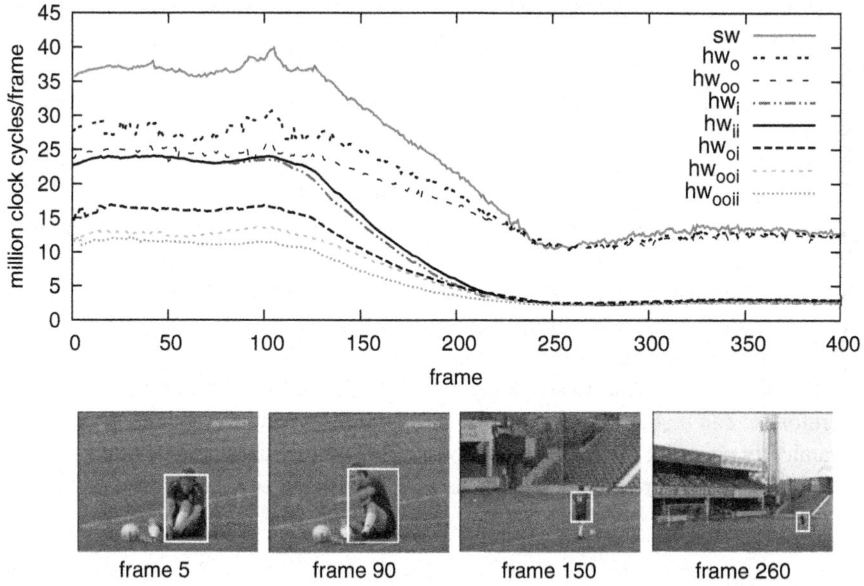

Fig. 13.8 Design space exploration for a video object tracker: the graphs show the execution times per frame for different hardware/software mappings for an example video (taken from [Hes13]). A pure software mapping is labeled with *sw* and hybrid hardware/software mappings with one to four hardware threads are labeled with *hw*. The index defines the number of observation (o) and importance (i) threads in hardware [AHL$^+$14b]

In a first step, we rewrite the monolithic C-code into a multi-threaded implementation that can execute on a desktop computer. The C-code can be naturally transformed into threads, where each filter stage is implemented as a software thread. In contrast to the initial C-code, we use four stages: sampling, observation, importance, and resampling. We have decided to introduce an additional observation stage, which computes the color histograms of the particles. The importance stage then only compares color histograms. Our multi-threaded video object tracker supports to instantiate multiple threads per filter stage. We implement our threads as POSIX threads (pthreads) under Linux and use message boxes and shared memory for synchronization and communication between threads.

In a second step, we port our multi-threaded software implementation to our target processor that is embedded on a Xilinx FPGA. We receive the video stream over an Ethernet interface using a TCP/IP socket. This transformation step usually requires only little effort. The desktop and system-on-chip platforms show many similarities. Both run Linux as operating system and support the same application programming interfaces (APIs).

In a third step, we profile the performance of the individual filter stages and decide which software threads should be also implemented as hardware threads. We identify that the particle filter spends most time to compute and compare color histograms. We therefore implement the observation (o) and importance (i) stages as ReconOS hardware threads. This step requires the highest implementation effort, because it requires the programmer to write VHDL code.

We can use ReconOS to perform a design space exploration of hardware/software mappings by simply instantiating the appropriate number of hardware and software threads. Figure 13.8 shows the tracking performance in time of several mappings for a video where a soccer player moves towards the background. The mapping hw_{ooii} uses four hardware threads and outperforms the other mappings (that have fewer hardware threads) while the player is still in the foreground. The computational demand of the filter decreases significantly when the player moves into the background. In this case, the mapping hw_i that uses just a single importance hardware thread shows comparable results. This simple scenario shows that the best hardware/software mapping might depend on the current workload. Therefore, it can be beneficial to update the mapping at runtime. ReconOS supports dynamically adapting the hardware/software mapping by using the dynamic partial reconfiguration feature of Xilinx FPGAs as explained in section "Dynamic Partial Reconfiguration".

13.7 Applications of ReconOS

Since its creation, ReconOS has been used in numerous research projects. In the following, we briefly present selected ongoing projects to show the wide applicability of the approach.

Dynamic Partial Reconfiguration ReconOS defines a standardized interface for hardware threads, which simplifies exchanging them, not only at design time but also during runtime using *dynamic partial reconfiguration* (DPR). DPR allows for exploiting FPGA resources in unconventional ways, for example, by loading hardware threads on demand, moving functionality between software and hardware, or even multi-tasking hardware slots by time-multiplexing. ReconOS supports DPR by dividing the architecture in a static and a dynamic part. The static part contains the processor, the memory subsystem, OSIFs, MEMIFs, and peripherals. The dynamic part is reserved for hardware threads, which can be reconfigured into the hardware slots. Our DPR tool flow builds on Xilinx PlanAhead and creates the static subsystem and the partial bitstreams for each desired hardware thread/slot combination. Time-multiplexing of hardware slots is supported through cooperative multi-tasking [LP09a] and preemptive hardware multi-tasking [HTK15].

Autonomous Network Architectures ReconOS has been used to implement *autonomous network architectures* that continuously optimize the network protocol stack on a per-application basis to cope with varying transmission characteristics, security requirements, and compute resources availability. The developed architecture [KBNH14] autonomously adapts itself by offloading performance-critical, network processing tasks to hardware threads, which are loaded at runtime using dynamic partial reconfiguration. The autonomous network architecture was evaluated on a distributed smart camera network that communicates over encrypted channels [HHK14].

Self-adaptive and Self-aware Computing Systems Another line of research also leverages the unified software/hardware interface and partial reconfiguration to create *self-adaptive and self-aware* computing systems that autonomously optimize performance goals under varying workloads. For example, we have created self-adaptive implementations of the particle filter presented in Sect. 13.2 that start and stop additional threads on worker CPUs and in reconfigurable hardware slots to keep the resulting frame rate for the video object tracker within a pre-defined band. In the EPiCS project[1] funded by the European Commission, we even advance the autonomy of computing systems and enable them to optimize for diverse goals such as performance, energy consumption and chip temperature based on the current quality-of-service requirements, workload characteristics and system state [AHL$^+$14a].

High-Performance Computing So far ReconOS has been used in embedded systems where the CPU and the hardware cores are implemented in Xilinx platform FPGAs. The general approach of ReconOS to provide a reconfigurable computing platform with a unified multi-threaded programming model, the associated structured development flow, and the portability between reconfigurable computing platforms, is equally attractive in a *high-performance computing* context. For example, ReconOS is currently being evaluated for use in high-speed data acquisition and particle physics applications.[2] In current work[3] we also are studying how ReconOS can be ported to x86-based server systems that attach FPGA accelerator cards via PCIe.

13.8 Availability of ReconOS

ReconOS has been actively developed since its inception in 2006 at the University of Paderborn. Since then it has gone through three major revisions and has been ported to several operating systems and hardware platforms.

- *Version 1* used the eCos operating system running on PowerPC CPUs embedded in Xilinx Virtex-2 Pro and Virtex-4 FPGAs.
- *Version 2* improved on the original version by providing FIFO interconnects between hardware threads, adding support for the Linux operating system, and offering a common virtual address space between hardware and software threads.

---

[1] http://www.epics-project.eu.

[2] http://openlab.web.cern.ch/ice-dip.

[3] http://sfb901.uni-paderborn.de.

- *Version 3*, which was released in early 2013, is a major overhaul that streamlines the hardware architecture towards a more lightweight and modular design. It brings ReconOS to the Microblaze/Linux and Microblaze/Xilkernel architectures and has been used extensively on Virtex-6 FPGAs.
- *Version 3.1* supports Xilinx Zynq platforms on ARM Cortex/Linux and ARM Cortex/Xilkernel architectures and utilizes the AXI bus in an improved architecture.

ReconOS is available as open source and follows a public development model that encourages contributions by third parties. The source code and further information is available at http://www.reconos.de.

13.9 Related Work on Operating Systems for Reconfigurable Computing

In this section, we present an overview on the initial motivations for and origins of operating systems for reconfigurable computing in the late 1990s and outline the subsequent and recent developments in the field.

The introduction of the partially reconfigurable Xilinx XC6200 FPGA series in the mid 1990s and, later on, the JBits software library for bitstream manipulation inspired researchers to investigate dynamic resource management for reconfigurable hardware. Early works, e.g., [Bre96, CCKH00, BKS00], drew an analogy between tasks in software and so-called virtual or swappable hardware modules and studied fundamental operations such as scheduling; placement, relocation and defragmentation; slot-based device partitioning and reconfiguration schemes; and inter-module routing. Although these works suggested to centralize resource management in a runtime layer for convenience, an integration with a software OS was not a predominant design goal. The very few projects that resulted in implementations used FIFOs or shared memory to interface reconfigurable hardware modules with other parts of an application running in software. However, the nature of these hardware modules was still that of a passive coprocessor, which was fed with data from software tasks.

After the development of more sophisticated prototypes, e.g., a multimedia appliance using multitasking in hardware [NCV$^+$03], several researchers, e.g., [ANJ$^+$04, SWP04, BWHC06], concurrently pushed the idea of treating hardware tasks as independent execution units, equipped with similar access to operating system functions as their software peers. Around 2004, these projects fundamentally changed the concept of reconfigurable hardware operating systems since the emerging prototypes turned hardware modules into threads or processes and offered them a set of operating system functions for inter-task communication and synchronization. These approaches can be considered the first operating systems directly dedicated to reconfigurable computing.

Soon after these first operating systems have been developed it was found that promoting hardware tasks to peers of software threads while carrying over a manually managed local memory architecture was too restrictive. Thus, researchers have studied how hardware tasks can autonomously access the main memory. For reconfigurable operating systems that build on general purpose OS such as Linux, this meant that virtual memory had to be supported. The first approaches, e.g., [VPI05, GC07], solve this challenge by creating a transparently-managed local copy of the main memory and modifying the host operating system to handle page misses on the CPU. To improve the efficiency of accessing main memory, especially for non-linear data access patterns, ReconOS has later pioneered a hardware memory management unit [APL11] for hardware modules that translates virtual addresses without the CPU.

Current research projects on operating systems for reconfigurable computing differ mainly with respect to whether a hardware module is turned into a process, a thread or a kernel module, and in the richness of OS services made available to reconfigurable hardware. While projects such as BORPH [SB08] choose UNIX processes, Hthreads [ASA+08] and ReconOS use a light-weight threading model to represent hardware modules. More recently, SPREAD [WYZ+12] started to integrate multi-threading and streaming paradigms, while FUSE [IS11] focuses on a closer, more efficient kernel integration of hardware accelerators.

Finally, the LEAP operating system, which is presented in detail in Chap. 14 in this book, is built on the idea of socket-like latency-insensitive communication channels between hardware and software modules. Through these channels, communication but also STDIO functions can be handled. Additionally, memory buffers that can be shared among modules are provided as part of the operating system. ReconOS provides a superset of these functions and offers additional synchronization and inter-process communication mechanisms.

Compared to other approaches leveraging the threading model, especially Hthreads that focuses on low-jitter hardware implementations of operating system services, ReconOS with its unified hardware/software interfaces allows us to offer an essentially identical and rich set of OS services to both software and hardware threads. ReconOS does not require any change to the host OS, which leads to a comparatively simple tool flow for building applications, to an improved portability and interoperability through standard OS kernels, and to a step-by-step design process starting with a fully functional software prototype on a desktop.

13.10 Summary

ReconOS is an open source operating system layer for FPGAs which is particularly attractive for developers with a software development background. Due to the deep semantic integration of hardware accelerators into an operating system environment and the use of standard operating system kernels, hardware threads can access a rich set of operating system functions. Since hardware threads and software threads share the same operating system interfaces they become interchangeable from the application's perspective. This allows ReconOS to support a step-by-step development process in which a multi-threaded application is ported to a hybrid hardware/software version, by porting the most profitable threads from software to hardware one by one. In addition the ReconOS approach allows for rapid design space exploration at design time and even migration of functions across the hardware/software border at run-time. Our experience from our research projects and numerous student projects conducted at our universities shows that these features can significantly lower the entry barrier for reconfigurable computing technology.

Acknowledgements This work was partially supported by the German Research Foundation (DFG) within the Collaborative Research Centre "On-The-Fly Computing" (SFB 901), the International Graduate School of Dynamic Intelligent Systems, and the European Union Seventh Framework Programme under grant agreement no. 257906 (EPiCS).

Chapter 14
The LEAP FPGA Operating System

Kermin Fleming and Michael Adler

In contrast to modern software, current FPGA programming environments remain primitive. Many of the abstractions typically available in general purpose systems: high-level languages, abundant libraries, code portability, and automatic resource management, do not exist for FPGAs. Creating a design for an FPGA often requires that programmers start from bare metal, bringing up not only their application, but significant operating infrastructure including memory controllers, I/O systems, and debugging facilities. All of these activities require large amounts of time, slowing the FPGA development process to the point that the relative simplicity of software becomes very attractive. Worse, getting an application to work once is insufficient. The lack of portability of FPGA programs between platforms and generations means that building the next generation implementation usually requires significant re-implementation. Although FPGAs offer attractive power and performance relative to traditional sequential processors, they are not commonly used in systems architecture *because they are painful to program.*

To address these development issues, we built the LEAP FPGA operating system. Like a software operating environment, LEAP provides both basic device abstractions for FPGAs and a collection of standard I/O and memory management services. These abstraction layers shield programs from the complex details of the underlying FPGA hardware while simultaneously providing design portability across all FPGAs supported by LEAP.

LEAP resembles general-purpose operating systems in function. However, because LEAP targets FPGAs, its implementation is necessarily very different. In general-purpose systems, user programs are organized into processes and threads that share a common execution substrate with the operating system: the processor and its memory system. Interaction between the program and the operating system

K. Fleming (✉) • M. Adler
Intel Corporation, Hudson, MA, USA
e-mail: Kermin.Fleming@intel.com

D. Koch et al. (eds.), *FPGAs for Software Programmers*,
DOI 10.1007/978-3-319-26408-0_14

generally occurs by storing data in memory and then changing contexts to the operating system by a function call. FPGAs differ from this general-purpose model in two ways, both of which profoundly affect the organization of LEAP.

In the FPGA, user programs and the operating system share neither a common execution substrate nor a common, global view of storage. Instead, FPGA programs are organized as spatially distributed modules, with portions of the FPGA fabric dedicated to each of the different functions of the program and the operating system. To accommodate this distributed programming model, LEAP's fundamental abstraction is *communication*. To enable portable, abstract communications, LEAP builds upon the concept of latency-insensitive design [CMS01]. LEAP formalizes latency-insensitive design as *latency-insensitive (LI) channels*, programming constructs that provide point-to-point, reliable communication but do not explicitly specify the timing or the implementation of the communication. By decoupling the semantics of communication from the physical and timing details of an implementation of data transmission, latency-insensitive channels slacken the requirements of synchronous timing while preserving the parallelism intrinsic to hardware designs. This timing relaxation is critical in providing operating system abstractions on FPGAs, since the timing behavior of operating system services must necessarily change between platform and program configurations. We will demonstrate, by example, that latency-insensitive channels can capture most important operating systems functionalities in FPGAs while providing a user-friendly programming interface.

In general-purpose systems, new instructions can be introduced into a program at any time during execution. As a result, general-purpose operating system activities occur at program run time: context-switching is lightweight and operating system instruction flows incur overhead only when those instructions are executed. FPGAs are typically programmed only once per execution. While possible, including dynamic management layers within a static FPGA program incurs fixed area and performance overhead that lasts for the life of the FPGA program. To avoid these penalties, most resource management decisions in LEAP, for example memory and clock management, are made statically at compile time. This decision results in a deeper coupling of compiler and operating system. LEAP's static incorporation of resources resembles early software operating systems, which required re-compilation to integrate new functionalities. A key contribution of this work is an extensible compiler interface that permits the integration of new operating systems services within the LEAP framework.

LEAP aims to allow developers to focus on core algorithms, while retaining the full flexibility of FPGAs. In building platforms for FPGAs, there is much to be learned from general-purpose operating systems. We will highlight the similarities and differences between LEAP and traditional software operating systems and the importance of communication to FPGA design. We will also explore the unification of operating-systems features, many of which have been previously evaluated individually, into a conceptual whole that has never been available, previously, for FPGAs.

14.1 LI Channels: The LEAP Design Philosophy

Operating systems are built on abstraction. In software, memory serves as the primary operating system abstraction layer. Core features like virtual memory, communication, I/O, and dynamic library linking all occur through memory. Even communication between the user program and the operating system, in the form of system calls, occurs through memory. Memory-as-abstraction is enabled in part by the dynamic time-multiplexing capabilities of the processor and in part by support for sharing within the processor memory system.

In contrast to general-purpose systems, FPGAs have no intrinsic concept of a globally shared storage or execution infrastructure. Moreover, constructing such an infrastructure in the FPGA as a fundamental operating system abstraction layer is inefficient. FPGAs are comprised of small state and logic elements which are typically dedicated to a single task and which communicate with each other using wires.

Communication of data, as opposed to the storage of data in memory, is fundamental to the FPGA. However the basic communication primitives provided by the FPGA, wires, are not sufficiently abstract for an operating system. Wires in the FPGA are synchronous, and clock edges have concrete semantic meaning. Synchronous semantics present a problem for operating systems: the timing behavior of operating system services necessarily changes depending on the user program and the platform targeted. Although it is possible to automatically abstract away synchronicity within an FPGA program [BTD$^+$97], the overhead of such abstraction is very high. Thus, communication over pure wires cannot be used as a basic interface into the operating system. To enable FPGA operating systems, a higher-level abstraction for communication is required.

LEAP introduces latency-insensitive channels as its primary communication abstraction. Latency-insensitive channels have operating behaviors and interfaces similar to the concurrent FIFO modules commonly available in hardware and software programming libraries—a simple enqueue and a simple dequeue operation along with some status methods, e.g. `notFull` and `notEmpty`, for use by the user program in determining when to send and receive data. Syntactically, latency-insensitive channels consist of named send and receive endpoint pairs [PACE09], instantiated with the following syntax:

```
module mkA (Empty);
  SEND(Bit(42)) toB = mkSend("AtoB");
endmodule
module mkB (Empty);
  RECV(Bit(42)) fromA = mkRecv("AtoB");
endmodule
```

Semantically, these pairs represent a reliable, in-order channel from the sender to the receiver. Unlike library FIFOs, which have a fixed implementation within a given library, the latency-insensitive channel denotes abstract communication and makes only two basic guarantees. First, the channel guarantees reliable, FIFO delivery of messages. Second, the channel guarantees that at least one message can be in flight at any point in time. Consequently, a latency-insensitive channel *may* have dynamically-variable transport latency and arbitrary, but non-zero, buffering. At compile time, LEAP's compilation infrastructure (Sect. 14.5.1) matches channel endpoints and produces physical implementations of the latency-insensitive channels.

LEAP makes use of the flexible semantics of the latency insensitive channel in three ways. First, LEAP leverages the abstract nature of latency insensitive channels as a means of orchestrating communications between FPGA and CPU and across FPGAs (Sect. 14.2). Second, LEAP relies on latency-insensitive channels to safely decouple its services, like memory, from the user program (Sects. 14.3 and 14.4). Finally, LEAP uses latency insensitive channels to abstract physical device interfaces, providing portability among platforms (Sect. 14.5).

14.2 LEAP Communications

When a programmer instantiates a latency-insensitive channel, he asserts that the potentially variable timing behavior of the channel *does not impact the functional correctness of the design*. Thus, the semantics of latency-insensitive channels permit LEAP to select *any* channel implementation that preserves the latency-insensitive channel semantics, e.g. reliable, FIFO message delivery. Much like the POSIX `socket` interface, LEAP uses latency-insensitive channels both for communication between hardware modules within an FPGA and for communication with other types of programs on external platforms. At compile time, LEAP analyzes the user program and discovers channels that terminate externally. These channels are tied, via a synthesized communications network, to physical devices capable of communicating externally.

In addition to managing communication between hardware and software, LEAP leverages latency-insensitive channels to map user programs automatically across arbitrary sets of FPGAs [FAP+12]. This operation is analogous to the mapping of threads to a multi-core processor by a general purpose operating system. Mapping an FPGA program to multiple platforms first requires a mechanism for partitioning the program. Again, LEAP leverages latency-insensitive channels. We define a latency-insensitive module to be a region of a program that interacts externally only by way of latency-insensitive channels. LEAP views all user programs as collections of latency-insensitive modules, a subdivision that can be thought of as the FPGA equivalent of operating system process or thread management. To map a program across FPGAs, LEAP allocates the latency-insensitive modules comprising the program to the available FPGAs in the system based on area consumption and communication. As in the case of FPGA-processor communications, LEAP synthesizes a networking layer to transport messages between connected FPGAs.

Figure 14.1b shows an example of a simple program that has been mapped to two FPGAs. As in software operating systems, no changes are necessary in the user program to make use of LEAP's partitioning.

Latency-insensitive modules offer a good balance between the abstraction needed for operating-system-level management and program expressivity. Moreover, this balance is achieved without a significant impact to program performance or area consumption: most latency-insensitive channels will be implemented as RTL FIFOs, preserving the performance of hand-coded RTLs. Latency-insensitive modules are already a common design paradigm in hardware systems, both at the system and micro-architectural levels. For example, SoCs targeting FPGAs are generally framed in terms of network-on-chip protocols, which are usually latency-insensitive. Within individual hardware blocks, it is a common design practice to decouple components with guarded FIFO interfaces. Because LEAP's latency-insensitive channel syntax and semantics resemble existing RTL structures, LEAP's latency-insensitive channel syntax can often be substituted directly into existing RTL.

Although LEAP primarily targets RTLs mapped to FPGAs with some support for software, latency-insensitive modules admit a diversity of programming substrates and languages. So long as the external latency-insensitive interface semantics of a module are maintained by the programmer, modules may have any implementation internally that programmers require. Such implementations include not only RTL, but also arbitrary software either on an external host processor or an internal soft processor.

14.3 LEAP Scalable FPGA Memory Hierarchies

Although we argue that abstract communication is the fundamental building block of FPGA programs, memory is critical to many applications. FPGAs offer a rich set of memory primitives: SRAM resources are available within the fabric, FPGA boards include off-chip DRAM, and host virtual memory may be accessible over a communications link. Unfortunately, these memory resources are often difficult to use, in part because FPGA programmers are fully exposed to the low-level details of the memory: the number of memory banks, the width of memory, timing, and the number of beats per memory access. Memory systems in general purpose machines are also complex, and large fractions of processor die area are consumed by memory management: caches, memory controllers, and virtualization hardware. However, most user programs in general purpose systems are insulated from the complexity of the memory hierarchy through a simple, abstract load-store interface to virtualized memory. The internal complexity of the memory hierarchy improves throughput dramatically, but the simple interface to memory allows programmers to focus on algorithms and enables program portability.

LEAP provides a software-like in-fabric memory abstraction for FPGA programs. LEAP's memories have a latency-insensitive interface with three methods: read-request, read-response, and write. A LEAP program may instantiate as many

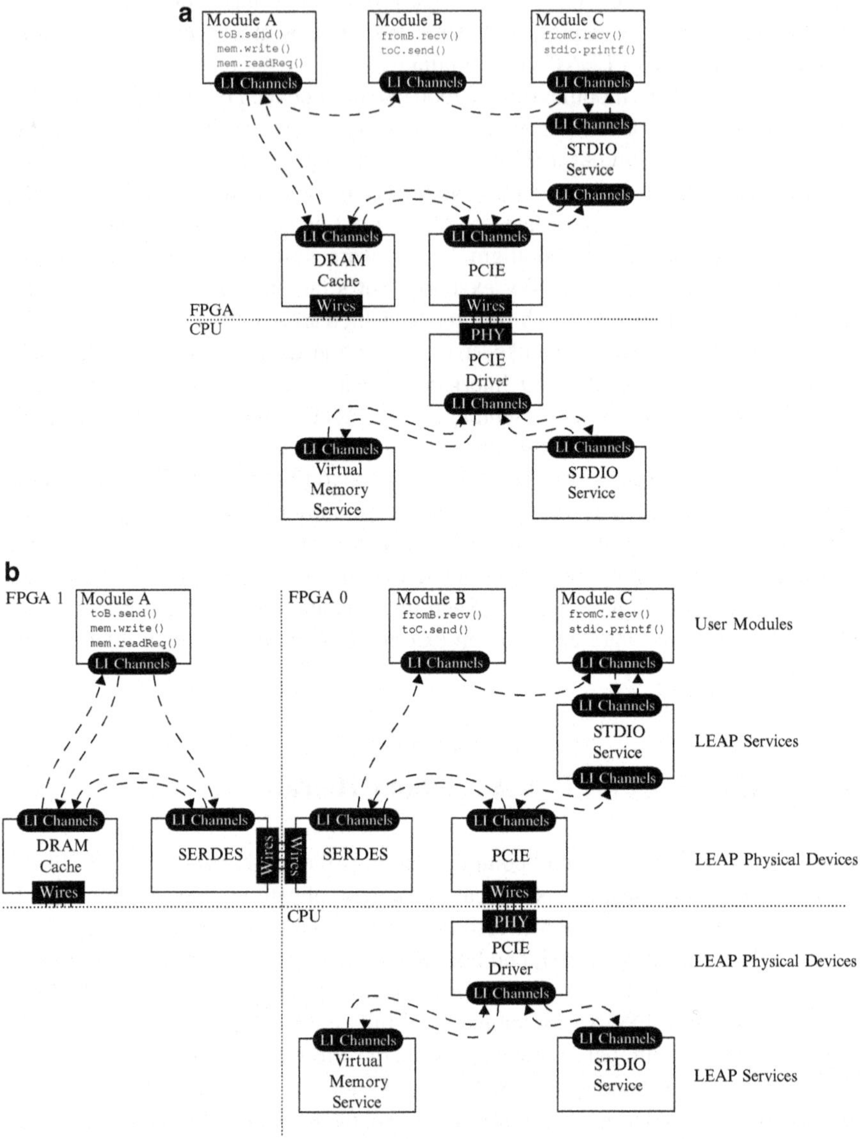

Fig. 14.1 Two implementations of a LEAP-based program. *Dotted lines* represent the logical latency insensitive channels. These channels are inferred at compile time from the combination of user source and platform devices. LEAP targets multiple FPGA platforms and communicates with software by stretching channels across chip boundaries. (**a**) A complete LEAP program. (**b**) The same program, mapped to two FPGAs

memories as needed and these memories may have arbitrary size, even if the target FPGA does not have sufficient physical memory to cover the entire requested memory space. Programmers instantiate LEAP memories using a simple, declarative syntax:

```
// Instantiate two memory interfaces with private
// address spaces
module mkModuleA();
  MEM_IFC(ADDR,DATA) priv1 = mkMem();
endmodule
module mkModuleB();
  MEM_IFC(ADDR,DATA) priv2 = mkMem();
endmodule
```

LEAP provides two basic interfaces to memory: private and shared [YFAE14]. Each declaration of a memory resource creates a latency-insensitive interface to memory. For shared memories, programmers declare a named coherence domain and instantiate a LEAP coherence controller in addition to the memory interface itself.

LEAP's memory primitives are portable among and across FPGAs. Like the load-store interface of general-purpose machines, LEAP's memory interface permits complex backing implementations. LEAP's memory interface does not explicitly state how many operations can be in flight, nor how quickly in-flight operations will be retired. As was the case in communications, this ambiguity provides significant freedom of implementation to the operating system. For example, a small memory could be implemented as a local SRAM, while a larger memory could be backed by a cache hierarchy and host virtual memory. LEAP leverages this freedom to build complex, optimized memory architectures on behalf of the user, bridging the simple user interface and complex physical hardware.

At compile time, LEAP aggregates the declared user memory interfaces into a cache hierarchy, making use of whatever physical memory resources are provided by the platform, as shown in Figs. 14.1 and 14.2. This hierarchy is optionally backed by an interface to host virtual memory, as shown in Fig. 14.1, which provides the illusion of an arbitrarily large address space. LEAP's backing memory hierarchy automatically multiplexes host virtual memory by explicitly assigning segments of virtual memory to each memory region declared in the user program, preventing independent memory regions from interfering with one another within the cache hierarchy.

14.4 LEAP Service Libraries

One of the most attractive features of software programming is that minimal programs, e.g. `hello world`, are small and easy to understand, even for a novice programmer. Contrast the conciseness and clarity of a basic software program

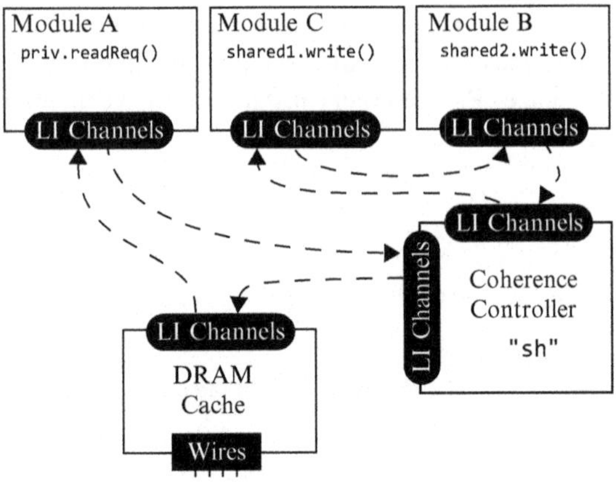

Fig. 14.2 The memory hierarchy inferred by LEAP for a program with a single private memory and two shared memories. Memories are organized as networks of latency-insensitive channels. All memory interfaces are able to take advantage of the large DRAM cache

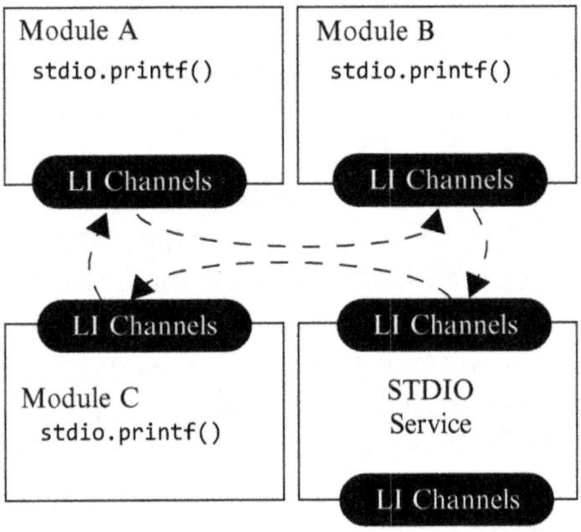

Fig. 14.3 STDIO, an example of a typical LEAP service library. Several user modules instantiate STDIO clients. Clients are aggregated by LEAP at compile time and attached to the STDIO server. In turn, this hardware server is connected to a software-side server, which implements the service library

with a basic RTL program running on an FPGA. The difference is stark! The chief distinction between the software and RTL program is not the complexity of the binary: the software binary is comprised of tens of thousands of unique instructions. Rather, the perceived simplicity of software arises from the availability of good libraries, which mask system complexity through intuitive APIs and strong composability. By combining communications and memory primitives with an extensible compilation infrastructure, LEAP is able to provide libraries that rival software in their scope and simplicity. Indeed, the programmer-supplied portion of LEAP's `hello world` application is only a dozen lines of code.

Although we argue that the LEAP libraries are as easy to use as software libraries, there are some obvious differences. Software libraries consist of instruction flows to which the user program gives control. LEAP service libraries take a distributed approach: service libraries consist of clients and servers which interact over latency-insensitive channels. The typical architecture of a service library, a central server and multiple clients, is shown in Fig. 14.3. To make use of the service library, programmers instantiate client modules in their code. Client modules provide a latency-insensitive local interface framed in terms of request and response methods,

```
interface STDIO(type data);
  // fopen request/response interface, returning a file handle
  Void fopen_req(GLOBAL_STRING_UID nameID,
                 GLOBAL_STRING_UID modeID);
  STDIO_FILE fopen_rsp();

  Void printf(GLOBAL_STRING_UID msgID, List(data) args);
  Void fprintf(STDIO_FILE file, GLOBAL_STRING_UID msgID,
               List(data) args);
endinterface
```

Fig. 14.4 A portion of the interface to an STDIO client. LEAP supports most common STDIO functions

as in Fig. 14.4. Pairwise, these methods resemble software function calls, but are decoupled to enable hardware pipelining. At runtime, clients issue requests to the controller over the network. These requests are serviced by the controller, often by invoking a software routine using a latency-insensitive channel which terminates in software. Direct composition with software libraries, for example, using software to allocate physical memory or handling printing to a terminal, greatly simplifies the FPGA-side server implementation and underscores the value of good hardware-software communications support.

To facilitate the aggregation of library clients, LEAP provides a broadcast communication primitive: the latency-insensitive ring. Participants in the broadcast instantiate named ring stops. At compile time, ring stops are aggregated by name and connected in a ring topology via latency-insensitive channels. Rings are advantageous for service libraries because they both carry a low area overhead and provide a convenient way to describe library implementations in which the number of clients is unknown prior to compilation. Indeed, the overhead is so low that LEAP's debugger service library can instantiate debugger clients at every latency-insensitive channel in a program, enabling programmers to monitor channel state at runtime.

LEAP provides many basic service libraries, including assertions, statistics collection, command line parameters, locking, and synchronization. For brevity, we describe only Standard I/O (STDIO) in detail.

STDIO is one of the most fundamental libraries available in C and often serves as an introduction to software programming. LEAP's STDIO service, part of which is listed in Fig. 14.4, provides the functional analog of STDIO in the FPGA. Like its software counterpart, LEAP STDIO provides a simple API which masks the complexity of actually writing formatted data to a file.

Programmers instantiate STDIO clients and invoke methods on them. Internally, STDIO clients marshal programmer commands into a packet format, which is then streamed over the service ring network to an STDIO server. The server transmits these packets to software on an attached processor. Software then invokes the appropriate C STDIO functions using the FPGA-supplied arguments.

One issue in Standard I/O is dealing with strings. Strings, which are logically unbounded, are expensive to manipulate directly in the FPGA. To avoid the area overhead of string manipulation in fabric, we borrow a tactic from C and cast strings as pointers. At compile time, string literals are replaced by pointers, the `GLOBAL_STRING_UID`. When FPGA programs manipulate strings, as in `snprintf`, they do so by passing string pointers to software. Software creates a new string and returns a pointer to the new string back to the FPGA. LEAP's string pointer management makes use of the SoftServices interface described in Sect. 14.5.1 to keep track of string pointers at compile time.

14.5 LEAP Platforms

Physical devices are hardly abstract, whether they are attached to a general purpose processor or to an FPGA. Interfacing to these devices requires a deep understanding of device behavior, the details of which can leak back into user code if the interface programmer is not careful. This is especially true in FPGAs, where device timing details are often absorbed by user logic. For example, a design targeting an FPGA with an attached SRAM may absorb and come to rely on the fixed-latency behavior of a particular SRAM. As a result of this dependence, the design becomes unportable.

To deal with physical devices, software operating systems utilize hardware abstraction layers. Each device provides some basic API which is common among all devices of that type. LEAP adopts this approach. In LEAP, classes of physical devices each provide a uniform, abstract device interface. Rather than a call-based API, LEAP devices provide a latency-insensitive-channel-based API, usually framed in terms of request and response channels. Internally, any implementation may be chosen by the driver writer, but the external driver interface is constrained by the LEAP compilation flow to use only latency-insensitive channels.

LEAP-enabled FPGA platforms may be viewed as a collection of driver modules for these low-level devices, in much the same way that `linux/arch` is used to differentiate low-level interfaces to different processor architectures. For each target platform, LEAP accepts a platform description file which includes abstract drivers for each physical device on the platform. LEAP uses this configuration file at compile time to instantiate those drivers required by the user program. As with software, a platform description must be written only once per platform and may be shared by all programs targeting the platform. To represent systems with multiple platforms, LEAP simply bundles together descriptions of single platforms with a description of platform interconnectivity.

Some FPGA platforms will not provide all of the resources required by all of the LEAP services. To maintain compatibility with such platforms, LEAP provides virtualized device implementations. These devices may rely on support from an external platform, either FPGA or processor, which can provide the service. In the case of LEAP Scratchpads, if no backing memory is available on the local FPGA, backing memory on a remote FPGA or on a remote processor may be used instead. LEAP virtual services resemble the memory-based virtualization layers typically found in software, such as `ramdisk`. LEAP's latency-insensitive approach directly enables platform virtualization: virtual devices are functionally equivalent to physical devices but their timing and performance may be radically different.

LEAP's FPGA platform abstraction directly enables its multiple FPGA partitioning capabilities. Because operating system interfaces are consistent across platforms, LEAP can map user modules to any platform irrespective of the services and resources provided by that platform.

14.5.1 LEAP Compilation

In software, applications typically access system resources by making a request to the operating system for an interface object. For example, `fopen` requests access to a file by way of a `FILE` handle. From the programmer's perspective, this interface is very clean: a request results in a simple object, which the program then manipulates. In this transaction, the operating system functions as an intermediary between the program and the underlying file system resource, and the operating system's responsibility is to provide an efficient implementation of the accessor object on behalf of the user program.

Resource access in LEAP is functionally analogous to software resources. Programmers request abstract resource accessor objects, which are then supplied by LEAP. The key difference between LEAP and software operating systems is when accessor objects are created. Since instructions incur low overhead, resource management decisions in general purpose systems can be made dynamically and the operating system has the freedom to optimize its provided implementation at runtime. Although LEAP makes FPGA resource management decisions dynamically wherever prudent, dynamic management of resources in an FPGA often requires both additional control and storage overhead. These overheads frequently outweigh the benefit of dynamism: for point-to-point communication, a FIFO is much cheaper to implement than a network router. Avoiding dynamism at runtime means that resource allocation decisions must be made at compile time, resulting in a tighter coupling of compiler and operating system than is typically found in modern general-purpose operating systems.

Consider, as an example, the problem of allocating clocking resources in an FPGA. On the surface, clocking seems like a trivial problem in FPGA design: the designer instantiates a new clock primitive and ties it to the design RTL. However, even a resource as simple as a clock benefits from abstraction. Clock primitives are finite resources within the FPGA fabric. If a program instantiates too many clocks, it will not fit in the FPGA, *even if these clocks all have the same frequency*. A second issue in clocking is portability. Since FPGA clocks are parameterized in terms of ratios, including a fixed clocking primitive in the RTL immediately renders a design unportable.

To address these issues, LEAP provides the SoftClocks service. Programs ask for a clock at a particular frequency. Internally, the service maintains a list of previously requested frequencies. In response to a user request for a clock, the service searches its list for a clock of that frequency. If found, the service returns the appropriate clock object, which may be used directly in the program. Otherwise, the service instantiates a new clock, inserts it in the clock list, and returns the new clock. All clocks generated by the service are derived from physical clocks provided by the target platform. User designs are thus portable, depending only on clocks derived from the SoftClocks service.

SoftClocks illustrates a general resource management paradigm: a resource management service gathers information at compile time about how a resource is used within an FPGA program, and then creates an efficient implementation of the resources required by the program. To manage general classes of FPGA resources at compile time, LEAP defines a programmer-extensible compiler interface, called *SoftServices*. LEAP uses SoftServices to manage many diverse functionalities, including clocking, the implementation of the latency-insensitive channels described in Sect. 14.1, and the construction of efficient scan chains for run-time debugging.

Conceptually, SoftServices are objects that contain arbitrary, service-specific state. Services provide two interfaces: a private compiler interface and a user interface, by which programmers interact with the service object and request resources. Service objects register with the LEAP compiler, and the compiler invokes the private interface at certain points during compilation. In a typical service implementation, the user program invokes the public service interface to request resource access. The service builds a representation, often a list, of these requests. At the end of program compilation, LEAP invokes a service-provided handler, which allows the service to examine its data and produce an efficient, program-specific service implementation.

SoftServices provide the following interface to the LEAP compilation flow:

```
interface SOFT_SERVICE(type service_state);
  service_state initService();
  Void finalizeService(service_state state);
  Void handleModule(service_state state);
endinterface
```

Each service may maintain whatever state information it requires, as represented by the structure `service_state`. Services must provide three functions to the compiler. `initService` is called at the start of compilation and `finalize-Service` is called at the end of compilation. This permits a service to view all requests for resource accessors for the entire program, thereby permitting the construction of globally efficient management hardware. `handleModule` is called at each latency-insensitive module, allowing services to make regionally-scoped implementation decisions.

SoftServices may make use of other SoftServices. For example, the implementation of named latency-insensitive channels interacts with the SoftClocks service by automatically inserting channels with clock domain crossings when necessary. To capture these inter-service dependencies, LEAP allows system programmers to specify the order in which the private service interfaces should be invoked by the compiler, in much the same way that `init.d` manages service startup in a general purpose operating system.

LEAP's SoftService infrastructure requires that the RTL compiler have strong support for static elaboration. Although legacy compilers are limited in this respect, recent hardware-oriented compilers [BVR+12, Blu04] provide sufficient infrastructure for LEAP.

14.6 Related Work

Previous work on FPGA operating systems has focused on adding communications support within existing general-purpose operating systems. BORPH [SB06] views FPGA programs as UNIX processes that can communicate externally by means of UNIX pipes. HThreads [ASA+08] takes a similar processor-centric approach, in which fabric-based accelerators are treated as threads that coordinate with other activities on a soft processor. FSMLanguage [Agr09] proposes a new domain-specific language for finite state machines. FSMLanguage abstracts communications between hardware and software FSM components, using channel constructs analogous to latency-insensitive channels. LEAP generalizes the concepts put forward in these works by supporting channel-based communication between arbitrary combinations of execution platforms, including multiple FPGAs. LEAP, BORPH, HThreads, and FSMLanguage all provide strong compilation support for mapping user programs to the FPGA.

In addition to communication, researchers have investigated memory abstractions for FPGAs. CoRAM [CHM11] proposes a cache interface similar to LEAP's memory [AFP+11] interface. Whereas LEAP allows the user RTL to control the memory interface directly, CoRAM advocates control by way of control threads programmed using a C-like language. Unlike LEAP, CoRAM does not address communication and provides no support for shared memory within or among FPGAs. FSMLanguage provides a basic memory abstraction, but its treatment of memory is limited to scheduling the ports of in-fabric SRAM resources.

Both Xilinx and Altera have produced OpenCL [Khr08] tool flows which simplify the coupling of a processor and FPGA. These flows allow a user to specify a C kernel, which the tool will then implement on the FPGA. We believe OpenCL is a good option for a class of kernel-based programs, and LEAP's latency-insensitive primitives can capture the kernel model of computation. However, OpenCL does not expose the full flexibility of FPGAs. For example, the kernel model proposed by OpenCL does not capture any design in which a kernel makes requests back to the host for service or in which kernels communicate with each other dynamically. LEAP's interfaces are intended to facilitate the expression of a more general class of parallel programs.

14.7 Conclusion

FPGAs have great potential as platforms for many kinds of computation. However, the difficulty of programming FPGAs hinders their adoption in general systems. In this work, we presented the general philosophy and implementation of LEAP, an operating system for FPGAs. Like a software operating system, LEAP reduces the burden of programming FPGAs by providing uniform, abstract interfaces to underlying hardware resources, automatic management of these resources, and powerful system libraries that aid program design. LEAP achieves these goals through formalizing the principle of latency-insensitive design and providing strong compiler support for automating implementation decisions.

We believe that LEAP, or at least the design principles embodied in LEAP, is applicable to a wide range of other systems. We have found LEAP particularly useful in describing programs partitioned across multiple, heterogeneous platforms. The rise of programmable accelerators has made heterogeneous systems increasingly attractive. Because programming heterogeneous systems fundamentally requires strong communication support and because accelerator-based systems share many characteristics with FPGAs, we see LEAP as being valuable in programming and managing these new architectures.

LEAP is open-sourced under a BSD-style license and may be freely downloaded at http://leap-fpga.org.

Part IV
SoC and Overlays on FPGAs

Chapter 15
Systems-on-Chip on FPGAs

Jeffrey Goeders, Graham M. Holland, Lesley Shannon, and Steven J.E. Wilton

This chapter provides an overview of SoCs on reconfigurable technology. SoCs are customized processing systems, typically consisting of one or more processors, memory interfaces, and I/O peripherals. FPGA vendors provide SoC design tools, which allow for rapid development of such systems by combining together different IP cores into a customized hardware system, capable of executing user-provided software. Using FPGAs to implement an SoC provides software designers a fabless methodology to create and tailor hardware systems for their specific software workloads. The vendor tools support a vast range of system architectures that can span from small embedded microcontroller-like systems to multiprocessor/Network-on-Chip architectures.

This chapter describes the advantages and limitations of designing SoCs on reconfigurable technology, what is possible with modern FPGA vendor SoC design tools, and the main steps in creating such systems. The SoC development tools for FPGAs are rapidly changing. As such, this chapter intentionally does not contain step-by-step instructions; instead, it focuses on the overarching concepts and techniques used in the latest SoC tools.

J. Goeders (✉) • S.J.E. Wilton
Department of Electrical and Computer Engineering, University of British Columbia, Vancouver, BC, Canada
e-mail: jgoeders@ece.ubc.ca

G.M. Holland • L. Shannon
School of Engineering Science, Simon Fraser University, Burnaby, BC, Canada

D. Koch et al. (eds.), *FPGAs for Software Programmers*,
DOI 10.1007/978-3-319-26408-0_15

15.1 Challenges and Opportunities for SoC on Reconfigurable Technology

A SoC is a complete hardware processing system, often including one or more processors, memory controllers, and I/O interfaces. Each processor executes user-provided software, which can perform computational tasks and manipulate the operation of many different I/O devices. By mapping these architectures to an FPGA, the software designer has the opportunity to create a *custom* system platform while remaining *fabless*. Furthermore, designing and deploying their system on an FPGA enables them to repeatedly modify the system architecture as needed. This flexibility empowers software designers to create hardware tailored specifically to the software workload that it will execute.

SoCs on FPGAs are typically designed as a set of hardware blocks, connected together, as shown in Fig. 15.1. The hardware blocks, commonly referred to as IP (intellectual property) cores, each provide a piece of functionality to the system, and the interconnect represents wires and buses that allow the blocks to communicate. The vendor design tools provide a library of IP cores, including processors, memory controllers, communication ports, multimedia interfaces, and more. Users are able to select and connect these components through a graphical interface, allowing them to quickly create a system with the key elements they require. This design abstraction allows users to rapidly build processing systems, ranging from simple designs with a light-weight processor and UART, to large systems with multiple processors, fast I/O, and multimedia capabilities.

FPGA-based SoCs offer many advantages over both custom digital hardware circuits, and traditional x86-based processing systems. They key strengths of these platforms are:

Rapid Fabless Design Vendor provided IP libraries allow users to create an SoC very quickly; blocks can be added with a few mouse-clicks, and the communication interconnect is automatically handled by the design tools. In a matter of

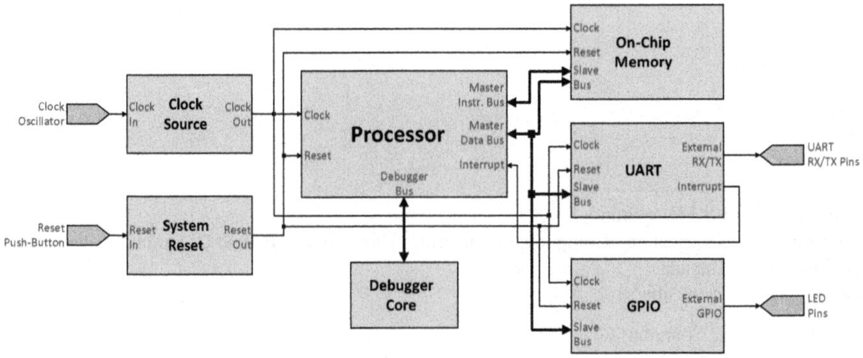

Fig. 15.1 Basic FPGA-based SoC

minutes, a complete hardware system, with processor, memory, and I/O can be created. This reduces design time over full custom hardware designs, lowering costs, and providing a fast time-to-market.

Flexibility FPGA-based SoCs can be made as small and simple or as large and complex as needed. Simple designs for embedded solutions can be developed and implemented on small, inexpensive FPGAs; or large systems with many processors can be created, targeting high-capacity FPGAs. This flexibility means the user only pays to deploy and power the minimum features required. This can reduce both production and operating costs over x86-based processing systems, while limiting non-recurring engineering costs if a redesign is needed.

Reconfigurability The reconfigurable nature of FPGAs means that the processor systems can be altered as needed, even after deployment. If software workloads change, the entire hardware system can be reconfigured. Designers can also incorporate hardware accelerators—custom compute units—into the SoC on the reconfigurable fabric of the same die, facilitating their integration. In fact, there is also the potential to create a Reconfigurable system-on-chip (RSoC), where the system's underlying architecture (e.g. the hardware accelerators) are reconfigured during the system's operation to support additional services.

These numerous advantages are offset by a set of limitations and challenges. For example, compared to high performance x86-based processor systems, the operating frequency, and thus performance, may be much lower. The flexibility that FPGAs offer comes at the cost of much lower clock speeds, and SoC designs implemented only using an FPGA's reconfigurable fabric generally operate in the range of 100 s of MHz. Although some modern FPGAs offer embedded ARM processors, running at 1–1.5 GHz, the processing performance will still be lower than modern x86-based systems. Despite this difference in clock speeds, it is still possible to obtain high performance SoCs on FPGAs, even beating the performance of x86-based systems, by leveraging the opportunities for a custom platform that better leverages spatial and temporal parallelism, through multiple processor systems and/or using hardware acceleration [TPD15].

Another potential drawback of using an FPGA-based SoC, versus a traditional x86 system, is the software environment. Although FPGA vendors make it easy to generate the underlying hardware for an SoC, the native software environment is generally "bare metal". In other words, there is no operating system, no filesystem, limited driver support, no TCP/IP support, and designers are often limited to programming in C or C++. Although all of these limitations can be overcome, by expanding the SoC, this may require greater expertise, and take significant time to design/integrate, implement and debug within the system. Conversely, x86 systems offer these features out of the box.

The remainder of this chapter describes the process for creating an FPGA-based SoC, comprising both the hardware system and the accompanying software. The hardware design is described in Sects. 15.2 and 15.3, with the former describing the process of creating a bare-bones, processor-based SoC using both Altera and Xilinx design tools, and the latter outlining how this basic system can be expanded to interface with the many peripherals that are often offered on FPGA boards. Section 15.4 describes how to program these systems, detailing the software development process from a basic bare-metal approach, to a full operating system. Section 15.5 outlines the process for expanding a system to contain multiple processors, and outlines the considerations and limitations of such an approach.

15.2 Basic Hardware System

This section describes the setup of a very basic FPGA-based SoC design, consisting of a processor running software, a UART to provide stdin/stdout communication with the system, and a GPIO to control LEDs.

15.2.1 Basic SoC Design

Figure 15.1 provides a conceptual diagram of a basic SoC design. The blue boxes represent instances of IP cores, and provide the main functionality of the design. The thin lines show single wire connections and are used to propagate signals between blocks, such as clock and reset signals. The wider lines represent buses, which provide communication between cores using standard bus protocols. The grey pentagons at the far left and right of the design represent external connections off the FPGA. They are used to connect to specific FPGA pins, which are wired to different devices on the printed circuit board on which the FPGA is mounted. The following describes the purpose of each IP core for the basic system shown in Fig. 15.1:

Processor At the heart of the system is a processor, which executes the software and is (usually) responsible for managing the flow of data throughout the system. This may be a soft processor that is implemented using reconfigurable logic on the FPGA; or a hard processor that is embedded into the FPGA's silicon (see Sect. 15.2.3).

Local Memory Connected directly to the processor is a local memory. This is an on-chip memory, meaning it uses memory resources within the FPGA, versus using a separate memory chip on the circuit board. This memory can be used for storing both the program instructions and data assuming they do not require too much memory. This option is preferential when possible as it has extremely low memory latency, often a single clock cycle. As shown in the figure, the instruction and data buses from the processor are connected to the memory.

Debug Module The debug module is connected to the processor and is necessary to enable interactive debugging of the software from a connected workstation.

Clock Source This core contains a clock input, which is driven from a pin on the FPGA, usually connected to a clock oscillator on the circuit board. The core is responsible for propagating the clock signals to the rest of the system, and ensuring that they are properly synchronized throughout the system, handling issues of clock skew and jitter.

Reset This core takes a reset input, often from a push-button on the circuit board, and forwards it to all components in the system.

UART This core provides logic to handle a serial UART connection, which can be used to provide stdin and stdout for the software running on the processor.

GPIO The General Purpose Input/Output (GPIO) core allows the processor to get and set the value of external pins. In Fig. 15.1, the core is connected to LEDs on the circuit board, allowing the processor to control the LEDs.

Bus Interconnect The processor communicates with IP cores through bus transactions. In the system shown, the bus connects the master device (the processor) to multiple slave devices (the UART, GPIO, and on-chip memory). This allows the processor to provide data to the other IP cores, such as sending UART data, or request information back, such as reading memory values in the on-chip memory. This is described in greater detail in Sect. 15.2.2.

Altera and Xilinx both offer SoC design tools to target their respective FPGAs. Although the tools differ in their visual style, they both offer similar core functionality. The design tools provide Graphical User Interfaces (GUIs) that allow a user to build a system that looks very similar to Fig. 15.1; IP blocks can chosen from vendor supplied IP libraries, and connections can be made with a few mouse clicks.

Creating the Basic System with Xilinx Vivado

This section describes creating the basic system from Fig. 15.1 using the Xilinx Vivado tool flow (version v2015.2). This system can be quickly set up by creating a new project, and selecting the *Base MicroBlaze* example project.

The created system will appear similar to Fig. 15.1, and should contain the same IP cores. There should also be one extra core, the *AXI Interconnect*; this core provides the logic for the bus transactions between the processor and slave devices (UART and GPIO). This functionality is implied in Fig. 15.1; however, the Vivado tools show this logic explicitly.

During project creation, the user must choose from a set of preconfigured FPGA boards. By doing so, the external connections are automatically bound to the appropriate pins on the FPGA, and the IP cores are configured as needed. For example, if the Virtex-7 VC707 board is chosen then the following will automatically be set: (1) the clock input will be connected to a 200 MHz oscillator on the circuit board, and the clock source IP core will be configured to scale this down to 100 MHz, (2) the reset input will be connected to a push-button on the circuit board, and (3) the I/O pins for the UART and LEDs will automatically be connected to appropriate pins. To use an FPGA board not in the preconfigured list, first choose a supported board to create the system, then alter the configuration options as appropriate.

Once the system is created and configured, it must be synthesized before it can be placed on the FPGA. The *Generate Bitstream* command in Vivado will perform the process of translating the SoC design to an FPGA bitstream. This process can take several hours for large, complex, SoC designs; however, for this simple example it should take only a few minutes. Once this is complete, the *Hardware Manager* tool can be used to connect to the FPGA and configure it to implement the SoC circuit. After this is complete, the FPGA will contain a full processor system; however, it will sit idle unless it is given software to execute, which is described later in Sect. 15.4. If at any point the SoC design is modified in Vivado, the synthesis and programming process will have to be repeated.

Creating the Basic System with Altera Qsys

This section describes creating the basic system using the Altera Qsys tool (version 15.0). The visual format differs slightly from Fig. 15.1; rather than a block diagram, the IP blocks are shown in a list format, each with a sub-list of available connections. Connections between blocks are shown in a matrix format, and connections can be made by clicking the intersections of the wires.

Although the tool does not include a pre-configured example system, the system shown in Fig. 15.1 can be created with a few mouse clicks. This is done by locating the *Clock Source*, *NIOS II Processor*, *On-Chip Memory*, *UART*, and *Parallel I/O* cores from the *IP Catalog* and adding them to a new (empty) design. After the cores are added, the connections should be made as shown in the figure, with the following exceptions: (1) in Qsys, the *Clock Source* core handles the connections for both the clock and reset signals, and (2) the debug module is automatically contained within the processor, and is not displayed as a separate block.

External connections are added using the *Export* column. In our example design, this includes the clock input, reset input, UART interface, and LED connections. In Qsys, these external connections are not automatically connected to the appropriate FPGA pins. Instead, the Qsys design must be imported into the Quartus II software in order to specify how these external connections are connected to the physical pins on the FPGA. This requires creating a hardware description language (HDL) wrapper around the Qsys system, and using the Quartus II software to add pin constraints. This process does require some familiarity with the hardware design tools, and further documentation can be found in the Quartus II handbook [Alt15c].

Performing synthesis using the Altera tools is a two-step process. First the *Generate HDL* command must be issued in Qsys, which will translate the SoC high-level description to a digital circuit description. Next, Quartus II is used to generate an FPGA bitstream. Again, complex designs may take hours to synthesize; however, this basic system should only take a few minutes. Once complete, the *Programmer* tool is used to program the FPGA with the hardware bitstream. If the user changes the design in Qsys, the entire synthesis process must be repeated.

Table 15.1 Address space configurator provided by FPGA Vendor SoC tools

Device	Address
Processor instruction bus	
↪ On-chip memory	`0x4000-0x7FFF`
Processor data bus	
↪ On-chip memory	`0x4000-0x7FFF`
↪ UART	`0x0000-0x001F`
↪ GPIO	`0x8000-0x8003`

15.2.2 Address Space Layout

In SoCs, the primary method for IP cores to communicate is through memory-mapped bus transactions. An IP core that initiates these transactions is referred to as a *master* on the bus, and a core that waits and responds to requests is referred to as a *slave*. A master can provide data to a slave via write operations, and can retrieve data via read operations. To facilitate this memory-mapped system, each device on the bus must be assigned its own unique region in the memory space.

Table 15.1 provides an example of an address space configuration, similar to the editors provided in the Vivado and Qsys tools, for the system in Fig. 15.1. The following provides a description of this example address space, assuming Altera IP cores are used:

On-Chip Memory This memory is a 16k on-chip memory, and thus requires a 16k address space. As shown in the Address Map, it is assigned the range 0x4000–0x7FFF. If the processor performs a write to address 0x4008, the bus logic will recognize that this request falls within the address space of the on-chip memory and propagate the request to the on-chip memory controller, which will write the specified value to address 0x0008 in the memory.

UART The communication interface with the UART consists of six 4-byte registers, so it is assigned a 32 byte address space (rounded up to the nearest power of 2), located at 0x0000–0x001F. The processor accesses data received by the UART by reading from the first register address 0x0000, and can send messages by writing to address 0x0004. The other four registers are used for status and configuration.

For most IP cores, this low-level communication is handled by the included software drivers; however, for some IP cores it may be necessary to interact at the register level (see Sect. 15.4). It does not matter where in the overall system address space each core is located; however, the designer must make sure that (1) no IP cores overlap in the address space, and (2) each IP core is assigned a large enough address space to handle its functionality. Both Vivado and Qsys provide menu options to automatically assign the addresses to the IP cores.

15.2.3 Hard vs. Soft Processors

The previous examples of SoCs on Xilinx and Altera FPGAs both used soft processors. A soft processor is implemented using the standard logic resources on an FPGA. The other option is to use hard processors, which are embedded into the silicon during fabrication of certain models of FPGAs. For example, Altera offers an *SoC* variant of their FPGAs, such as the *Cyclone V SoC*, *Arria V SoC*, *Arria 10 SoC*, and *Stratix 10 SoC*, which includes a hardened ARM multicore processor. Likewise, Xilinx offers the *Zynq* and *Zynq Ultrascale* line of FPGAs, also with a multicore ARM processor. In both Vivado and Qsys, the hard processor is added to the system using the IP catalog, in the same manner as a soft processor or other IP blocks. When using a hard processor, the user must provide a boot ROM that is used to initialize the hard processor and start up the system. This is typically connected to the system using a flash medium, such as an SD card.

There are several trade-offs between using hard and soft processors, primarily regarding performance vs. configurability, that is now discussed.

Performance

Soft processors are implemented using the highly reconfigurable fabric of FPGAs, which introduces overheads versus a hardened implementation of a processor. The operating frequency of a softcore processor depends on many factors, including FPGA generation, model, processor configuration, and utilization of the FPGA fabric. For the latest generation of FPGAs, the *maximum* operating frequency of the soft processors provided by the Qsys and Vivado tools is in the range of 165–469 MHz [Alt15b, Xil15a]. In contrast, the ARM cores provided on the latest FPGAs operate in the 1.0–1.5 GHz range [Xil15g, Xil15f, Alt14b]. In addition to operating at a higher frequency, the hard processors offer architectural advantages, providing greater performance, even when scaled to the same frequency. The soft processors offer performance in the 0.15–1.44 DMIPs/MHz range [Xil15b, Alt15b], while the hard ARM processors provide about 2.5 DMIPs/MHz per core [Xil15f].

Although hard processors offer superior performance, they are limited to the quantity that are fabricated onto the chip; whereas soft processors can be continually added to the design until there is no more room on the FPGA. This can provide performance advantages for a soft processor system, provided that the software workload can be sufficiently parallelized.

Configurability

The main advantage that soft processors offer is configurability. Options can be added or removed to trade off between performance and FPGA resource requirements. Removing options from the processor will likely lower performance.

For example, if the processor is configured without a floating point unit, but the user's software includes floating point instructions, the software versions of floating point operations will incur a significant penalty. However, it has added benefits, such as fitting on a smaller and cheaper FPGA, requiring less power, or being able to fit more processor instances on the same chip. Some of the configuration options available with the Altera Nios II and Xilinx MicroBlaze processors include:

- Adding instruction and/or data caches, and changing cache sizes.
- Adding specialized computation units, including hardware multipliers and dividers, and floating-point units.
- Adding a memory management unit (MMU), memory protection unit (MPU), or exception handler.
- Configuring the resources allocated for debugging, including the number of hardware breakpoints or memory watchpoints.
- Adding branch prediction, and configuring the branch prediction cache size.

The benefit of configurability is not just in the initial design phase; the reconfigurable nature of FPGAs allow these options to be changed even after production roll-out. All that is required is that the FPGA be reprogrammed with the new bitstream.

15.3 Expanding the Hardware System

This section describes how to expand the basic system from the previous section to include additional functionality, such as memory systems, I/O interfaces, and hardware accelerators.

15.3.1 Expanding Using IP Cores

SoC hardware designs are expanded by adding IP cores from the vendor supplied libraries. Behind the scenes, IP cores typically consist of two parts: a hardware description language (HDL) circuit, which implements the functionality of the core, and a software driver that facilitates the user's software communications with the hardware (described in Sect. 15.4). When the hardware system is synthesized, the HDL circuits from all IP cores are bundled together into a single large design, which is implemented on the FPGA using the programmable fabric.

Some FPGAs may contain permanent hard logic for a processor, DDR memory controller, PCIe interface, or other I/O interfaces; this is done to provide higher performance for commonly-used IP cores, and to meet timing requirements that would be difficult to obtain using soft logic. In these cases, the IP cores are still added to the design in the same fashion, but instead of using the FPGA fabric, the SoC circuit will incorporate the existing hard cores on the FPGA. In many cases, the same version of the IP core from the library will be used for both hard and soft logic implementations; the soft logic version will automatically be used if the FPGA lacks a specialized hard core.

The vendor provided IP library is sufficient for most designs; however, if needed, IP cores can be obtained from other sources, such as:

- 3rd party IP cores. These can be obtained from online open-source repositories, or purchased as licenses from other companies.
- User-created hardware accelerator IP cores generated using the HLS tools or traditional HDL flow.
- Packaging the user-created SoC design into an IP core, for use in a hierarchical SoC design.

Some cores provided in the vendor libraries are not available in all editions of the design tools. For example, the lowest performance configuration of the soft processors are available with the most basic editions of the design tools, while the higher performance configurations require upgraded editions of the design tools.

To add an IP core to the design, first select it from the IP library, and then add it to the layout. Next, the connections need to be made to the rest of the system. Figure 15.2 shows the common connections that are available; note that not all cores will have all connection types. The connections should be made as follows:

Clock and Reset Most cores contain clock and reset signals, and usually these should all be connected to the system wide clock and reset signals. In some cases IP cores will require a specific clock rate, which may be different from the system clock (see *Clocks* in Sect. 15.3.3).

Buses Most IP cores have one or more bus connections to communicate with and transfer data to other IP cores in the design. Most commonly, the core will contain a memory-mapped slave bus connection that should be connected to the master port of the processor (for Xilinx Vivado, rather than connecting directly to the processor, the connection is made to the AXI Interconnect core, which is in turn connected to the processor). This allows the processor to communicate

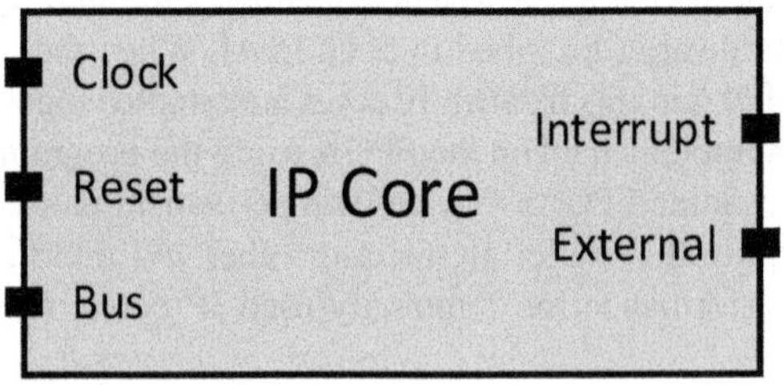

Fig. 15.2 IP core connections

with and control the IP core (usually through the provided software drivers). Although memory-mapped buses are common, there are also other bus types, such as data streaming, or first-in first-out (FIFO) style buses. For example, when processing video content, it is common to use streaming connections to feed video frame data from one processing IP core to the next.

Interrupt Some IP cores contain interrupt connections, which should be connected to the processor (or interrupt IP core, if separate from the processor). These connections allow the core to generate an interrupt request, which triggers the processor to halt execution and run the interrupt handler routine.

External External connections are used to connected to signals outside of the SoC. Typically these are connected to FPGA pins that are physically connected to various devices on the board (clock oscillator, Ethernet PHY, multimedia codec chips, etc.)

The connection types listed above, and their descriptions, are not exhaustive, nor without exception. The user should consult the documentation for each IP core when adding it to the system. The vendor tools make this easy, and the documentation can usually be accessed by simply right-clicking on the IP core and selecting the appropriate menu option.

IP Core Configuration

Most IP cores contain several configuration options. As examples, a UART core will usually allow you to change the baud rate or error detection schemes, and on-chip memory will allow you to configure the memory width and depth, number of ports, latency, and more. The configuration pages for each IP core are usually opened within the GUI when you add the IP to the design, but can also be accessed later if modifications are desired. Again, it is best to consult the documentation for each IP core for full details on the configuration options.

The IP core configuration options are not just on/off switches in a static circuit; rather, changing the options will actually alter the circuit that is implemented on the FPGA. If a feature is disabled, it will usually be removed entirely from the circuit; thus, changes to the configuration options will often affect the number of FPGA resources required by the IP core. For example, enabling large data and instruction caches on the processor will consume large amounts of on-chip memory, leaving less available for the main program memory. If trying to fit many processors on the same FPGA, one will have to decide between having *fewer* processors with large caches and specialized multiply/divide or floating-point units, versus having *many* simple processors.

15.3.2 The Memory System

In the bare-bones system from Sect. 15.2, on-chip memory resources are used for storing the program instruction and data. On-chip memory has very low latency, typically providing read and write operations with only one or two cycle delay. However, FPGA on-chip memory is relatively scarce, ranging from 100 s of KBs in the smallest FPGAs, to 10 s of MBs in the largest. In some cases, the memory may not even be large enough to store the program executable, especially when large software libraries are linked in. Even if the executable does fit, there may not be enough remaining memory to provide a sufficiently large stack and heap for the application.

DDR Memory

To provide a larger memory for both the program and its data, most SoCs contain an SDRAM memory controller that can access the DDR SDRAM memory module(s) on the FPGA board. When adding a memory controller, the user will be prompted to provide timing information for the SDRAM chip; this information can be found in the datasheet for the memory module. The FPGA vendor SoC tools already contain preconfigured timing information for certain FPGA boards and memory modules, so it is not always necessary to manually look up the information.

The SDRAM controller will also require a clock signal, running at a specific frequency range for the type of memory module used. Some tools offer a wizard when adding an SDRAM controller, which will also automatically add a clock generator IP core to your design, while other tools require you to do this manually. Refer to the documentation for the memory controller IP core for more detailed information.

Once the SDRAM controller is added to the system, the system needs to be modified so that the software executable is run from the newly added memory module. This is done by configuring the software linker, as described in Sect. 15.4. Finally, although SDRAM is the most commonly used memory type, the vendor libraries also contain IP cores for interfacing with other types, such as SRAM and Flash.

15.3.3 Commonly Used IP Cores

The vendor libraries offer IP cores for interfacing with a wide variety of I/O peripherals. These vary greatly in their complexity; some can be added to the system with a few clicks and include simple software drivers as their interface; others are much more complex and may require expert configuration. The IP cores also vary greatly in software driver support, ranging from no software driver to large, complex libraries.

Basic I/O General purpose I/O (GPIO) cores can be used to get and set values of FPGA pins. In the basic example (Sect. 15.2), this was used to control LEDs; however, there are many other uses, such as reading the value of switches and push-buttons, or interacting with simple devices in an embedded environment.

Timers Timer IP cores can be added to the system to provide very high resolution (cycle-accurate) timers, which are useful for profiling and system scheduling.

Clocks Certain IP cores may require a clock input with a specific frequency; this is common for DDR memory controllers, audio and video interfaces, and other I/O devices. In such a case, a Digital clock manager (DCM) or Phase-locked loop (PLL) can be added to the system. These IP cores are able to derive a new clock signal with a specified frequency, using the system clock as input.

Ethernet Ethernet is a commonly-used method for providing high-speed communication between the SoC and other devices. The Ethernet IP cores are designed to support a very wide range of FPGA boards and system types and, as such, can be quite complex with many configuration options. However, since it is a widely used feature, one can often find online step-by-step tutorials for a specific FPGA board. Included with these cores is software to support the full TCP/IP stack; allowing the SoC to communicate with standard IP networks as both client and server.

DMA DMA cores perform memory transfer operations, either within the same memory, or between two different memories. The processor can request a DMA operation by providing source and destination memory addresses and transfer size, and the DMA core will carry out the memory copy operation while the processor continues on with other tasks. This allows the processor to execute other code in parallel with the memory operation, increasing performance.

Persistent Storage Many FPGA boards offer SATA, USB or Flash connections that can be used to interface with persistent storage devices. It may be straightforward to add IP cores to provide a hardware interface with such systems; however, often the complexity lies in the software. The software drivers may only provide primitive interfaces with the devices, leaving the complex task of file systems and communication protocols up to the user.

Multimedia FPGA boards offer a wide range of multimedia, including audio and video input and output through a variety of formats and connections. IP cores are provided to interface with many of these protocols, and in many cases basic processing (video scaling, cropping, etc.) can be performed using existing hardware cores without any intervention from the processor. This allows for high-performance media streaming and processing, while the processor performs other tasks in parallel.

Hardware Accelerators Most IP cores are designed to provide interaction with external I/O; however, another class of IP cores, hardware accelerators, use the FPGA fabric to implement hardware circuits that can perform a task faster than the processor (or in parallel with it). Creating such IP cores usually requires hardware expertise; a HDL circuit that performs the acceleration is wrapped in logic that provides bus protocols for communication with the rest of the SoC. However, many modern high-level synthesis (HLS) tools (see Chap. 3), provide the ability to automatically create a hardware accelerator IP core from a software description of an algorithm, which could then be imported into the users SoC design.

PCIe Some FPGA boards offer PCI-Express (PCIe) interfaces so that the board can be inserted into a server system or workstation. PCIe allows a high-speed connection between the host (commonly an x86 processor) and the SoC; however, such systems are very complex compared to the simple SoC designs presented here.

15.4 Programming Environment

This section gives an overview of the tools available to software developers that allow them to write applications targeting SoC platforms, both in the case of bare metal and with the use of an operating system. Both Altera and Xilinx provide Integrated Development Environments (IDEs) that integrate with their SoC design tools to aid the developer in configuring a cross compilation toolchain, using libraries and vendor supplied IP core drivers, as well as downloading their code to the target platform and debugging.

15.4.1 Bare Metal Software

Depending on the requirements of an application, the performance overhead and memory requirements of running an entire operating system kernel may not justify the benefits gained. This is especially true of soft-processor-based SoCs, which trade performance per core for configurability. In these cases, a bare metal application development workflow may be more suitable.

In a bare metal environment, the abstractions commonly provided by an operating system, including process management, virtual address spaces, and file systems are unavailable. As a result, the user's programs have complete access to the underlying physical memory of the system, without any protection mechanisms. Since SoC platforms use memory mapped I/O, writing software to control different I/O hardware in the system is simply a matter of reading and writing appropriate memory addresses that get mapped to device registers. For this reason, the languages of choice for bare metal programming are those that support direct access to memory, such as C and C++.

To aid software developers, both Altera and Xilinx provide software IDEs (Altera EDS and Xilinx SDK) that integrate with their respective hardware design tools. While there are differences between the IDEs, both are Eclipse-based and provide developers with tools for application development including: source code editing, build tools based on GCC, cross-compiler configuration, organization of applications into projects, board support package generation and customization, device programming and debugging. Since the soft processor targets within an SoC are highly configurable, compiler options must be correctly set to match the hardware configurations in order to build binaries that are compatible with a given configuration. For example, a soft processor may be configured to use a floating point unit, in which case it will be able to execute floating point instructions. The compiler must be passed an option to insert these floating point instructions into the output binary. This setting of compiler options is just one example of the automation that these vendor IDEs provide.

Board Support Package

After completing the hardware design for an FPGA-based SoC, the first step toward developing a bare metal application is to export a hardware description file from the hardware design tool. In the Altera flow, this is a *.sopc file*, and in the Xilinx flow this is a *.hdf file*. While the exact semantics vary between vendors, this file generally contains information about the hardware design relevant for software, including a list of IP cores in the design, their configurations and address space mappings. This hardware description file is read by the IDE and used to generate and customize a *board support package*.

A board support package contains a collection of libraries and device drivers for peripheral IP, and can be thought of as a minimal low level software layer. A board support package also contains a header file (system.h in Altera flow, xparameters.h in Xilinx flow), that defines constants related to the hardware system, for example interrupt constants, IP core base addresses, etc. These values may be required as parameters to IP driver Application Programming Interface (API) calls. Since these values vary dependent on the system configuration, the header file is automatically generated from the hardware description file for each system.

Both Altera and Xilinx allow a board support package to be customized after it is first created. Examples of the customizations this enables the user incorporate include: adding libraries, selecting which software drivers and which respective versions are included, selecting a device to use for stdin and stdout (e.g. UART or JTAG), and setting compiler flags. We now discuss the included libraries and drivers in more detail.

Libraries

The board support package contains at minimum an implementation of the C standard library, but other libraries can be included as needed. By default, the board support package will also contain some library code for accessing basic processor features including caches, exceptions, and interrupts. The libraries that are available vary between vendors, but some examples include, bare metal TCP/IP stacks and file system libraries. More information about the available libraries for bare metal platforms can be found in [Xil15c, Alt15a].

Drivers

The task of writing software that interacts with various IP cores in the SoC design is simplified by the availability of drivers that exist for the majority of vendor library IP. The API level of these drivers varies for each IP block and between vendors. Lower level drivers typically export functions to access individual device registers and the programmer must handle manipulating register bits to enable the desired device functions, via masking and other means. Higher level drivers will usually provide a more programmer-friendly API that hides the hardware register interface from the programmer. As an example, a higher level UART driver may provide functions to set the baud rate and modify serial settings, while a low level driver would merely export functions to read and write the baud rate bits within a device register.

Documentation for the Altera and Xilinx IP driver APIs can be accessed by clicking on the IP block in Qsys and Xilinx SDK respectively.

Application Development

In order to create a new application, a new project can be created in the IDE. This encapsulation within a project allows the IDE to handle Makefile generation and setup paths to the various parts of the compiler toolchain. Upon application project creation, the programmer must select a board support package that the application will reference, as this allows application code to call into the libraries and drivers included therein. Multiple applications may share a common board support package. The Altera and Xilinx IDEs provide a number of sample applications varying from simple "hello world" programs, to more complex memory and peripheral test programs.

SoCs often contain multiple different memory regions, including local on-chip memory, external flash memory and DDR, for example. A programmer can decide which parts of different code sections (stack, heap, text and data, etc.) map to which memory regions by configuring the linker with a script. This linker script also defines the maximum sizes of the stack and heap memories.

Debugging

The Altera and Xilinx IDEs contain device programming features that allow a user to both configure an FPGA with a bitstream and download executable code to the

SoC conveniently from within the same interface. Additionally, these IDEs provide debugging capability, that will likely be familiar to users of Eclipse or other software IDEs. Software breakpoints, stepping through both C and assembly code, stack tracing, inspecting variables and viewing memory are all fully supported.

15.4.2 Operating Systems

While developing programs that run on bare metal is sufficient for many application domains, sometimes it is desirable to make use of the abstractions afforded by an operating system. This is particularly true if significant code is being reused that already targets an OS such as Linux, or makes use of standard libraries (POSIX threads, sockets, etc.). Building an operating system to run on a specific FPGA board/SoC platform is a more complex task than setting up a bare metal environment. However, a number of vendor supplied and third party tools exist to make the process of building a working kernel image for a specific FPGA board easier and less error prone. While a complete tutorial on configuring and building an OS image is beyond the scope of this book, we provide an overview of the steps involved and provide references to additional resources.

Linux is the operating system of choice for use on many embedded computing platforms. Both Altera and Xilinx have added architectural support for their soft processors (Nios II and MicroBlaze respectively) to the mainline Linux kernel within the last few years. However, this support is currently limited to single processor configurations only. More details on multiprocessor SoCs are given in Sect. 15.5.

Hardware Requirements

In order for an SoC platform to run a Linux operating system, certain IP cores must be present in the hardware system. These required IP cores, along with some commonly included optional cores, are described below:

Processor with MMU Since Linux provides a virtual memory system, the processor must be configured to use a memory management unit (MMU) to perform translation of virtual memory address to physical addresses.

Timer A hardware timer is required for the OS to manage software timers.

UART A UART is needed to provide a serial console to the Linux kernel for printing of messages and to provide the user with a command line shell.

External Memory An external memory such as DDR is required to store the OS binary and optionally the root filesystem. As a result, this memory must be large enough to fit these components.

Ethernet (Optional) While not required, including Ethernet in the hardware system enables the use of the network stack within Linux. If Ethernet is present, the bootloader running on the SoC can be configured to download the OS image from a development host over a local network. Since Ethernet allows for high-speed data transfer this can save significant time during development.

Non-volatile Memory (Optional) Optionally, a non-volatile memory such as NAND flash, or an SD card controller may be included in the hardware system. The root filesystem can then be mounted to this memory to provide persistent storage. If this hardware is not included, the filesystem may be mounted in external memory.

Building and Booting an Operating System Image

The process of generating a bootable Linux operating system image for an SoC target is similar in many aspects to generating an operating system image for any other embedded CPU target. The main steps are outlined below:

- Setting up a toolchain
- Configuring and building a bootloader
- Configuring and building the Linux kernel
- Configuring and building a root filesystem
- Specifying a device tree

The first step is to obtain a cross-compiler toolchain for the target CPU that will be used to compile the various components of the operating system. If the target is a soft-processor based SoC, these can be obtained from Altera and Xilinx for the Nios II and MicroBlaze, respectively. Usually the easiest way to setup these toolchains is by installing the appropriate vendor IDEs, which are also used for bare metal application development.

U-boot is the de-facto standard bootloader for embedded targets, including FPGA-based SoC platforms. U-boot must be configured for the appropriate target and compiled as a bare metal application. In order to boot Linux, u-boot requires the kernel binary, a device tree blob (i.e. a binary description of the device tree), and a root filesystem.

While Nios II and MicroBlaze support is present in the mainline kernel repository, both Altera and Xilinx maintain their own kernel source repositories, which typically also include Linux device drivers for their library IP. After downloading an appropriate kernel tree, the kernel must be then configured and compiled.

As with other embedded targets, FPGA-based SoC platforms use a *device tree* data structure to specify the hardware configuration, including information about the CPU, physical address space, interrupts, etc. This device tree must be converted from its text format to a binary format (blob) using the device tree compiler from the toolchain. Lastly, one must build and populate a root filesystem.

In order to boot the system, the u-boot binary must be downloaded to the target device, and the kernel binary, device tree and root filesystem should be placed in the board's external memory. This can be achieved via commands at the u-boot prompt. Once a Linux image has been built and booted successfully, users may wish to write the u-boot and kernel binaries, root filesystem and device tree blob to non-volatile memory, such as a NAND Flash or an SD card, if available on the platform. This way the system can be booted in the future without being connected to a development host PC.

Additional Resources

Building and booting an operating system for a particular FPGA board can be a complicated task. Fortunately there is a large amount of information available online, usually in the form of wikis and user guides, that provide step by step instructions on how to obtain, configure and build a Linux image for Altera and Xilinx development boards. Additionally, there are some vendor-supplied and third party toolkits that aim to make entire embedded Linux development process easier. These tools include Xilinx's PetaLinux Tools [Xil15d], Altera's SoC EDS suite [Alt14a], and Wind River Linux [Win15].

15.5 Multiprocessor Systems

In this section, we discuss multiprocessor systems for reconfigurable hardware. We describe how to construct an SoC with multiple processors, including the hardware required for interprocessor communication. We also give an overview of some of the unique aspects of writing software for multiprocessor systems including cache coherency and mutual exclusion.

As stated previously, the Nios II and MicroBlaze soft processors are simple RISC processors, with limited instructions per cycle (IPC) due to their relatively low operating frequencies because they are implemented using the logic resources of the FPGA. To overcome this limitation, a designer can often improve overall system performance by adding more processors to the system and dividing software tasks among them. Using a soft processor based system has an advantage in that the number of processors and each of their configurations can be tuned exactly to the application task requirements they will execute. For these types of systems, designers are limited in their choices only by the available logic resources of the target FPGA device.

15.5.1 Building a Multiprocessor System

Creating a hardware system with multiple processors is a fairly straightforward task in both Qsys and Vivado. Beginning with a basic single processor system, as described in Sect. 15.2, additional processors may be added to the design from the IP core library in the same manner as any other IP core. At a minimum, connections must be made for each processor to valid clock and reset signals and at least one memory. In order to save resources, common peripherals, as well as connections to external memory are often shared between processors in a multiprocessor system. To implement this sharing in hardware, multiple processors can connect to the same bus interconnect, which in turn is connected to the shared peripheral/memory. When connecting multiple processors to the same bus, the design tools are generally capable of automatically adding the required bus arbitration logic, but this can be dependent on the type of bus interface being used.

Users can create symmetric multiprocessor systems by configuring each processor in the system identically. Alternatively, asymmetric topologies can be created by modifying the configurations for different processors in the system. For example, if a user knows that only one software task will require floating point operations, only one of the processors needs to be configured to include a hardware floating point unit, while the remaining processors will only support integer arithmetic.

Note that these multiprocessor systems are symmetric in that they have the same processor configuration. They would also share any off-chip memory system. However, their on chip local memories and caches have no coherency mechanisms that ensure that if one processor writes data to off-chip memory, the other processors will evict the same data from their caches as invalid.

Interprocessor Communication

Processors within a multiprocessor system need to communicate for the purposes of data sharing, synchronization and coordination of tasks. One method for achieving this interprocessor communication is through the use of a shared memory, as described above.

Another technique is to provide direct communication between processors with the use of a FIFO buffer memory. Both Qsys and Vivado, include FIFO IP cores in their libraries that may be added to a user design in the same manner as any other IP. FIFOs are directional, and have ports for connecting to a master and slave, for writing and reading to/from the FIFO respectively. When connecting two processors with a FIFO, one processor will connect as a master, allowing it to write data into the FIFO, and the other processor will connect as a slave, allowing it to read data from the FIFO. To enable bidirectional communication between processors, two FIFOs must be used.

Coherency and Mutual Exclusion

As suggested previously, two problems that arise when dealing with a shared memory multiprocessor environment are memory coherency and exclusive access to memory. While creating a shared memory multiprocessor system for reconfigurable hardware is generally straightforward by following the steps in Sect. 15.5.1, ensuring cache coherency and exclusive access to memory is not easily achieved.

Implementing hardware to enable cache coherency protocols is a complex task that is beyond the scope of our discussion. Software developers must take this lack of coherency into consideration when writing software, as failing to do so so can lead to bugs that are difficult to diagnose and correct. As such, while an individual processor's cache maintains coherency with itself, software designers are better off viewing these caches as scratchpads that support hardware prefetching and writebacks to memory and use other mechanisms to maintain coherency system-wide when there are multiple processors.

For ensuring mutual exclusion, both Qsys and Vivado contain mutex IP cores in their libraries, that provide implementations of hardware mutexes. Such IP can be used to ensure exclusive access to memory and other peripherals if required.

Hard Multiprocessors

Although this discussion of multiprocessors focuses largely on soft processor based systems, as discussed in Sect. 15.2.3, both vendors also ship FPGA devices with hardened multicore ARM processors. Specifically, Xilinx's Zynq family of parts and the Altera *SoC* FPGA part variants all include embedded ARM cores. Generally speaking these hardened multiprocessors do have cache coherency protocols implemented in hardware as well as exclusive memory instructions.

15.5.2 Software for Multiprocessors

In the bare metal software environment, each processor in a multiprocessor system requires its own application binary that is specifically compiled to account for its specific processor configuration. As mentioned in Sect. 15.4.1, an application's referenced board support package encapsulates this processor configuration and sets up the compiler toolchain as appropriate. For this reason, the board support package must target a particular processor in the system. In the case of a single core system, this defaults to the only available processor, but in a multiprocessor system it may be necessary to use multiple board support packages, with each one targeting a different processor. Note that if the targeted processors have identical configurations, as would likely be the case in a symmetric multiprocessor system, then applications can share board support packages and all compiled program binaries should be portable between processors.

If applications targeting different processors within a system are to be located in a shared memory, developers must ensure that the code sections do not overlap unless this is specifically intended. This can be achieved by modifying the linker script which specifies address ranges for each of the code sections within an application binary. For example, each processor can use their local memory to store their individual instruction code and then store any shared data in the shared system memory.

In a bare metal environment, programmers can write "multi-threaded" style code, where applications targeting different CPUs run in parallel and synchronize and share data through global variables. It is often necessary for a single processor to setup these shared variables and pass references to the other processors in the system. For such a task, interprocessor communication only through shared memory may be insufficient and FIFOs may be required to pass memory references between processors.

Operating Systems

Due to the lack of hardware to allow cache coherency and exclusive memory accesses, vendor Linux support is currently not implemented for multiprocessor systems built using either Nios II or MicroBlaze. Multiprocessor support may be added in future versions of the Linux kernel, however, this would also necessitate a change in the hardware IP used as well [MSF12].

Conversely, the hardened multicore ARM processors available in certain devices, are supported in Linux, since they have exclusive memory instructions and coherency protocols implemented in hardware.

15.6 Conclusions

This chapter provides an overview of how reconfigurable technology can be used to create custom SoC designs. It describes how the vendor tool flows support the quick generation of a basic hardware system through a graphical user interface. We have outlined the varied levels of software support available, ranging from bare metal systems to those with an operating system as well as third party support.

Currently, the main benefit of creating an FPGA-based SoC is the ability to create a custom hardware system for a specific embedded application. Software designers can start with a very basic system and then update and customize it as needed to include hardware accelerators and/or multiple processors with minimal non-recurring engineering costs. The FPGA provides a configurable substrate that allows software designers to remain fabless, while selecting between low cost, low power, and low performance soft processor based systems to ARM based systems, to systems that combine both hard and soft processors.

Looking forward, there are some significant opportunities for designing SoCs on reconfigurable technology. Although there currently is not vendor O/S support for SMP multicore systems, with the inclusion of multicore ARM processors in commercial products, there is an increased possibility for this extension. Even now, it is possible to update the open source Linux kernel for these platforms for those comfortable with editing kernel code. Another unique opportunity for FPGA-based SoCs is to leverage the dynamic partial reconfiguration support available from both

Altera and Xilinx. Dynamic partial reconfiguration is the ability to dynamically reconfigure the hardware substrate *while* the application is running. As such, there exists the possibility to incorporate this functionality into SoC designs on FPGAs to create *Reconfigurable* SoCs (RSoCs). While the hardware infrastructure needed to create RSoCs exists, the necessary vendor tool support still needs to be developed to make this accessible to non-FPGA experts.

Chapter 16
FPGA Overlays

Hayden Kwok-Hay So and Cheng Liu

Developing applications that run on FPGAs is without doubt a very different experience from writing programs in software. Not only is the hardware design process fundamentally different from that of software development, software programmers also often find themselves constantly battling with the much lower *design productivity* in developing hardware designs.

To begin, first time FPGA designers are often put off by the steep learning curve of the complex labyrinth of tools involved. Beyond that, instead of spending merely seconds to compile and debug a quick-and-dirty proof-of-concept design, software programmers soon also discover that implementing even the simplest FPGA design can consume at least tens of minutes of their development time. As the size of their designs increase, the run time of these implementation tools quickly increase to hours or even days, greatly limiting the number of possible debug-edit-implement cycles per day. Indeed, to debug, they may turn to the use of a cycle-accurate simulator. Unfortunately, tracing the behaviors of even just a handful of signals through tens of thousands of cycles soon becomes a slow and intractable process.

In this chapter, we explore how the concept of FPGA overlay may be able to alleviate some of these burdens. We will look at how by using an overlay architecture, designers are able to compile applications to FPGA hardware in merely seconds instead of hours. We will also look at how overlays are able to help with design portability, as well as to improve debugging capabilities of low-level designs. Finally, we will explore the challenges and opportunities for future research in this area.

H.K.-H. So (✉) • C. Liu
Department of Electrical and Electronic Engineering, The University of Hong Kong, pokfulam, Hong Kong
e-mail: hso@eee.hku.hk

D. Koch et al. (eds.), *FPGAs for Software Programmers*,
DOI 10.1007/978-3-319-26408-0_16

16.1 Overview

The concept of overlaying a virtual architecture over a physical system is not entirely new—in networking, for example, virtual overlay networks are routinely constructed on top of the physical infrastructure in modern systems. However, the concept of overlaying a virtual architecture over a physical FPGA is only recently gaining traction among researchers, but is already generating a lot of excitements because of their potentials.

So what is an FPGA overlay exactly? Interestingly, because of the unique nature of FPGAs as a flexible configurable hardware, drawing up a precise definition for an FPGA overlay may not be as straightforward as it may seem. In this chapter, we will use the following definition as a starting point.

> An FPGA overlay is a virtual reconfigurable architecture that overlays on top of the physical FPGA configurable fabric.

From this definition, we can see that an FPGA overlay is a machine architecture that is able to carry out certain computation. Furthermore, this architecture is *virtual* because the overlay may not necessarily be implemented physically in the final design. Finally, it is *reconfigurable* because the overlay must be able to support customization or be reprogrammed to support more than one application.

In other words, an FPGA overlay is a virtual layer of architecture that conceptually locates between the user application and the underlying physical FPGA similar to that shown in Fig. 16.1. With this additional layer, user applications will no longer be implemented onto the physical FPGA directly. Instead, the application will be

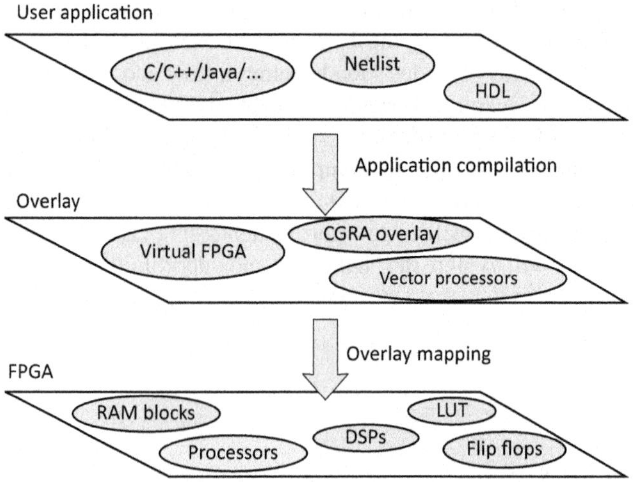

Fig. 16.1 Using an overlay to form a 2-layer approach to FPGA application development. Overlay may be designed as a virtual FPGA, or it may implement an entirely different compute architecture such as a coarse-grained reconfigurable array (CGRA), vector processor, multi-core processor, or even a GPU

targeted toward the overlay architecture regardless of what the physical FPGA may be. A separate step will subsequently translate this overlay architecture, together with the application that runs on it, to the physical FPGA.

The easiest way to understand how an FPGA overlay operates in practice, is to consider building a *virtual* FPGA (VA) using the configurable fabric of a physical FPGA (PA). By doing so, you now have an FPGA overlay architecture in the form of VA overlaying on top of PA. We say that VA is essentially "virtual" because architectural features of VA such as its muxes or I/O blocks may not necessarily be present in PA. Yet, if the overlay is constructed correctly, any design that was originally targeting VA may now execute unmodified on PA without knowing its details.

However, the power of employing FPGA overlay is not limited to making virtual FPGAs only. On the contrary, taking advantage of FPGA's general-purpose configurable fabric, many researchers have demonstrated the benefits of overlays that implement entirely different computing architectures such as multi-core processor systems, coarse-grained reconfigurable arrays (CGRAs), or even general-purpose graphic processing units (GPUs).

Now, using a multi-core processor overlay as an example, it should be apparent how overlays are able to improve a software programmer's design productivity. Instead of working with unfamiliar hardware-centric tools and design methodologies, software programmers are now able to utilize FPGAs as accelerators simply by writing programs that target a familiar architecture. In general, one benefit of using FPGA overlay is that it is able to bridge between the often software-inclined user and the low-level FPGA hardware fabric.

Of course, if an application can be readily accelerated on a multi-core processor overlay implemented using FPGAs, then it is understandably begging the question: why not simply run the design on an actual high-performance multi-core processor instead? The answer to this question, indeed, is the key challenge that will guide the future research on overlay designs—While an overlay offers many desirable features to software programmers such as improved design productivity, the additional layer on top of the physical FPGA inevitably introduces additional performance penalty to the system. A good overlay design must therefore ensure that despite the performance penalty introduced, the overall acceleration offered by FPGA must remain competitive for the accelerator system to be worthwhile. Furthermore, it must provide added value beyond a solution with a fixed general-purpose architecture. One such added value is the ability to customize the overlay for the particular application or group of applications concerned for sake of performance or power-efficiency.

16.1.1 Coarse-Grained Architectures

In most cases, FPGA overlays are essentially coarse-grained architectures that are built on top of the physical fine-grained configurable fabric. As their names suggest, the basic idea of a coarse-grained architecture is to reduce the configuration granularity of an FPGA from its physical fine-grained configurable fabric such as LUTs to one with coarser reconfiguration granularity. In some cases, such coarse-grained blocks may refer simply to arithmetic blocks of moderate sizes such as adders, multipliers, or digital signal processing blocks of some sort. In other cases, such coarse-grained blocks may refer to very complex microprocessors connected with sophisticated network-on-chip. Regardless of the implementation, the central goal of any coarse-grained architecture remains the same: to improve power-performance of the system by trading off design flexibility.

In addition, because of their reduced configuration flexibility, coarse-grained architectures can also enable design methodologies that are more productive than traditional hardware design flow. In particular, coarse-grained architectures improve a designer's productivity in two important ways. First, by constraining the flexibility of an FPGA, a coarse-grained architecture reduces the design space significantly, which has a net effect of reducing implementation tool flow run time considerably [LPL+11]. In addition, many coarse-grained architectures implement compute models that are more familiar to software designers, considerably lowering the barrier-to-entry to employ such designs. For instance, many recent overlays are implemented as coarse-grained reconfigurable arrays (CGRAs), where computation is carried out by a connected array of processing elements (PEs). Instead of implementing the user application using low-level configurable logic of the FPGA, these operations are translated into computational tasks that take place in the PEs. To many software programmers, programming a parallel processor array, while a daunting task in its own right, is arguably a lot more approachable than to implement designs on the native FPGA configurable fabric. As such, it is not surprising that many recent overlay designs are built on top of an underlying coarse-grained reconfigurable array [KHKT06, FVM+11, SBB06, LS12, CA13, JFM15].

16.2 Benefits of Overlays

As a virtualization layer that sits between a user application and the physical configurable fabric, an FPGA overlay inherits many of the benefits that software programmers have learned to expect from their CPU virtualization experience—portability, compatibility, manageability, isolation, etc. On top of that, employing FPGA overlays has also been demonstrated as a good way to improve a designer's productivity through improved compilation speed and better debugging support. Along the same line, by carefully partitioning the complex hardware-software design flow around an intermediate overlay layer, it is also possible to provide separation of concerns between software and hardware engineers in the design team. The overlay essentially acts as a bridge between the two teams, while allowing the overall system to take advantage of the FPGA resources efficiently.

16.2.1 Virtualization

Virtualization of FPGA resources has long been an active area of research since the early days of reconfigurable computing. These pioneering works have demonstrated many of the possibilities as well as challenges associated with virtualizing hardware resources that are not designed to be time-multiplexed. A common trend among these early works was that virtualization can be used as a mean to provide the designers and/or tools the illusion of having infinite hardware resources. Early works by Trimberger et al. [TCJW97], virtual wire [BTA93] and SCORE [DMC$^+$06], for instances, gave the users the illusion of a system with unlimited FPGA resources through carefully structured hardware/CAD system. Others have studied the problem of time-sharing of FPGA resources from an operating system's perspective as a way to provide shared accelerator resources among users/processes [LP09b, SB08, FC05].

As the concept of FPGA overlay continues to mature, the idea of virtualizing FPGAs has taken on a new focus. As an overlay, virtualizing FPGAs allows an additional benefit of providing a compatibility and portability layer for FPGA designs. In the work of Zuma [BL12], for instance, virtual, embedded FPGAs were proposed. By providing a virtual FPGA layer, the authors demonstrated that it is possible to execute the same netlist on multiple FPGAs from competing vendors using multiple different design tools.

16.2.2 Reduced Compilation Time

A key difference in design experience between software compilation and implementing FPGA designs rests on the drastically different run time of the involved tools. With modern compiler technologies, software compilation has already become a straightforward, predictable, and most importantly, very rapid process. Compiling even a relatively complex piece of software application rarely takes longer than a few minutes on a reasonably fast computer. On the other hand, implementing applications for FPGAs involves a complex labyrinth of low-level tools that are convoluted, unpredictable, and takes a long time to complete. Compiling even the smallest design may take tens of minutes, while spending hours or even days on some of the largest designs are not unheard of. Unfortunately, this 2 orders of magnitude difference in run time, together with the unpredictable nature and the often-mystical error reporting mechanisms, are all contributing to a very high barrier-to-entry that shies away most first time software programmers. Technically, this much longer run time of the tools also significantly reduces the number of possible debug-edit-implement cycles per day, causing project delays as well as lowered productivity of the designers.

By using an overlay as intermediate compilation target, together with careful crafting of the design process, researchers have demonstrated how such lengthy hardware development process can be reduced significantly. For example, in the work of Intermediate Fabric [CS10], an intermediate coarse-grained reconfigurable fabric was introduced as an overlay. With the overlay architecture, the average place and route times for the tested benchmark were significantly reduced by more than 500 times.

Similarly in the work of QuickDough, a coarse-grained reconfigurable array was used as overlay to accelerate compute intensive loop kernels. Loops are scheduled to execute on the CGRA instead of compiling to the reconfigurable fabric. As a result, when compared to manually generating custom hardware for the loops using standard hardware tools, up to 3 orders of magnitude reduction in compilation time was demonstrated.

16.2.3 Improved Debugging Capabilities

Unlike many software development frameworks where a range of debugging facilities and methodologies are readily available, tools for debugging applications that target FPGA-based systems are still in their infancy. Traditional FPGA design methodologies rely heavily on cycle-accurate simulations for application development and debugging. While such simulations are invaluable to understand the low-level operation of the FPGA, they are slow, tedious and provide only limited information about the run-time behavior of the design. To monitor run-time behavior of a design, users must rely on even more complex in-system emulation facilities or even external testing hardware. Taking advantage of FPGA overlays, researchers have demonstrated some promising results addressing the need for better debugging tools.

For instance, in a series of work by Hung and Wilton [HW14, HW13], an overlay network was incorporated into FPGA design to facilitate insertion of trace buffers after a design has been placed and routed. By carefully controlling the signal routes to utilize only unoccupied resources in the FPGA, they have demonstrated efficient ways to dynamically select and monitor signals during run time.

In other cases, since the overlay layer enables a virtual computing paradigm for the application developer that is different from the underlying FPGA, they also enable new debugging strategies that are more suitable to the designer. For instance in MARC, debugging the user-specified OpenCL applications that run on the generated multi-core architecture can follow traditional software debugging methodologies instead of relying on low level FPGA tools. Not only can it greatly increase the abstraction level, but it can also allow a debugging strategy that matches the user's expectation.

16.2.4 Separation of Hardware and Software Concerns

With careful planning, it is also possible to take advantage of the 2-layer design approach offered by FPGA overlays as a natural division point between hardware and software development efforts. Recall from Fig. 16.1 that implementing designs via an FPGA overlay involves two steps: user applications must first be mapped to the overlay architecture, which is subsequently implemented to the physical fabric in a second step. In many cases, the overlay architectures are designed so they can efficiently support the computational model expected by the programmer. As a result, mapping of software applications to the overlay is usually more intuitive to software programmers than to map the same application to the physical FPGA fabric in one step. On the other hand, mapping the overlay and the user application to the physical fabric involves intimate knowledge about the FPGA hardware implementation process. This task is best left to a separate hardware team. Consequently, one benefit of having an overlay is that the hardware team may now devote their efforts exclusively on implementing the relatively well-structured overlay on the fabric, rather than to implement many individual applications. The hardware team can be more focused, and can perceivably create a better, highly optimized hardware design.

For example in the work of MARC [LPL$^+$11], a multi-core processor-like architecture was used as an intermediate compilation target. In this project, user applications are expressed as OpenCL programs. To implement these applications, they are first compiled as an application-specific multi-core processor, which is subsequently implemented on the physical FPGA in a separate process. With this set up, users no longer need to understand the detailed implementation of their algorithm on FPGAs. Instead, they write essentially standard OpenCL programs, and focuses exclusively on writing the best code for the accelerated architecture assumed. The task to map their OpenCL code into custom core on the target overlay, and to implement that overlay on the FPGA fabric, was left to a separate team.

In their case studies, the resulting application achieves about one third of the speedup when compared to fully custom designs. Yet with the 2-layer approach using OpenCL, the design effort with MARC is significantly lower when compared to making a custom design of the same application.

16.3 Types of Overlays

Over the years, quite a few works on FPGA overlays have been developed. In this section, we will take a quick walk through some of them and also look at various types of FPGA overlays that have been explored in the past decade.

16.3.1 Virtual FPGAs

To begin, one of the most easiest to understand categories of overlay are virtual FPGAs [LMVV05, BL12, GWL11, CS10, KBL13]. They are built either virtually or physically on top of off-the-shelf FPGA devices. These overlays have different configuration granularity but typically feature coarser configuration granularity than a typical FPGA device. Similar to virtual machines running on a typical computer, such virtual FPGA provides an additional layer that improves application portability and compatibility. Furthermore, because of the coarser-grained configurable fabric, implementing designs on such overlay is relatively easier than on a fine-grained device. However, the additional layer imposes restrictions on the underlying fabrics' capability and usually results in moderate hardware overhead and timing degradation.

In one of the earlier works, Lysecky et al. developed a relatively fine-grained virtual FPGA as firm cores expressed as structural VHDL [LMVV05]. The virtual layer provides effective portability yet incurs relatively high performance and hardware overhead. In [GWL11], Grant et al. proposed a time-multiplexed virtual FPGA CAD framework MALIBU. The virtual FPGA used in MALIBU has both fine-grain and coarse-grain processing elements integrated into each logic cluster and can be used to reduce the compilation time significantly with moderate timing penalty. Around the same time, Coole and Stitt also proposed another island-style coarse-grained overlay called Intermediate Fabric [CS10]. It uses coarse-grained operators such as adders instead of logic clusters and routes data through 8–32 bit buses achieving both portability and fast compilation. Finally, Koch et al. developed a fine-grained FPGA overlay in [KBL13] to implement customized instructions on FPGAs from different vendors providing a portable application consisting of a program binary and an overlay configuration in a completely heterogeneous environment.

16.3.2 Coarse-Grained Reconfigurable Arrays

Another category of overlay architecture commonly employed is in the form of coarse-grained reconfigurable arrays (CGRAs) [KHKT06, FVM$^+$11, SBB06, LS12, CA13, JFM15]. The use of CGRAs provides an efficient tradeoff between flexibility of software and performance of hardware especially for compute intensive applications as demonstrated by numerous earlier works [TB01, CH02].

In one of the earlier works in the area, Kissler et al. developed WPPA (weakly programmable processor array), a VLIW architecture based parameterizable CGRA overlay [KHKT06]. It featured an interconnection wrapper unit for each processing element (PE) that could be used for dynamic CGRAs topology customization. Around the same time in [SBB06], a customized CGRA overlay called QUKU was developed for DSP algorithms. It had a two-level configuration capability, while the high-speed configuration was used for operator reuse within an application and low-speed reconfiguration was used for optimization between different applications. In [FVM$^+$11], Ferreira et al. proposed a heterogeneous CGRA overlay with a global multi-stage interconnection on FPGA. Compiling applications onto the overlay took only milliseconds for smaller DFGs. In [LS12], Lin and So also proposed a soft CGRA overlay for rapid compilation. In addition, they demonstrated that by customizing the overlay connection between PEs on a per-application basis, improvement in energy-efficiency could be obtained in the expense of longer tool run time. The authors in [CA13] built a generic high speed mesh CGRA overlay using the elastic pipeline technique to achieve the maximum throughput. It adopted a data-driven execution flow and was suitable for smaller pipelined DFG execution. Recently in [JFM15], Jain et al. also proposed an overlay that is constructed around the primitive FPGA DSP blocks to achieve high-frequency implementation and high throughput result. Also, in [CS15], Coole and Stitt proposed to provide the overlay with limited flexibility instead of full configurability specifically to a group of design. With this customization, the area overhead was reduced significantly.

16.3.3 Processor-Like Overlays

A third category of overlay moves away from the traditional FPGA architectures and instead explores using processor-like designs as an intermediate layer. The main concern for works in this category are usually compatibility and usability of the overlay from a user's perspective. To provide the necessary performance, these overlay architectures usually feature a high degree of control and provide ample of data parallelism to make them suitable for FPGA accelerations. As an early attempt, Yiannacouras et al. explored the use of a fine-grained scalable vector processor for code acceleration in embedded systems [YSR09]. Later in [SL12], Severance and Lemieux proposed a soft vector processor named VENICE to allow easy FPGA application development. It accepts simple C program as input and execute the code on the highly optimized vector processor on the FPGA for performance. In the work of MARC, Lebedev et al. explored the use of a many-core processor template as an intermediate compilation target [LCD+10]. In that work, they have demonstrated improved usability with the model while also highlighting the need for customizing computational cores for sake of performance. To explore the integration between processor and FPGA accelerators, a portable machine model with multiple computing architectures called MURAC was explored in [HIS14]. Finally, a GPU-like overlay was proposed in [KS11] that demonstrate good performance while maintaining a compatible programming model for the users.

16.4 Case Study: QuickDough

As a case study to illustrate how an FPGA overlay works in practice, the design and implementation of the research project QuickDough will be examined in this section.

In short, QuickDough is a nested loop accelerator generation framework that is able to produce hardware accelerators rapidly [LS12, LNS15]. Given a user-designated loop for acceleration, QuickDough automatically generates and

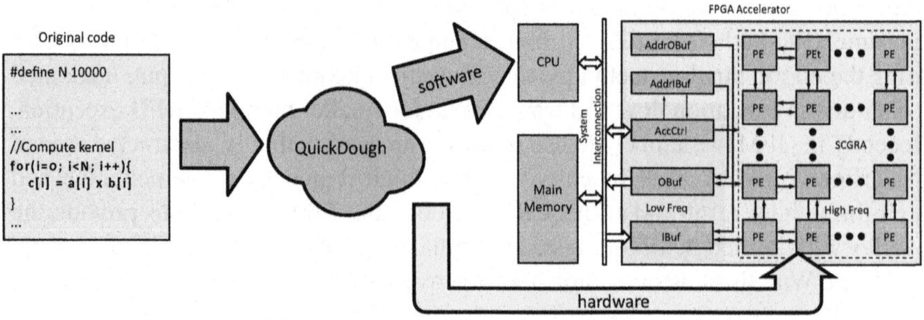

Fig. 16.2 QuickDough takes a user-designated loop as input and generate the corresponding hardware accelerator system using a soft coarse-grained reconfigurable array overlay

optimizes the corresponding hardware accelerator and its associated data I/O facilities with the host software (Fig. 16.2).

The overall design goal of QuickDough is to enhance designer's productivity by greatly reducing the hardware generation time and by providing automatic optimization of the data I/O between the host software and the accelerator. Instead of spending hours on conventional hardware implementation tools, QuickDough is capable of producing the targeted hardware-software system in the order of seconds. By doing so, it provides a rapid development experience that is compatible with that expected by most software programmers.

To achieve this compilation speed, while maintaining a reasonable accelerator performance, QuickDough avoids the creation of custom hardware directly for each application. Instead, the compute kernel loop bodies are scheduled to execute on a CGRA overlay, which is selected from a library of pre-implemented hardware library. By sidestepping the time-consuming low-level hardware implementation tool flow, the time to implementing an accelerator in QuickDough is reduced to essentially just the time spent on overlay selection and scheduling compute operations on the resulting overlay. In addition, taking advantage of the overlay's softness and regularity, QuickDough allows users to perform tradeoff between compilation time and performance by selecting and customizing the overlay on a per application basis. The result is a unified design framework that seamlessly produces the entire hardware-software infrastructure with a design experience similar to developing conventional software.

Through these facilities and through carefully partitioning the design process, QuickDough strives to improve design productivity of software programmers utilizing FPGA accelerators in three aspects:

1. It automates most of the hardware accelerator generation process, requiring only minimum input from the application designer;
2. It produces functional hardware designs at software compilation speed (order of seconds), greatly increasing the number of debug-edit-implement cycles per day achievable;

3. It allows software programmers to progressively improve performance of the generated accelerator through subsequent optimization phases, essentially separating the functional verification and optimization process of application development.

In the following subsections, an overview of QuickDough is presented to illustrate how it achieves the above goals. For details, please refer to [LS12, LNS15].

16.4.1 Generation Framework

Figure 16.3 shows an overview of the QuickDough compilation flow. The key to rapid accelerator generation in QuickDough is to partition the complex hardware-software compilation flow into a fast and a slow path.

The fast path of QuickDough consists of the steps necessary to generate a *functional* loop accelerator, which are shown in orange for hardware and in green for software generation in Fig. 16.3. To begin generating a loop accelerator,

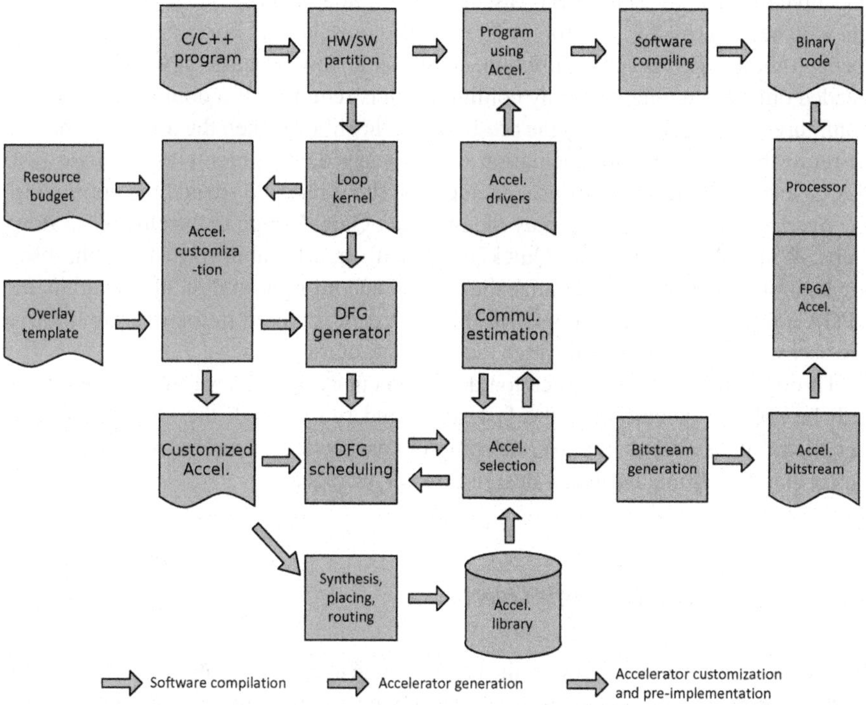

Fig. 16.3 QuickDough: FPGA loop accelerator design framework using SCGRA overlay. The compute intensive loop kernel of an application is compiled to the SCGRA overlay based FPGA accelerator while the rest is compiled to the host processor

QuickDough first partially unroll the compute kernel loop and extract the loop body into its corresponding data flow graph (DFG). Subsequently, a suitable overlay configuration is selected from the pre-built implementation library on which the DFG is scheduled to execute. Optionally, the user may choose to iteratively refine the selection by feeding back scheduling performance and estimated communication cost at each iteration. The selected pre-built overlay implementation is then updated with the corresponding scheduling result to create the final FPGA configuration bitstream. Finally, this updated bitstream is combined with the rest of the software to form the final application that will be executed on the target CPU-FPGA system.

While there may seem to be a lot of steps involved in this fast path, all of them run relatively quickly. Furthermore, the only loop in this compilation flow is the accelerator selection process, which as explained is an optional step that can be bypassed if the user opts for speed over quality. The result is that the run time of this fast path can be kept within the order of tens of seconds. It allows users to perform rapid design iterations, which is particularly important during early application development phases.

On the other hand, the slow path of QuickDough consists of all the remaining time-consuming steps in the flow from Fig. 16.3. These steps are responsible for implementing the overlay in hardware, optimizing the CGRA to the user application, and updating the overlay library as needed. Although these steps are slow, they are not necessarily by run for every design compilation. For example, running through the low-level FPGA implementation is a slow process, however, they are needed only when a new overlay configuration is required. If a compatible overlay configuration already exists in the pre-built overlay library, then the user may choose to reuse the existing implementation during early developments to facilitate fast design turn-around. When the user decides that she is ready to spend time optimizing the overlay configuration, she may then instruct QuickDough to execute these slow steps. With the slower steps, QuickDough will then be able to analyze the user application requirement, customize the overlay accordingly, and finally generate the FPGA configuration bitstream. Once this bitstream is stored in the overlay library, the user will not need to go through these slow process again.

Throughout the entire application development cycle, most of the time the user will be able to execute only the fast steps, and run through the slow steps only occasionally. As a result, the overlay compilation speed remains orders of magnitude better than traditional hardware design flow on average.

16.4.2 The QuickDough Overlay

Figure 16.4 shows an overview of the QuickDough overlay together with its data I/O infrastructure. The QuickDough overlay as shown in the right hand side consists of an array of simple processing elements (PEs) connected by a direct network. Each PE computes and forwards data to their neighbors synchronously according to a static schedule. This schedule is stored in the instruction ROM associated with each

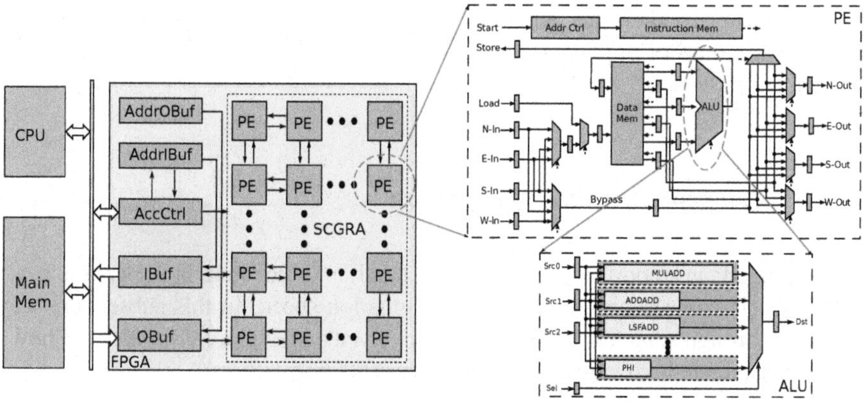

Fig. 16.4 Overview of the QuickDough generated system. Implemented on the FPGA is the QuickDough overlay, which is an array of PEs connected with a direct network, and the communication infrastructure with the host

PE and controls the action of each PE's action in every cycle. Finally, each PE contains a scratchpad data memory for run-time data that may be reused in the same PE or be forwarded in subsequent steps.

Communication between the accelerator and the host processor is carried through a pair of input/output buffers. Like with the rest of the PE array, accesses to these I/O buffers from the array also take place in lock step with the rest of the system. The location to access in each cycle is controlled by a pair of address buffers, which contains address information generated from the QuickDough compiler.

Figure 16.4 also shows the connection within a processing element. Each PE is constructed surrounding an ALU, with multiplexors connecting the input/output of the ALU to either the internal memory or the neighboring PEs. In this particular version, the optional load/store path is also shown, which is presented only at dedicated PE connected to the I/O buffer. Within the PE, the ALU is supported by a multi-port data memory and an instruction memory. Three of the data memory's read ports are connected to the ALU as inputs, while the remaining ports are sent to the output multiplexors for connection to neighboring PEs and the optional store path to output buffer (`OBuf`) external to the PE. At the same time, this data memory takes input from the ALU output, data arriving from neighboring PEs, as well as from the optional loading path from the data input buffer (`IBuf`).

The ALU itself has a simple design that supports up to 16 fully pipelined operations. It is constructed also as a template and may be customized to support any user-defined 3-input operations. Regardless of its function, each operation must have a deterministic pipeline depth. With that, the QuickDough scheduler will ensure there is never any output conflict and allow the operations to execute in parallel as needed.

Finally, note that the overlay is designed as a soft template. Many parameters of the overlay, including the SCGRA array size, I/O buffer size, as well as the operations supported by the ALU, are configurable. They can be optimized for a particular group of application upon user's request as part of the framework's customization steps.

16.4.3 Loop Accelerator Generation

As mentioned, an important goal of QuickDough is to generate high-performance FPGA accelerator systems in software compilation speed. In this subsection, we will walk through some of the major steps involved and explore the details on how they help generate hardware accelerators very efficiently.

DFG Generation

The top-level inputs to QuickDough are loops that the user has designated for acceleration. To accelerate a loop, the straightforward way is to treat each loop iteration as an individual acceleration target and rely on the host processor for loop control. However, it is not ideal because (1) the overhead for data and control transfer between the host and FPGA will be too high, (2) the amount of parallelism encapsulated in a single iteration of the loop body is likely to be too low for acceleration, and (3) there will be no data reuse between loop iterations to amortize the transfer overhead between the host and the FPGA.

Therefore, just like most other acceleration frameworks, QuickDough begins the accelerator generation process by partially unrolling the input loop U times to increase the amount of parallelism available. The body of this partially unrolled loop subsequently forms the basic unit (BU) for acceleration in later steps. This is the unit that is implemented on the FPGA accelerator using the SCGRA overlay. With a single copy of BU implemented in the accelerator, the original loop is completed by executing the accelerator N/U times, where N is the original loop bound.

Furthermore, to amortize the communication cost between the host processor and the accelerator, input/output data are transferred in groups for G invocations of BU as shown in Fig. 16.5. By buffering I/O and intermediate data on the accelerator, this grouping strategy also allow reusing data between loop iterations, further enhancing performance.

In our current implementation, the unrolling factor U is specified by the user, while the grouping factor G is determined by the framework automatically. G is determined based on the on-chip buffer size.

At the end of this process, the unrolled loop body BU will form the core of the accelerator and the corresponding data flow graph (DFG) will be extracted, which will drive the rest of the generation flow.

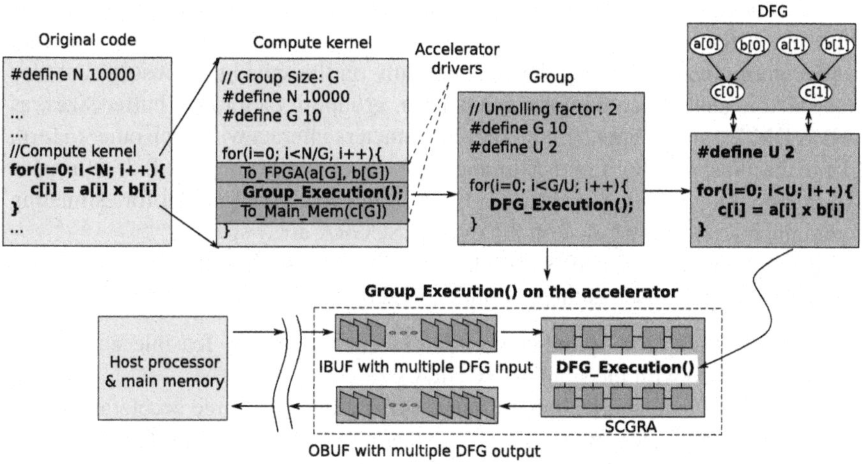

Fig. 16.5 Loop execution on an SCGRA overlay based FPGA accelerator

Accelerator Selection

As an accelerator generation framework, an important task of QuickDough is without doubt to generate the physical implementation of the accelerator on the FPGA. With its SCGRA serving as an intermediate layer, this translates into physically implementing the overlay on the FPGA. Unfortunately, no matter how simple the overlay is, running through the low-level hardware implementation tools is going to be a lengthy process, contradicting the original goal of employing the overlay in the first place. In order to sidestep this lengthy hardware implementation process, QuickDough instead transform it into a selection process from a library of pre-implemented overlay.

When compared to directly implementing the overlay using standard hardware implementation tools, this accelerator selection process is much faster. In return, the performance of the resulting overlay configuration may not be optimal for the given user application. Despite the suboptimal performance, the selected overlay implementation is functionally correct and should give a good sense of the overall hardware-software codesign process to the software programmer, which is necessary for rapid early development. The user may subsequently opt for the slower optimization and customization steps explained in the next subsection to progressively improve performance.

As such, the goal of this process is to select the *best* and *functional* overlay configuration from the pre-implemented library *rapidly*. While functionality is easy to determine, QuickDough must also be able to estimate the resulting performance swiftly for this selection process. Now, the performance of the accelerator depends mainly on two factors: (1) computational latency, and (2) communication latency. As the SCGRA overlay is regular and computes with a deterministic schedule, its exact computational latency of carrying out the user DFG can readily be obtained from the scheduler. Of all the configurable parameters of the overlay, the SCGRA array size is the dominant factor affecting this computational latency. On the other hand, communication latency depends not only on the scheduling results, but also other factors such as communication pattern, grouping factor, I/O buffer sizes, as well as DMA transfer latency, etc. These parameters interact with each other to form a large design space. For rapid estimation, instead of performing a full design space exploration, QuickDough is able to rely on simple analytical model for estimating communication latency based on the overlay SCGRA array size.

Finally, QuickDough leaves this choice of effort in finding the best accelerator configuration to the user as three optimization effort levels:

- Level 0 (`O0`)—No optimization. QuickDough selects a feasible accelerator configuration with the smallest SCGRA size.
- Level 1 (`O1`)—QuickDough estimates performances of three accelerators with different array sizes and select the one that results in the best performance.
- Level 2 (`O2`)—QuickDough performs an exhaustive search on all accelerators in the library and searches for the best accelerator configuration.

Obviously, the more effort is being put into the selection process, the longer the process will take, and the resulting performance is improved most of the time. The choice on how much effort to pay is up to the user.

DFG Scheduling

This is the step where the DFG extracted from the unrolled loop body is scheduled to execute on the selected SCGRA overlay. For each operation in the DFG, the scheduler must decide the cycle and the PE in which the operation should be carried out. The scheduler also determines the routing of data between the producing and consuming PEs, as well as to/from the I/O buffer.

As QuickDough tends to target DFGs with close to thousands of node, to keep the scheduling time short, a classical list scheduling algorithm was adopted [Sch96]. A scheduling metric proposed in [LS12] that considers both load balancing and communication cost was used in the scheduler.

At the end of this scheduling step, a schedule with operation of each PE and data I/O buffer in every cycle is produced. This schedule essentially turns the generic overlay implementation into the specific accelerator for the user application.

Accelerator Bitstream Generation

The final step of the accelerator generation process is to produce the instructions for each PE and the address sequences for the I/O buffers according to the scheduler's result.

As a way to reduce area overhead of the overlay, both the instruction memory of each PE as well as the I/O address buffers are implemented as read-only memories (ROMs) on the physical FPGA. To update the content of these memories, the QuickDough framework must update the FPGA implementation bitstream directly. To do that, the memory organization and placement information of the target overlay implementation is obtained from an `XDL` file corresponding to the overlay implementation [BKT11]. The resulting memory organization information is encoded in the corresponding `BMM` file, which is combined with the scheduler results to update the overlay implementation bitstream with the `data2mem` tool from Xilinx [Xil11].

While it may sound complicated, the whole process is automated and consumes only a few seconds of tools run time. The result of this process is an updated bitstream with all application-specific instructions embedded. This updated bitstream is the final configuration file used to program the physical FPGA, and it concludes the QuickDough flow.

16.4.4 Customization and Optimization

The true power of utilizing an FPGA overlay rests on the fact that the overlay is virtual, soft, and easily customizable for the need of the application. In this part, we will examine the various ways QuickDough enables a user to customize and optimize the overlay to improve power-performance of the generated accelerators. Unlike the fast generation path explained above, these customization steps are much more time consuming. As such, these customization steps are considered part of the slow path in QuickDough's design philosophy, and are expected to execute only occasionally as needed.

There are a number of reasons why a user may want to spend the time on customization despite the slow process. To begin, the user may want to construct a prebuilt implementation library specific to the target application domain. It is useful as the result can be memorized in the implementation library and be reused by all other target applications, amortizing the initial implementation effort. Once the project development has gone through the initial debugging phase, a user may also opt for creating a customized overlay for the particular application that needs to be accelerated. This per-application optimization will help further improve the performance of the accelerator but the process will undoubtedly be slower than to simply select a prebuilt implementation. Luckily, the result of this per-application customization can be stored in the prebuilt library such that it can be reused on subsequent compilations.

Customization in Action

For the purpose of customization, the QuickDough overlay was designed from the beginning to be a template from which a family of overlay instances could be generated. The SCGRA overlay consists of a regular array of simple PEs connected with direct connections. Depending on the application requirements, a range of design parameters of this overlay can then be customized, including the array size, on-chip scratchpad data memory capacity, instruction memory capacity and I/O buffer capacity. On top of that, as a hardware-software system generator for loop acceleration, QuickDough may also optimize the communication between the accelerator and the host software by varying the *grouping factor*, which determines the amount of data transferred between the two in each transaction. Finally, it may also optionally optimize the *loop unrolling factor* on behalf of the user depending on the computational capability of the overlay as a function of its array size.

Obviously, a major challenge when optimizing this set of parameters is that they tend to interact with one another in fairly complex ways. For example, increasing the array size increases the compute capability of the array, potentially improving the accelerator performance. However, the increased size is beneficial only if the input DFG has enough parallelism presented to take advantage of the increased capability. To increase the amount of available computation, QuickDough may optionally increase the amount of loop unrolling in the user-supplied kernel. However, that also increases the amount of data I/O required, as well as the on-chip instruction and data buffer requirement, which in turn limits the size of the array in the first place. A full discussion of the detailed customization and optimization process is beyond the scope of this chapter. Yet it is worthwhile to realize the potential of customization here.

As an illustration, Fig. 16.6 shows the results of customization on performance and resource consumption of a QuickDough generated accelerator. In this example, accelerators for a 49-tap FIR filter targeting the Zedboard were generated using

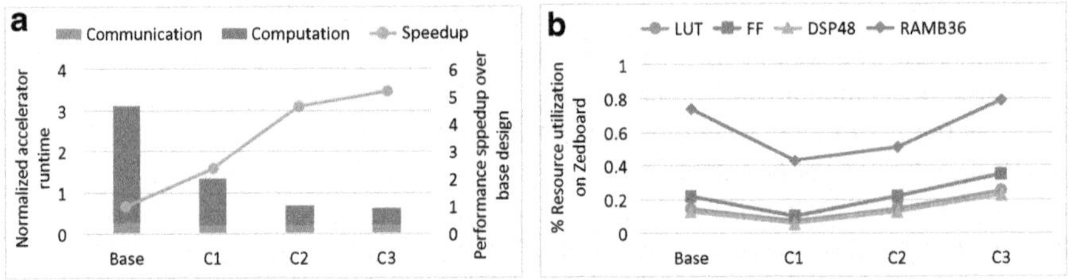

c

	base	C1	C2	C3
Loop unrolling factor	10×50	50×50	50×50	50×50
Grouping factor	100×50	2500×50	1250×50	5000×50
SCGRA array size	3×3	2×2	3×3	4×4
Instruction Memory Depth	4k	4k	2k	1k
IO Buffer Depth	1k	4k	2k	8k

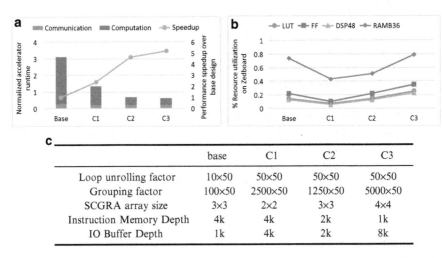

Fig. 16.6 Effect of overlay customization on performance and resource consumption. Example here shows the result of an FIR filter using QuickDough with three different overlay configurations over a base design. (**a**) Performance. (**b**) Resource consumptions. (**c**) Overlay configurations

QuickDough. Three different customized overlay configurations were generated with details shown in Fig. 16.6c.

A few interesting observations can be made from the figures. In terms of performance, comparing to the sub-optimal baseline configuration, the best configuration (C3) results in more than five times improvement in performance. Furthermore, comparing the baseline and C2 that features the same array size (3×3), 4.7× improvement in performance can be achieved by optimizing the unrolling and grouping factor in combination with the on-chip memory configuration. Interestingly, considering the resource consumptions, C2 in fact consumes 31 % less on-chip memory and the same amount of LUTs, flip-flops, and DSP blocks as the baseline. This demonstrates how customization can result in designs with improved performance while reducing resource consumptions.

16.4.5 Summary

Using QuickDough as a case study, we have illustrated how the use of an FPGA overlay has enabled a very different experience for software programmers when designing hardware-software systems. The 2-layer approach to hardware design allows for a very rapid design experience that sharply contrasts the lengthy low-level hardware tool flow. Furthermore, the softness of the overlay provides opportunities for further customization and optimization as the development effort progresses. Together, it creates for novice users a "software-like" experience to designing complex accelerator systems while maintaining considerable overall acceleration performance promised by FPGAs.

16.5 Research Challenges and Opportunities

While the early results of using FPGA overlay are encouraging, there remains many challenges that must be overcome before its full potential can be unleashed. In particular, for FPGA overlays to be useful, future research must be able to address two important challenges: reduce overhead and enhance customization.

16.5.1 Reducing Overhead

By introducing an additional architectural layer between a user application and the physical fabric, an FPGA overlay inevitably incurs additional performance and area overhead to the system. It is especially important when the FPGA is used as an accelerator in a processor-based system. If the overhead is so large that the

FPGA is no longer offering any significant performance enhancement, then software programmers will have little incentive to devote effort into using them.

From a hardware implementation's point of view, the additional layer must also incur additional area overhead. Such overhead usually manifests as additional usage of configurable fabric and on-chip memory. In some cases where spare resources are available, the effect of such area overhead may not be apparent. In fact, as illustrated earlier, researchers have deduced ways to make sure of such spare resources for debugging purposes [HW13]. However, in other cases where the area of FPGA overlay results in a reduction of resources available to implement a user's design, then it is likely that this area overhead will translate into performance overhead. On a circuit level, such area overhead may also manifest as additional delays impose on the critical path, resulting in overall performance loss. Therefore in short, area overhead, if not well controlled, may easily translate into performance overhead.

Luckily, as mentioned in some early works, despite such overhead, the use of an overlay may still be worthwhile from a performance perspective. For example in [CA13], Capalija and Abdelrahman has demonstrated that by taking advantage of the regularity of its overlay and with the help of detailed floor planning, they were able to control the overhead while maintaining reasonable throughput performance when compared to a simple push-button synthesis flow. The very use of overlay allows amortization of the optimization effort in the long run as the overlay structure may be reused many times.

16.5.2 Enhancing Customization

Another major challenge faced by overlay designers is the very notion of using FPGA overlay in the first place: "If by overlaying a different architecture over an FPGA may provide all sort of nice properties, then why not implement the overlay architecture directly on silicon to enjoy all the benefits instead?" For example, if a GPU overlay provides good performance and good programmability, then maybe the design should be implemented on an actual GPU instead of a GPU overlaying on top of an FPGA.

Indeed, if the so called "overlay" in question is simply a fixed reimplementation of another architecture on an FPGA, then the only incentive to so are probably to provide compatibility through virtualization, or simply to save cost on silicon implementation. In the latter case, this intermediate architecture may hardly be called an "overlay" any more.

However, the true power of FPGA overlays is that they can be adapted to the application through *customization*. As a virtual architecture, many aspects of an overlay architecture can be customized to the target application to improve power-performance. In [LCD+10], for example, the computational core in the multi-core overlay may be customized to the specific targeted application. By customizing the core to their targeted application, almost 10× improvement in the overlay performance can be observed, bring the overlay performance to be within factor of

3 of the reference custom FPGA design. In [LS12], the direct connection topology among the process elements in coarse-grained reconfigurable array overlay was customized against the input application. When compared to the best predefined topology, the application-specific interconnect provides up to 28 % improvement in resulting energy-delay product.

The improved results should come at no surprise: the extreme case of an overlay customization is simply a full-custom design of the application on the target FPGA, which is supposed to have the best possible performance if designed correctly. What is challenging is therefore to ability to fine-tune the tradeoff among design productivity, virtualization and performance of the resulting system. In other words, the research question to ask in the future should therefore be: "How to improve performance of an overlaying system through customization without significantly sacrificing the benefits of using an overlay?" The answer to this question is going to open up a wide field of exciting research in FPGA-based reconfigurable systems.

16.5.3 Closing Thoughts

FPGA overlay is without doubt going to be an important part of future FPGA-based reconfigurable systems. The potential is plentiful and we anticipate that overlaying technology will benefit all aspects of future reconfigurable systems, especially on the important aspect of design productivity. With silicon technologies continue to evolve, the amount of available on-chip configurable resource is going to increase. The abundance of high-performance configurable resources on FPGA will enable a new generation of systems that relies heavily on advanced overlay architectures for virtualization and to improve design productivity.

References

[AHS14] M. Abdelfattah, A. Hagiescu, D. Singh, Gzip on a chip: high performance lossless data compression on FPGAs using OpenCL, in *Proceedings of the International Workshop on OpenCL (IWOCL)*, May 2014, pp. 4:1–4:9

[AK08] S. Aditya, V. Kathail, *Algorithmic Synthesis Using PICO: An Integrated Framework for Application Engine Synthesis and Verification from High Level C Algorithms*, Chap. 4, pp. 53–74; in Coussy, Morawiec [CM08], 1st edn. (2008)

[AFP+11] M. Adler, K. Fleming, A. Parashar, M. Pellauer, J. Emer, LEAP scratchpads: automatic memory and cache management for reconfigurable logic, in *Proceedings of the 19th ACM/SIGDA International Symposium on Field Programmable Gate Arrays (FPGA)*, February 2011 (ACM, New York, 2011), pp. 25–28

[AHL+14a] A. Agne, M. Happe, A. Lösch, C. Plessl, M. Platzner, Self-awareness as a model for designing and operating heterogeneous multicores. ACM Trans. Reconfigurable Technol. Syst. **7**(2), 13:1–13:18 (2014)

[AHL+14b] A. Agne, M. Happe, E. Lübbers, B. Plattner, M. Platzner, C. Plessl, ReconOS – an operating system approach for reconfigurable computing. IEEE Micro **34**(1), 60–71 (2014)

[APL11] A. Agne, M. Platzner, E. Lübbers, Memory virtualization for multithreaded reconfigurable hardware, in *Proceedings of the International Conference on Field Programmable Logic and Applications (FPL)*, September 2011 (IEEE Computer Society, Los Alamitos, 2011), pp. 185–188

[Agr09] J. Agron, Domain-specific language for HW/SW co-design for FPGAs, in *Proceedings of the IFIP TC 2 Working Conference on Domain-Specific Languages*, July 2009. Lecture Notes in Computer Science (LNCS), vol. 5658 (Springer, Berlin, 2009), pp. 262–284

[AHA14] AHA Products Group, AHA3642 (2014), http://www.aha.com/DrawProducts.aspx?Action=GetProductDetails&ProductID=38. Accessed 4 Aug 2015 [Online]

[ALSU06] A. Aho, M. Lam, R. Sethi, J. Ullman, *Compilers: Principles, Techniques, and Tools*, 2nd edn. (Addison-Wesley Longman Publishing Co., Inc., Boston, 2006)

[AABC11] M. Aldham, J. Anderson, S. Brown, A. Canis, Low-cost hardware profiling of run-time and energy in FPGA embedded processors, in *Proceedings of the 22nd IEEE International Conference on Application-Specific Systems, Architectures and Processors (ASAP)*, September 2011 (IEEE Computer Society, Los Alamitos, 2011), pp. 61–68

[Alt10a] Altera Corp., San Jose, CA, DE4 Development Board (2010)

[Alt10b] Altera Corp., San Jose, CA, Stratix-IV Data Sheet (2010)

D. Koch et al. (eds.), *FPGAs for Software Programmers*,
DOI 10.1007/978-3-319-26408-0

[Alt13a] Altera Corp., Altera SoCs: When Architecture Matters (2013), http://www.altera.com/devices/processor/soc-fpga/overview/proc-soc-fpga.html. Accessed 4 Aug 2015 [Online]

[Alt14a] Altera Corp., San Jose, CA, *Altera SoC Embedded Design Suite User Guide (ug-1137)* (2014)

[Alt14b] Altera Corp., San Jose, CA, *Comparing Altera SoC Device Family Features* (2014) UF-1005

[Alt15a] Altera Corp., San Jose, CA, *Nios II Classic Software Developer's Handbook (NII5V2)* (2015)

[Alt15b] Altera Corp., San Jose, CA, *Nios II Core Implementation Details (NII51016)* (2015)

[Alt15c] Altera Corp., San Jose, CA, *Quartus II Handbook Volume 1: Design and Synthesis (QII5V1)* (2015)

[Alt13b] Altium Limited, *C-to-Hardware Compiler User Manual (GU0122)* (2013)

[AK98] H. Andrade, S. Kovner, Software synthesis from dataflow models for G and LabVIEW, in *Proceedings of the IEEE Asilomar Conference on Signals, Systems, and Computers*, November 1998, pp. 1705–1709

[ANJ+04] D. Andrews, D. Niehaus, R. Jidin, M. Finley, W. Peck, M. Frisbie, J. Ortiz, E. Komp, P. Ashenden, Programming models for hybrid FPGA-CPU computational components: a missing link. IEEE Micro **24**(4), 42–53 (2004)

[ASA+08] D. Andrews, R. Sass, E. Anderson, J. Agron, W. Peck, J. Stevens, F. Baijot, E. Komp, Achieving programming model abstractions for reconfigurable computing. IEEE Trans. Very Large Scale Integr. VLSI Syst. **16**(1), 34–44 (2008)

[ANS+14] O. Arcas-Abella, G. Ndu, N. Sonmez, M. Ghasempour, A. Armejach, J. Navaridas, W. Song, J. Mawer, A. Cristal, M. Luján, An empirical evaluation of high-level synthesis languages and tools for database acceleration, in *Proceedings of the 24th International Conference on Field Programmable Logic and Applications (FPL)*, September 2014, pp. 1–8

[ALJ15] J. Arram, W. Luk, P. Jiang, Ramethy: reconfigurable acceleration of bisulfite sequence alignment, in *Proceedings of the 23rd ACM/SIGDA International Symposium on Field-Programmable Gate Arrays (FPGA)*, February 2015 (ACM, New York, 2015), pp. 250–259

[BTA93] J. Babb, R. Tessier, A. Agarwal, Virtual wires: overcoming pin limitations in FPGA-based logic emulators, in *Proceedings of the IEEE Workshop on FPGAs for Custom Computing Machines (FCCM)*, April 1993, pp. 142–151

[BTD+97] J. Babb, R. Tessier, M. Dahl, S. Hanono, D. Hoki, A. Agarwal, Logic emulation with virtual wires. IEEE Trans. Comput. Aided Des. **16**(6), 609–626 (1997)

[BVR+12] J. Bachrach, H. Vo, B. Richards, Y. Lee, A. Waterman, R. Avizienis, J. Wawrzynek, K. Asanovic, Chisel: constructing hardware in a Scala embedded language, in *Proceedings of the 49th ACM/EDAC/IEEE Design Automation Conference (DAC)*, June 2012, pp. 1212–1221

[Bar76] M. Barbacci, The symbolic manipulation of computer descriptions: ISPL compiler and simulator, Technical report, Department of Computer Science, Carnegie Mellon University, Pittsburgh, 1976

[BS73] M. Barbacci, D. Siewiorek, Automated exploration of the design space for register transfer (RT) systems. ACM SIGARCH Comput. Archit. News **2**(4), 101–106 (1973)

[Bas04] C. Bastoul, Code generation in the polyhedral model is easier than you think, in *Proceedings of the 13th International Conference on Parallel Architectures and Compilation Techniques (PACT)* (IEEE Computer Society, Washington, DC, 2004), pp. 7–16

[Bat68] K. Batcher, Sorting networks and their applications, in *Proceedings of the Spring Joint Computer Conference (AFIPS)* (ACM, New York, 1968) pp. 307–314

[BKS00] K. Bazargan, R. Kastner, M. Sarrafzadeh, Fast template placement for reconfigurable computing systems, IEEE Des. Test Comput. **17**(1), 68–83 (2000)

[BBZT14] A. Becher, F. Bauer, D. Ziener, J. Teich, Energy-aware SQL query acceleration through FPGA-based dynamic partial reconfiguration, in *Proceedings of the 24th International Conference on Field Programmable Logic and Applications (FPL)* (IEEE, New York, 2014), pp. 1–8

[BKT11] C. Beckhoff, D. Koch, J. Torresen, The Xilinx design language (XDL): tutorial and use cases, in *Proceedings of the 6th International Workshop on Reconfigurable Communication-centric Systems-on-Chip (ReCoSoC)*, June 2011 (IEEE, New York, 2011) pp. 1–8

[BWHC06] N. Bergmann, J. Williams, J. Han, Y. Chen, A process model for hardware modules in reconfigurable system-on-chip, in *Workshop Proceedings of the International Conference on Architecture of Computing Systems (ARCS)*, March 2006. Lecture Notes in Informatics (LNI), vol. 81 (Gesellschaft für Informatik (GI), Bonn, 2006), pp. 205–214

[BRM99] V. Betz, J. Rose, A. Marquardt, *Architecture and CAD for Deep-Submicron FPGAs* (Kluwer Academic, Norwell, 1999)

[Bis06] C. Bishop, *Pattern Recognition and Machine Learning (Information Science and Statistics)* (Springer, New York, Secaucus, 2006)

[Blu04] Bluespec Inc., *Bluespec SystemVerilog Version 3.8 Reference Guide* (2004)

[Bob07] C. Bobda, *Introduction to Reconfigurable Computing: Architectures, Algorithms, and Applications* (Springer, Netherlands, 2007)

[Bol08] T. Bollaert, *Catapult Synthesis: A Practical Introduction to Interactive C Synthesis*, Chap. 3, pp. 29–52; in Coussy, Morawiec [CM08], 1st edn. (2008)

[BL12] A. Brant, G. Lemieux, ZUMA: an open FPGA overlay architecture, in *Proceedings of the IEEE 20th Annual International Symposium on Field-Programmable Custom Computing Machines (FCCM)* (2012), pp. 93–96

[Bre96] G. Brebner, A virtual hardware operating system for the Xilinx XC6200, in *Proceedings of the International Workshop Fiel-Programmable Logic and Applications (FPL)* (1996), pp. 327–336

[BR96] S. Brown, J. Rose, FPGA and CPLD architectures: a tutorial. IEEE Des. Test Comput. **13**(2), 42–57 (1996)

[BCVN10] B. Buyukkurt, J. Cortes, J. Villarreal, W. Najjar, Impact of high-level transformations within the ROCCC framework. ACM Trans. Archit. Code Optim. **7**(4), 17:1–17:36 (2010)

[CBA14a] N. Calagar, S. Brown, J. Anderson, Source-level debugging for FPGA high-level synthesis, in *Proceedings of the 24th International Conference on Field Programmable Logic and Applications (FPL)*, September 2014 (IEEE, New York, 2014), pp. 1–8

[CW91] R. Camposano, W. Wolf (eds.), *High-Level VLSI Synthesis* (Kluwer Academic, Norwell, 1991)

[CAB13] A. Canis, J. Anderson, S. Brown, Multi-pumping for resource reduction in FPGA high-level synthesis, in *Proceedings of the Conference on Design, Automation and Test in Europe (DATE)* (2013), pp. 194–197

[CBA14b] A. Canis, S. Brown, J. Anderson, Modulo SDC scheduling with recurrence minimization in high-level synthesis, in *Proceedings of the 24th International Conference on Field Programmable Logic and Applications (FPL)*, September 2014 (IEEE, New York, 2014), pp. 1–8

[CCA+13] A. Canis, J. Choi, M. Aldham, V. Zhang, A. Kammoona, T. Czajkowski, S. Brown, J. Anderson, LegUp: an open-source high-level synthesis tool for FPGA-based processor/accelerator systems. ACM Trans. Embed. Comput. Syst. **13**(2), 24:1–24:27 (2013)

[Can15] Canterbury Corpus, *Descriptions of the Corpora* (2015), http://corpus.canterbury.ac.nz/descriptions/. Accessed 4 Aug 2015 [Online]

[CA13] D. Capalija, T. Abdelrahman, A high-performance overlay architecture for pipelined execution of data flow graphs, in *Proceedings of the 23rd International Conference on Field Programmable Logic and Applications (FPL)*, September 2013, pp. 1–8

[CDW10] J. Cardoso, P. Diniz, M. Weinhardt, Compiling for reconfigurable computing: a survey. ACM Comput. Surv. **42**(4), 13:1–13:65 (2010)

[CMS01] L. Carloni, K. McMillan, A. Sangiovanni-Vincentelli, Theory of latency-insensitive design. IEEE Trans. Comput. Aided Des. **20**(9), 1059–1076 (2001)

[CO14] J. Casper, K. Olukotun, Hardware acceleration of database operations, in *Proceedings of the ACM/SIGDA International Symposium on Field-Programmable Gate Arrays (FPGA)* (2014), pp. 151–160

[CCP06] D. Chen, J. Cong, P. Pan, FPGA design automation: a survey. Found. Trend. Electron. Des. Autom. **1**(3), 139–169 (2006)

[CBA13] J. Choi, S. Brown, J. Anderson, From software threads to parallel hardware in high-level synthesis for FPGAs, in *Proceedings of the IEEE International Conference on Field-Programmable Technology (FPT)* (2013), pp. 270–277

[CHM11] E. Chung, J. Hoe, K. Mai, CoRAM: an in-fabric memory abstraction for FPGA-based computing, in *Proceedings of the 19th ACM/SIGDA International Symposium on Field Programmable Gate Arrays (FPGA)* (ACM, New York, 2011), pp. 97–106

[CG15] A. Cilardo, L. Gallo, Improving multibank memory access parallelism with lattice-based partitioning. ACM Trans. Archit. Code Optim. **11**(4), 45:1–45:25 (2015)

[CCKH00] K. Compton, J. Cooley, S. Knol, S. Hauck, Configuration relocation and defragmentation for reconfigurable computing, in *Proceedings of the International Symposium on Field-Programmable Custom Computing Machines (FCCM)* (2000), pp. 279–280

[CH02] K. Compton, S. Hauck, Reconfigurable computing: a survey of systems and software. ACM Comput. Surv. **34**(2), 171–210 (2002)

[CD94] J. Cong, Y. Ding, FlowMap: an optimal technology mapping algorithm for delay optimization in lookup-table based FPGA designs. IEEE Trans. Comput. Aided Des. **13**(1), 1–12 (1994)

[CHL+12] J. Cong, M. Huang, B. Liu, P. Zhang, Y. Zou, Combining module selection and replication for throughput-driven streaming programs, in *Proceedings of the Conference on Design, Automation and Test in Europe (DATE)* (EDA Consortium, San Jose, 2012), pp. 1018–1023

[CHZ14] J. Cong, M. Huang, P. Zhang, Combining computation and communication optimizations in system synthesis for streaming applications, in *Proceedings of the ACM/SIGDA International Symposium on Field-Programmable Gate Arrays (FPGA)* (ACM, New York, 2014), pp. 213–222

[CJLZ09] J. Cong, W. Jiang, B. Liu, Y. Zou, Automatic memory partitioning and scheduling for throughput and power optimization, in *Proceedings of the International Conference on Computer-Aided Design (ICCAD)* (ACM, New York, 2009), pp. 697–704

[CJLZ11] J. Cong, W. Jiang, B. Liu, Y. Zou, Automatic memory partitioning and scheduling for throughput and power optimization. ACM Trans. Des. Autom. Electron. Syst. **16**(2), 15:1–15:25 (2011)

[CLN+11] J. Cong, B. Liu, S. Neuendorffer, J. Noguera, K. Vissers, Z. Zhang, High-level synthesis for FPGAs: from prototyping to deployment. IEEE Trans. Comput.-Aided Des. **30**(4), 473–491 (2011)

[CZ06] J. Cong, Z. Zhang, An efficient and versatile scheduling algorithm based on SDC formulation, in *Proceedings of the IEEE/ACM Design Automation Conference (DAC)* (2006), pp. 433–438

[CZZ12] J. Cong, P. Zhang, Y. Zou, Optimizing memory hierarchy allocation with loop transformations for high-level synthesis, in *Proceedings of the 49th Annual Design Automation Conference (DAC)* (ACM, New York, 2012), pp. 1233–1238

[Con15] Convey Computers (2015), http://www.conveycomputer.com/. Accessed 2 April 2015 [Online]

[CS10] J. Coole, G. Stitt, Intermediate fabrics: virtual architectures for circuit portability and fast placement and routing, in *Proceedings of IEEE/ACM/IFIP International Conference on Hardware/Software Codesign and System Synthesis (CODES+ISSS)*, October 2010, pp. 13–22

[CS15] J. Coole, G. Stitt, Adjustable-cost overlays for runtime compilation, in *Proceedings of the IEEE 23rd Annual International Symposium on Field-Programmable Custom Computing Machines (FCCM)*, May 2015, pp. 21–24

[CCB+08] P. Coussy, C. Chavet, P. Bomel, D. Heller, E. Senn, E. Martin, *GAUT: A High-Level Synthesis Tool for DSP Applications*, Chap. 9, pp. 147–169; in Coussy, Morawiec [CM08], 1st edn. (2008)

[CM08] P. Coussy, A. Morawiec (eds.), *High-Level Synthesis: From Algorithm to Digital Circuit*, 1st edn. (Springer, New York, 2008)

[Cra98] D. Craft, A fast hardware data compression algorithm and some algorithmic extensions. IBM J. Res. Dev. **42**(6), 733–746 (1998)

[DMC+06] A. DeHon, Y. Markovsky, E. Caspi, M. Chu, R. Huang, S. Perissakis, L. Pozzi, J. Yeh, J. Wawrzynek, Stream computations organized for reconfigurable execution. Microprocess. Microsyst. **30**, 334–354 (2006)

[DGY+74] R. Dennard, F. Gaensslen, H.-N. Yu, V. Rideout, E. Bassous, A. LeBlanc, Design of ion-implanted MOSFET's with very small physical dimensions. IEEE J. Solid-State Circuits **9**(5), 256–268 (1974)

[DBG+14] P. Dlugosch, D. Brown, P. Glendenning, M. Leventhal, H. Noyes, An efficient and scalable semiconductor architecture for parallel automata processing. IEEE Trans. Parallel Distrib. Syst. **25**(12), 3088–3098 (2014)

[DLYS13] Z. Du, X. Li, X. Yang, K. Shen, A parallel multigrid poisson PDE solver for Gigapixel image editing, in *High Performance Computing*. Communications in Computer and Information Science, vol. 207 (Springer, Berlin, 2013)

[Edw06] S. Edwards. The challenges of synthesizing hardware from C-like languages. IEEE Des. Test Comput. **23**(5), 375–386 (2006)

[ESK07] M. El Ghany, A. Salama, A. Khalil, Design and implementation of FPGA-based systolic array for LZ data compression, in *Proceedings of the IEEE International Symposium on Circuits and Systems (ISCAS)*, May 2007, pp. 3691–3695

[Exa13] Exar, *GX 1700 Series* (2013), http://www.exar.com/common/content/document.ashx?id=21282&languageid=1033. Accessed 4 Aug 2015 [Online]

[Fea92] P. Feautrier, Some efficient solutions to the affine scheduling problem. Part II. Multidimensional time. Int. J. Parallel Prog. **21**(6), 389–420 (1992)

[FVLN15] E. Fernandez, J. Villarreal, S. Lonardi, W. Najjar, FHAST: FPGA-based acceleration of BOWTIE in hardware. IEEE/ACM Trans. Comput. Biol. Bioinform. **12**(5), 973–981 (2015)

[FVM+11] R. Ferreira, J. Vendramini, L. Mucida, M. Pereira, L. Carro, An FPGA-based heterogeneous coarse-grained dynamically reconfigurable architecture, in *Proceedings of the 14th International Conference on Compilers, Architectures and Synthesis for Embedded Systems (CASES)* (ACM, New York, 2011), pp. 195–204

[Fin10] M. Fingeroff, *High-Level Synthesis Blue Book* (Xlibris Corporation, Bloomington, 2010)

[FAP+12] K. Fleming, M. Adler, M. Pellauer, A. Parashar, Arvind, J. Emer, Leveraging latency-insensitivity to ease multiple FPGA design, in *Proceedings of the ACM/SIGDA international symposium on Field Programmable Gate Arrays (FPGA)* (2012), pp. 175–184

[FPM12] M. Flynn, O. Pell, O. Mencer, Dataflow supercomputing, in *Proceedings of the 22nd International Conference on Field Programmable Logic and Applications (FPL)* (2012), pp. 1–3

[FVM+08] F. Franchetti, Y. Voronenko, P. Milder, S. Chellappa, M. Telgarsky, H. Shen, P. D'Alberto, F. de Mesmay, J. Hoe, J. Moura, M. Püschel, Domain-specific library generation for parallel software and hardware platforms, in *IEEE International Symposium on Parallel and Distributed Processing (IPDPS)*, April 2008, pp. 1–5

[FRC90] R. Francis, J. Rose, K. Chung, Chortle: a technology mapping program for lookup table-based field programmable gate arrays, in *Proceedings of the 27th ACM/IEEE Design Automation Conference (DAC)* (ACM, New York, 1990), pp. 613–619

[Fre96] M. Freidlin, Diffusion processes and PDE's in narrow branching tubes, in *Markov Processes and Differential Equations*. Lectures in Mathematics ETH Zurich (Birkhäuser, Basel, 1996), pp. 79–89

[FC05] W. Fu, K. Compton, An execution environment for reconfigurable computing, in *Proceedings of the 13th Annual IEEE Symposium on Field-Programmable Custom Computing Machines (FCCM)*, April 2005, pp. 149–158

[GDWL92] D. Gajski, N. Dutt, A. Wu, S. Lin, *High-level Synthesis: Introduction to Chip and System Design* (Kluwer Academic, Norwell, 1992)

[GFY+14] L. Gan, H. Fu, C. Yang, W. Luk, W. Xue, O. Mencer, X. Huang, G. Yang, A highly-efficient and green data flow engine for solving Euler atmospheric equations, in *Proceedings of the 24th International Conference on Field Programmable Logic and Applications (FPL)*, September 2014 (IEEE, New York, 2014), pp. 1–6

[GC07] P. Garcia, K. Compton, A reconfigurable hardware interface for a modern computing system, in *Proceedings of the International Symposium on Field-Programmable Custom Computing Machines (FCCM)*, April 2007 (IEEE Computer Society, Los Alamitos, 2007), pp. 73–84

[GW14] J. Goeders, S. Wilton, Effective FPGA debug for high-level synthesis generated circuits, in *Proceedings of the International Conference on Field Programmable Logic and Applications (FPL)*, September 2014 (IEEE, New York, 2014), pp. 1–8

[GW15] J. Goeders, S. Wilton, Using dynamic signal-tracing to debug compiler-optimized HLS circuits on FPGAs, in *Proceedings of the IEEE International Symposium on Field-Programmable Custom Computing Machines (FCCM)*, May 2015, pp. 127–134

[GGF+11] V. Gopal, J. Guilford, W. Feghali, E. Ozturk, G. Wolrich, *High Performance DEFLATE on Intel Architecture Processors* (2011), http://www.intel.com/content/dam/www/public/us/en/documents/white-papers/ia-deflate-compression-paper.pdf. Accessed 4 Aug 2015 [Online]

[GA13] M. Gort, J. Anderson, Range and bitmask analysis for hardware optimization in high-level synthesis, in *Proceedings of the IEEE/ACM Asia and South Pacific Design Automation Conference (ASP-DAC)* (2013), pp. 773–779

[GHSV+11] N. Goulding-Hotta, J. Sampson, G. Venkatesh, S. Garcia, J. Auricchio, P. Huang, M. Arora, S. Nath, V. Bhatt, J. Babb, S. Swanson, M. Taylor, The GreenDroid mobile application processor: an architecture for silicon's dark future. IEEE Micro **31**(2), 86–95 (2011)

[GWL11] D. Grant, C. Wang, G. Lemieux, A CAD framework for Malibu: An FPGA with time-multiplexed coarse-grained elements, in *Proceedings of the 19th ACM/SIGDA International Symposium on Field Programmable Gate Arrays (FPGA)* (ACM, New York, 2011), pp. 123–132

[GBC+08] Z. Guo, A. Buyukkurt, J. Cortes, A. Mitra, W. Najjar, A compiler intermediate representation for reconfigurable fabrics. Int. J. Parallel Prog. **36**(5), 493–520 (2008)

[GBN04] Z. Guo, B. Buyukkurt, W. Najjar, Input data reuse in compiling window operations onto reconfigurable hardware, in *Proceedings of the ACM SIGPLAN/SIGBED Conference on Languages, Compilers and Tools for Embedded Systems*, June 2004 (ACM, New York, 2004), pp. 249–256

[GNB08] Z. Guo, W. Najjar, B. Buyukkurt, Efficient hardware code generation for FPGAs. ACM Trans. Archit. Code Optim. **5**(1), 6:1–6:26 (2008)

[GB08] R. Gupta, F. Brewer, *High-Level Synthesis: A Retrospective*, Chap. 2, pp. 13–28; in Coussy, Morawiec [CM08], 1st edn. (2008)

[GGDN04a] S. Gupta, R. Gupta, N. Dutt, A. Nicolau, *SPARK: A Parallelizing Approach to the High-Level Synthesis of Digital Circuits* (Kluwer Academic, Norwell, 2004)

[GGDN04b] S. Gupta, R. K. Gupta, N. D. Dutt, A. Nicolau, Coordinated parallelizing compiler optimizations and high-level synthesis. ACM Trans. Des. Autom. Electron. Syst. **9**(4), 441–470 (2004)

[GL91] B. Gustafsson, P. Lötstedt, Analysis of multigrid methods for general systems of PDE, in *Multigrid Methods III*. International Series of Numerical Mathematics, vol. 98 (Birkhäuser Basel, 1991), pp. 223–234

[HCS+15] S. Hadjis, A. Canis, R. Sobue, Y. Hara-Azumi, H. Tomiyama, J. Anderson, Profiling-driven multi-cycling in FPGA high-level synthesis, in *Proceedings of the Conference on Design, Automation and Test in Europe (DATE)* (2015), pp. 31–36.

[HWBR09] A. Hagiescu, W.-F. Wong, D. Bacon, R. Rabbah, A computing origami: Folding streams in FPGAs, in *Proceedings of the 46th ACM/IEEE Design Automation Conference (DAC)* (IEEE, New York, 2009), pp. 282–287

[HANT15] R. Halstead, I. Absalyamov, W. Najjar, V. Tsotras, FPGA-based multithreading for in-memory hash joins, in *Online Proceedings of the Seventh Biennial Conference on Innovative Data Systems Research (CIDR)* (2015), pp. 1–9

[HVN14] R. Halstead, J. Villarreal, W. Najjar, Compiling irregular applications for reconfigurable systems. Int. J. High Perform. Comput. Netw. **7**(4), 258–268 (2014)

[HIS14] B. Hamilton, M. Inggs, H. So, Mixed-architecture process scheduling on tightly coupled reconfigurable computers, in *Proceedings of the 24th International Conference on Field Programmable Logic and Applications (FPL)*, September 2014, pp. 1–4

[HLB+14] F. Hannig, V. Lari, S. Boppu, A. Tanase, O. Reiche, Invasive tightly-coupled processor arrays: a domain-specific architecture/compiler co-design approach. ACM Trans. Embed. Comput. Syst. **13**(4s), 133:1–133:29 (2014)

[HRDT08] F. Hannig, H. Ruckdeschel, H. Dutta, J. Teich, PARO: synthesis of hardware accelerators for multi-dimensional dataflow-intensive applications, in *Proceedings of the Fourth International Workshop on Applied Reconfigurable Computing (ARC)*, London, United Kingdom. Lecture Notes in Computer Science (LNCS), vol. 4943 (Springer, Berlin, 2008), pp. 287–293

[HRT08] F. Hannig, H. Ruckdeschel, J. Teich, The PAULA language for designing multi-dimensional dataflow-intensive applications, in *Proceedings of the GI/ITG/GMM-Workshop* (Shaker, Dresden, 2008), pp. 129–138

[HHK14] M. Happe, Y. Huang, A. Keller, Dynamic protocol stacks in smart camera networks, in *Proceedings of the International Conference on Reconfigurable Computing and FPGAs (ReConFig)*, December 2014 (IEEE, New York, 2014), pp. 1–6

[HLP11] M. Happe, E. Lübbers, M. Platzner, A self-adaptive heterogeneous multi-core architecture for embedded real-time video object tracking. J. Real-Time Image Proc. **8**(1), 1–16 (2011)

[HTK15] M. Happe, A. Traber, A. Keller, Preemptive hardware multitasking in ReconOS, in *Proceedings of the Symposium on Applied Reconfigurable Computing (ARC)*, March 2015. Lecture Notes in Computer Science (LNCS), vol. 9040 (Springer, Berlin, 2015), pp. 79–90

[HTHT09] Y. Hara, H. Tomiyama, S. Honda, H. Takada, Proposal and quantitative analysis of the CHStone benchmark program suite for practical C-based high-level synthesis. J. Inf. Process. **17**, 242–254 (2009)

[HS88] C. Harris, M. Stephens, A combined corner and edge detector, in *Proceedings of the 4th Alvey Vision Conference* (1988), pp. 147–151

[HD07] S. Hauck, A. DeHon, *Reconfigurable Computing: The Theory and Practice of FPGA-Based Computation* (Morgan Kaufmann Publishers, San Francisco, 2007)

[Hes13] R. Hess, *Particle Filter Object Tracking – C code* (2013), http://blogs.oregonstate.edu/hess/code/particles

[HA00] J. Hoe, Arvind, Hardware synthesis from term rewriting systems, in *VLSI: Systems on a Chip*, ed. by L. Silveira, S. Devadas, R. Reis. IFIP – The International Federation for Information Processing, vol. 34 (Springer, Berlin, 2000), pp. 595–619.

[Hof13] P. Hofstee, The big deal about big data, in *Keynote Talk at the 8th IEEE International Conference on Networking, Architecture, and Storage (NAS)*, July 2013

[Hog15] S. Hogg, *What is LabVIEW?* (2015), http://www.ni.com/newsletter/51141/en/. Accessed 4 Aug 2015 [Online]

[HLC+15] Q. Huang, R. Lian, A. Canis, J. Choi, R. Xi, N. Calagar, S. Brown, J. Anderson, The effect of compiler optimizations on high-level synthesis-generated hardware. ACM Trans. Reconfigurable Technol. Syst. **8**(3), 14:1–14:26 (2015)

[HW13] E. Hung, S. Wilton, Towards simulator-like observability for FPGAs: a virtual overlay network for trace-buffers, in *Proceedings of the ACM/SIGDA International Symposium on Field Programmable Gate Arrays (FPGA)* (ACM, New York, 2013), pp. 19–28

[HW14] E. Hung, S. Wilton, Incremental trace-buffer insertion for FPGA debug. IEEE Trans. Very Large Scale Integr. VLSI Syst. **22**(4), 850–863 (2014)

[HB06] M. Hutton, V. Betz, *FPGA Synthesis and Physical Design*, Chap. 13 (CRC Press, Boca Raton, 2006), pp. 13:1–13:30

[Ino12] Inomize, *GZIP HW Accelerator* (2012), http://www.inomize.com/index.php/content/index/gzip-hw-accelerator. Accessed 4 Aug 2015 [Online]

[Int13] Intel Corp., *Scaling Acceleration Capacity from 5 to 50 Gbps and Beyond with Intel QuickAssist Technology* (2013), http://www.intel.com/content/dam/www/public/us/en/documents/solution-briefs/scaling-acceleration-capacity-brief.pdf. Accessed 4 Aug 2015 [Online]

[IS11] A. Ismail, L. Shannon, FUSE: front-end user framework for O/S abstraction of hardware accelerators, in *Proceedings of the International Symposium on Field-Programmable Custom Computing Machines (FCCM)* (IEEE, New York, 2011), pp. 170–177

[IBE+10] X. Iturbe, K. Benkrid, A. Erdogan, T. Arslan, M. Azkarate, I. Martinez, A. Perez, R3TOS: a reliable reconfigurable real-time operating system, in *Proceedings of the NASA/ESA Conference on Adaptive Hardware and Systems (AHS)* (2010), pp. 99–104

[JFM15] A. Jain, S. Fahmy, D. Maskell, Efficient overlay architecture based on DSP blocks, in *Proceedings of the IEEE 23rd Annual International Symposium on Field-Programmable Custom Computing Machines (FCCM)*, May 2015, pp. 25–28

[KAS+02] V. Kathail, S. Aditya, R. Schreiber, B. Rau, D. Cronquist, M. Sivaraman, PICO: automatically designing custom computers. Computer **35**(9), 39–47 (2002)

[KBNH14] A. Keller, D. Borkmann, S. Neuhaus, M. Happe, Self-awareness in computer networks. Int. J. Reconfigurable Comput. (2014). Article ID 692076

[KR88] B. Kernighan, D. Ritchie, *The C programming language*, 2nd, ANSI-C edn. (Prentice-Hall, Englewood Cliffs, 1988)

[Khr08] Khronos OpenCL Working Group, *The OpenCL Specification, version 1.0.29* (2008)

[KS11] J. Kingyens, J. Steffan, The potential for a GPU-Like overlay architecture for FPGAs. Int. J. Reconfigurable Comput. (2011). Article ID 514581

[KHKT06] D. Kissler, F. Hannig, A. Kupriyanov, J. Teich, A dynamically reconfigurable weakly programmable processor array architecture template, in *Proceedings of the 2nd International Workshop on Reconfigurable Communication Centric System-on-Chips (ReCoSoC)* (2006), pp. 31–37

[Koc13] D. Koch, *Partial Reconfiguration on FPGAs – Architectures, Tools and Applications* (Springer, Berlin, 2013)

[KBL13] D. Koch, C. Beckhoff, G. Lemieux, An efficient FPGA overlay for portable custom instruction set extensions, in *Proceedings of the 23rd International Conference on Field Programmable Logic and Applications (FPL)*, September 2013, pp. 1–8

[KBT08] D. Koch, C. Beckhoff, J. Teich, ReCoBus-Builder: a novel tool and technique to build statically and dynamically reconfigurable systems for FPGAs, in *Proceedings of the International Conference on Field Programmable Logic and Applications (FPL)*, September 2008, pp. 119–124

[KT11] D. Koch, J. Torresen, FPGASort: a high performance sorting architecture exploiting run-time reconfiguration on FPGAs for large problem sorting, in *Proceedings of the ACM/SIGDA International Symposium on Field Programmable Gate Arrays (FPGA)* (2011), pp. 45–54

[KR06] I. Kuon, J. Rose, Measuring the gap between FPGAs and ASICs, in *Proceedings of the ACM/SIGDA 14th International Symposium on Field Programmable Gate Arrays (FPGA)* (ACM, New York, 2006), pp. 21–30

[LA04] C. Lattner, V. Adve, LLVM: a compilation framework for lifelong program analysis & transformation, in *Proceedings of the International Symposium on Code Generation and Optimization (CGO)* (IEEE Computer Society, Los Alamitos, 2004), pp. 75–88

[LPL+11] C. Lavin, M. Padilla, J. Lamprecht, P. Lundrigan, B. Nelson, B. Hutchings, HMFlow: accelerating FPGA compilation with hard macros for rapid prototyping, in *Proceedings of the IEEE 19th Annual International Symposium on Field-Programmable Custom Computing Machines (FCCM)*, May 2011, pp. 117–124

[LCD+10] I. Lebedev, S. Cheng, A. Doupnik, J. Martin, C. Fletcher, D. Burke, M. Lin, J. Wawrzynek, MARC: a many-core approach to reconfigurable computing, in *Proceedings of the International Conference on Reconfigurable Computing and FPGAs (ReConFig)*, December 2010, pp. 7–12

[LS12] C. Lin, H. So, Energy-efficient dataflow computations on FPGAs using application-specific coarse-grain architecture synthesis. ACM SIGARCH Comput. Archit. News **40**(5), 58–63 (2012)

[LCP+11] O. Lindtjorn, R. Clapp, O. Pell, H. Fu, M. Flynn, O. Mencer, Beyond traditional microprocessors for geoscience high-performance computing applications. IEEE Micro **31**(2), 41–49 (2011)

[LNS15] C. Liu, H. Ng, H. So, Automatic nested loop acceleration on FPGAs using soft CGRA overlay, in *Proceedings of the Second International Workshop on FPGAs for Software Programmers (FSP)*, September 2015, pp. 13–18

[LLV15] LLVM, *LLVM - Low Level Virtual Machine* (2015), http://www.llvm.org. Accessed 1 April 2015 [Online]

[LP09a] E. Lübbers, M. Platzner, Cooperative multithreading in dynamically reconfigurable systems, in *Proceedings of the IEEE International Conference on Field Programmable Logic and Applications (FPL)* (IEEE, New York, 2009), pp. 1–4

[LP09b] E. Lübbers, M. Platzner, ReconOS: multithreaded programming for reconfigurable computers. ACM Trans. Embed. Comput. Syst. **9**(1), 8:1–8:33 (2009)

[LMVV05] R. Lysecky, K. Miller, F. Vahid, K. Vissers, Firm-core virtual FPGA for just-in-time FPGA compilation, in *Proceedings of the 2005 ACM/SIGDA 13th International Symposium on Field-Programmable Gate Arrays (FPGA)* (ACM, New York, 2005), p. 271

[MSY+15] G. Malazgirt, N. Sonmez, A. Yurdakul, O. Unsal, A. Cristal, High level synthesis based hardware accelerator design for processing SQL queries, in *Proceedings of the FPGAworld Conference*, September 2015, pp. 1–6

[MJA13] A. Martin, D. Jamsek, K. Agarwal, FPGA-based application acceleration: case study with GZIP compression/decompression streaming engine, in *Proceedings of the International Conference on Computer-Aided Design (ICCAD)* (2013)

[MS09] G. Martin, G. Smith, High-level synthesis: Past, present, and future. IEEE Des. Test Comput. **26**(4), 18–25 (2009)

[MSF12] E. Matthews, L. Shannon, A. Fedorova, Polyblaze: from one to many bringing the MicroBlaze into the multicore era with linux SMP support, in *International Conference on Field Programmable Logic and Applications*, August 2012, pp. 224–230

[ME95] L. McMurchie, C. Ebeling, Pathfinder: a negotiation-based performance-driven router for FPGAs, in *Proceedings of the ACM third International Symposium on Field-Programmable Gate Arrays (FPGA)* (1995), pp. 111–117

[MVG+12] W. Meeus, K. Van Beeck, T. Goedemé, J. Meel, D. Stroobandt, An overview of today's high-level synthesis tools. Des. Autom. Embed. Syst. **16**(3), 31–51 (2012)

[MHT+12] R. Membarth, F. Hannig, J. Teich, M. Körner, W. Eckert, Generating device-specific GPU code for local operators in medical imaging, in *Proceedings of the 26th IEEE International Parallel & Distributed Processing Symposium (IPDPS)*, May 2012 (IEEE, New York, 2012), pp. 569–581

[MRHT14] R. Membarth, O. Reiche, F. Hannig, J. Teich, Code generation for embedded heterogeneous architectures on Android, in *Proceedings of the Conference on Design, Automation and Test in Europe (DATE)*, March 2014, pp. 1–6

[MRH+16] R. Membarth, O. Reiche, F. Hannig, J. Teich, M. Körner, W. Eckert, HIPAcc: a domain-specific language and compiler for image processing. IEEE Trans. Parallel Distrib. Syst. **27**(1), 210–224 (2016)

[Mer08] M. Meredith, *High-Level SystemC Synthesis with Forte's Cynthesizer*, Chap. 5, pp. 75–97; in Coussy, Morawiec [CM08], 1st edn. (2008)

[Mic94] G. Micheli, *Synthesis and Optimization of Digital Circuits*, 1st edn. (McGraw-Hill Higher Education, New York, 1994)

[Moo65] G. Moore, Cramming more components onto integrated circuits. Electronics **38**(8), 114–117 (1965)

[Muc97] S. Muchnick, *Advanced Compiler Design and Implementation* (Morgan Kaufmann Publishers, San Francisco, 1997)

[MTA10] R. Mueller, J. Teubner, G. Alonso, Glacier: a query-to-hardware compiler, in *Proceedings of the ACM SIGMOD International Conference on Management of Data* (2010), pp. 1159–1162

[NBD+03] W. Najjar, A. Böhm, B. Draper, J. Hammes, R. Rinker, J. Beveridge, M. Chawathe, C. Ross, High-level language abstraction for reconfigurable computing. IEEE Comput. **36**(8), 63–69 (2003)

[NI14] W. Najjar, P. Ienne, Reconfigurable computing. IEEE Micro **34**(1), 4–6 (2014)

[Nat15a] National Instruments, *LabVIEW Communications System Design Suite Overview* (2015), http://www.ni.com/white-paper/52502/en/. Accessed 4 Aug 2015 [Online]

[Nat15b] National Instruments, *LabVIEW FPGA* (2015), http://www.ni.com/fpga/. Accessed 4 Aug 2015 [Online]

[Nat15c] National Instruments, *LabVIEW RIO Architecture* (2015), http://www.ni.com/white-paper/10894/en/. Accessed 4 Aug 2015 [Online]

[Nat15d] National Instruments, *NI myRIO* (2015), http://www.ni.com/myrio/. Accessed 4 Aug 2015 [Online]

[Nat15e] National Instruments, *NI Scan Engine Advanced I/O Access* (2015), http://www.ni.com/white-paper/8071/en/. Accessed 4 Aug 2015 [Online]

[Nik04] R. Nikhil, Bluespec system verilog: efficient, correct RTL from high level specifications, in *Proceedings of the Second ACM and IEEE International Conference on Formal Methods and Models for Co-Design (MEMOCODE)*, June 2004, pp. 69–70

[Nik08] R. Nikhil, *Bluespec: A General-Purpose Approach to High-Level Synthesis Based on Parallel Atomic Transactions*, Chap. 8, pp. 129–146; in Coussy, Morawiec [CM08], 1st edn. (2008)

[NCV+03] V. Nollet, P. Coene, D. Verkest, S. Vernalde, R. Lauwereins, Designing an operating system for a heterogeneous reconfigurable SoC, in *Proceedings of the International Parallel and Distributed Processing Symposium (IPDPS)* (2003), pp. 1–7

[OSV11] M. Odersky, L. Spoon, B. Venners, *Programming in Scala: A Comprehensive Step-by-Step Guide*, 2nd edn. (Artima Incorporation, Walnut Creek, 2011)

[Ope15a] OpenACC, *OpenACC directives for accelerators* (2015), http://www.openacc-standard.org/. Accessed 4 Aug 2015 [Online]

[Ope15b] OpenMP, *The OpenMP API Specification for Parallel Programming* (2015), http://openmp.org/. Accessed 4 Aug 2015 [Online]

[Pap03] I. Papaefstathiou, Titan II: an IPComp processor for 10Gbit/sec networks, in *Proceedings of the IEEE Computer Society Annual Symposium on VLSI*, February 2003, pp. 234–235

[PPA+13] A. Parashar, M. Pellauer, M. Adler, B. Ahsan, N. Crago, D. Lustig, V. Pavlov, A. Zhai, M. Gambhir, A. Jaleel, R. Allmon, R. Rayess, S. Maresh, J. Emer, Triggered instructions: a control paradigm for spatially-programmed architectures. ACM SIGARCH Comput. Archit. News **41**(3), 142–153 (2013)

[PK89] P. Paulin, J. Knight, Force-directed scheduling for the behavioral synthesis of ASICs. IEEE Trans. Comput. Aided Des. **8**(6), 661–679 (1989)

[PACE09] M. Pellauer, M. Adler, D. Chiou, J. Emer, Soft connections: addressing the hardware-design modularity problem, in *Proceedings of the 27th ACM/IEEE Design Automation Conference (DAC)* (ACM, New York, 2009), pp. 276–281

[PSKK15] N. Pham, A. Singh, A. Kumar, M. Khin, Exploiting loop-array dependencies to accelerate the design space exploration with high level synthesis, in *Proceedings of the Conference on Design, Automation and Test in Europe (DATE)* (EDA Consortium, San Jose, 2015), pp. 157–162

[Pou10] L.-N. Pouchet, Iterative optimization in the polyhedral model, PhD thesis, University of Paris-Sud 11, Orsay, France, 2010

[PZSC13] L.-N. Pouchet, P. Zhang, P. Sadayappan, J. Cong, Polyhedral-based data reuse optimization for configurable computing, in *Proceedings of the ACM/SIGDA International Symposium on Field-Programmable Gate Arrays (FPGA)* (ACM, New York, 2013), pp. 29–38

[PBKE12] K. Pulli, A. Baksheev, K. Kornyakov, V. Eruhimov, Real-time computer vision with OpenCV. Commun. ACM **55**(6), 61–69 (2012)

[PCC+14] A. Putnam, A. Caulfield, E. Chung, D. Chiou, K. Constantinides, J. Demme, H. Esmaeilzadeh, J. Fowers, G. Gopal, J. Gray, M. Haselman, S. Hauck, S. Heil, A. Hormati, J.-Y. Kim, S. Lanka, J. Larus, E. Peterson, S. Pope, A. Smith, J. Thong, P. Xiao, D. Burger, A reconfigurable fabric for accelerating large-scale datacenter service, in *Proceedings of the ACM/IEEE 41st International Symposium on Computer Architecture (ISCA)* (IEEE, New York, 2014), pp. 13–24

[Rau94] B. Rau, Iterative modulo scheduling: an algorithm for software pipelining loops, in *Proceedings of the 27th Annual International Symposium on Microarchitecture (MICRO)* (ACM, New York, 1994), pp. 63–74

[Rau96] B. Rau, Iterative modulo scheduling, Technical report, HP, USA, 1996, http://www.hpl.hp.com/techreports/94/HPL-94-115.html

[RSH+14] O. Reiche, M. Schmid, F. Hannig, R. Membarth, J. Teich, Code generation from a domain-specific language for C-based HLS of hardware accelerators, in *Proceedings of the International Conference on Hardware/Software Codesign and System Synthesis (CODES+ISSS)*, October 2014 (ACM, New York, 2014), pp. 1–10

[RBK07] S. Rigler, W. Bishop, A. Kennings, FPGA-based lossless data compression using Huffman and LZ77 algorithms, in *Proceedings of the Canadian Conference on and Computer Engineering (CCECE)*, April 2007, pp. 1235–1238

[RVS+10] J. Robinson, S. Vafaee, J. Scobbie, M. Ritche, J. Rose, The Supersmall soft processor, in *Proceedings of the Southern Programmable Logic Conference (SPL)*, March 2010, pp. 3–8

[RNPL15] A. Rodchenko, A. Nisbet, A. Pop, M. Luján, Effective barrier synchronization on Intel Xeon Phi coprocessor, in *Proceedings of the 21st International Conference on Parallel Processing (Euro-Par)*, August 2015. Lecture Notes in Computer Science (LNCS), vol. 9233, pp. 588–600

[San12] Sandgate Technologies, Inc., *GZIP/GUNZIP Silicon IP Family* (2012), http://www.sandgate.com/new/static/QuickZIP%20Family%20Product%20Brief%20%28V1.2a%29.pdf. Accessed 4 Aug 2015 [Online]

[SW10] B. Schafer, K. Wakabayashi, Design space exploration acceleration through operation clustering. IEEE Trans. Comput. Aided Des. **29**(1), 153–157 (2010)

[SW12] B. Schafer, K. Wakabayashi, Divide and conquer high-level synthesis design space exploration. ACM Trans. Des. Autom. Electron. Syst. **17**(3), 29:1–29:19 (2012)

[SAHT14] M. Schmid, N. Apelt, F. Hannig, J. Teich, An image processing library for C-based high-level synthesis, in *Proceedings of the 24th International Conference on Field Programmable Logic and Applications (FPL)*, September 2014 (IEEE, New York, 2014), pp. 1–4

[STH+14] M. Schmid, A. Tanase, F. Hannig, J. Teich, V. Bhadouria, D. Ghoshal, Domain-specific augmentations for high-level synthesis, in *Proceedings of the 25th IEEE International Conference on Application-specific Systems, Architectures and Processors (ASAP)*, June 2014 (IEEE, New York, 2014), pp. 173–177

[SKH+14] C. Schmitt, S. Kuckuk, F. Hannig, H. Köstler, J. Teich, ExaSlang: a domain-specific language for highly scalable multigrid solvers, in *Proceedings of the 4th International Workshop on Domain-Specific Languages and High-Level Frameworks for High Performance Computing (WOLFHPC)*, November 2014 (IEEE Computer Society, Los Alamitos, 2014), pp. 42–51

[SSH+15] C. Schmitt, M. Schmid, F. Hannig, J. Teich, S. Kuckuk, H. Köstler, Generation of multigrid-based numerical solvers for FPGA accelerators, in *Proceedings of the 2nd International Workshop on High-Performance Stencil Computations (HiStencils)*, January 2015, pp. 9–15

[Sch96] J. Schutten, List scheduling revisited. Oper. Res. Lett. **18**(4), 167–170 (1996)

[SL12] A. Severance, G. Lemieux, VENICE: a compact vector processor for FPGA applications, in *Proceedings of the International Conference on Field-Programmable Technology (FPT)*, December 2012, pp. 261–268

[SBB06] S. Shukla, N. Bergmann, J. Becker, QUKU: a two-level reconfigurable architecture, in *Proceedings of the IEEE Computer Society Annual Symposium on Emerging VLSI Technologies and Architectures*, March 2006, pp. 1–6

[Sin11] S. Singh, Computing without processors. Commun. ACM **54**(8), 46–54 (2011)

[SB06] H. So, R. Brodersen, Improving usability of FPGA-based reconfigurable computers through operating system support, in *Proceedings of the International Conference on Field Programmable Logic and Applications (FPL)*, August 2006, pp. 1–6

[SB08] H. So, R. Brodersen, A unified hardware/software runtime environment for FPGA-based reconfigurable computers using BORPH. ACM Trans. Embed. Comput. Syst. **7**(2), 14:1–14:28 (2008)

[SS15] A. Soltani, S. Sharifian, An ultra-high throughput and fully pipelined implementation of AES algorithm on FPGA. Microprocess. Microsyst. **39**(7), 480–493 (2015)

[SWP04] C. Steiger, H. Walder, M. Platzner, Operating systems for reconfigurable embedded platforms: online scheduling of real-time tasks. IEEE Trans. Comput. **53**(11), 1392–1407 (2004)

[Ste04] F. Stein, Efficient computation of optical flow using the census transform, in *Pattern Recognition*. Lecture Notes in Computer Science (LNCS), vol. 3175 (Springer, Berlin, 2004), pp. 79–86

[ST82] K. Stüben, U. Trottenberg, Multigrid methods: fundamental algorithms, model problem analysis and applications, in *Multigrid Methods*. Lecture Notes in Mathematics, vol. 960 (Springer, Berlin, 1982), pp. 1–176

[Sum08] T. Summers, *Hardware based GZIP Compression, Benefits and Applications* (2008), http://www.comtechaha.com/Uploads/GZIP-Benefits-Apps.pdf. Accessed 4 Aug 2015 [Online]

[TMK10] M. Tahghighi, M. Mousavi, P. Khadivi, Hardware implementation of a novel adaptive version of deflate compression algorithm, in *Proceedings of the 18th Iranian Conference on Electrical Engineering (ICEE)*, May 2010 (IEEE, New York, 2010), pp. 566–569

[TB01] R. Tessier, W. Burleson, Reconfigurable computing for digital signal processing: a survey. J. VLSI Signal Process. Syst. Signal Image Video Technol. **28**(1–2), 7–27 (2001)

[TPD15] R. Tessier, K. Pocek, A. DeHon, Reconfigurable computing architectures. Proc. IEEE **103**(3), 332–354 (2015)

[TBU00] P. Thevenaz, T. Blu, M. Unser, Interpolation revisited. IEEE Trans. Med. Imaging **19**(7), 739–758 (2000)

[Tra08] Transaction Processing Performance Council, *TPC-H Benchmark Specification* (2008), http://www.tpc.org/tpch/spec/tpch2.6.0.pdf

[TCJW97] S. Trimberger, D. Carberry, A. Johnson, J. Wong, A time-multiplexed FPGA, in *Proceedings of the 5th IEEE Symposium on Field-Programmable Custom Computing Machines (FCCM)*, April 1997 (IEEE Computer Society, Washington, DC, 1997), pp. 22–28

[TGP07] J. Tripp, M. Gokhale, K. Peterson, Trident: from high-level language to hardware circuitry. Computer **40**(3), 28–37 (2007)

[Van94] E. Van de Velde, Poisson solvers, in *Concurrent Scientific Computing*. Texts in Applied Mathematics, vol. 16 (Springer New York, 1994), pp. 183–216

[VFHB14] E. Vermij, L. Fiorin, C. Hagleitner, K. Bertels, Exascale radio astronomy: can we ride the technology wave? in *Proceedings of the 29th International Conference on Supercomputing (ISC)*, June 2014, ed. by J. Kunkel, T. Ludwig, H. Meuer. Lecture Notes in Computer Science (LNCS), vol. 8488 (Springer, Berlin, 2014), pp. 35–52

[VPI05] M. Vuletic, L. Pozzi, P. Ienne, Seamless hardware-software integration in reconfigurable computing systems. IEEE Des. Test Comput. **22**(2), 102–113 (2005)

[WO00] K. Wakabayashi, T. Okamoto, C-based SoC design flow and EDA tools: an ASIC and system vendor perspective. IEEE Trans. Comput. Aided Des. Integr. Circuits Syst. **19**(12), 1507–1522 (2000)

[WS08] K. Wakabayashi, B. Schafer, *"All-in-C" Behavioral Synthesis and Verification with CyberWorkBench*, Chap. 7, pp. 113–127; in Coussy, Morawiec [CM08], 1st edn. (2008)

[WC91] R. Walker, R. Camposano (eds.), *A Survey of High Level Synthesis Systems* (Kluwer Academic, Norwell, 1991)

[WLC14] Y. Wang, P. Li, J. Cong, Theory and algorithm for generalized memory partitioning in high-level synthesis, in *Proceedings of the 2014 ACM/SIGDA International Symposium on Field-Programmable Gate Arrays (FPGA)* (ACM, New York, 2014), pp. 199–208

[WLZ+13] Y. Wang, P. Li, P. Zhang, C. Zhang, J. Cong, Memory partitioning for multidimensional arrays in high-level synthesis, in *Proceedings of the 50th Annual Design Automation Conference (DAC)* (ACM, New York, 2013), pp. 12:1–12:8

[WYZ+12] Y. Wang, J. Yan, X. Zhou, L. Wang, W. Luk, C. Peng, J. Tong, A partially reconfigurable architecture supporting hardware threads, in *Proceedings of the International Conference on Field-Programmable Technology (FPT)* (2012), pp. 269–276

[WZCC12] Y. Wang, P. Zhang, X. Cheng, J. Cong, An integrated and automated memory optimization flow for FPGA behavioral synthesis, in *Proceedings of the 17th Asia and South Pacific Design Automation Conference (ASP-DAC)* (IEEE, New York, 2012), pp. 257–262

[Wei84] R. Weicker, Dhrystone: a synthetic systems programming benchmark. Commun. ACM **27**(10), 1013–1030 (1984)

[WSRM12] S. Weston, J. Spooner, S. Racanière, O. Mencer, Rapid computation of value and risk for derivatives portfolios. Concurr. Comput. **24**(8), 880–894 (2012)

[Wik15] Wikipedia, *AirPlay — Wikipedia, The Free Encyclopedia* (2015), https://en.wikipedia.org/w/index.php?title=AirPlay&oldid=663198848. Accessed 4 Aug 2015 [Online]

[WFW+94] R. Wilson, R. French, C. Wilson, S. Amarasinghe, J. Anderson, S. Tjiang, S.-W. Liao, C.-W. Tseng, M. Hall, M. Lam, J. Hennessy, SUIF: an infrastructure for research on parallelizing and optimizing compilers. ACM SIGPLAN Not. **29**(12), 31–37 (1994)

[Win15] Wind River, *Wind River Linux* (2015), http://www.windriver.com/products/linux. Accessed 26 July 2015 [Online]

[WBC14] F. Winterstein, S. Bayliss, G. Constantinides, Separation logic-assisted code transformations for efficient high-level synthesis, in *Proceedings of the IEEE 22nd Annual International Symposium on Field-Programmable Custom Computing Machines (FCCM)* (IEEE, New York, 2014), pp. 1–8

[WFY+15] F. Winterstein, K. Fleming, H.-J. Yang, S. Bayliss, G. Constantinides, MATCHUP: Memory abstractions for heap manipulating programs, *Proceedings of the ACM/SIGDA International Symposium on Field-Programmable Gate Arrays (FPGA)* (ACM, New York, 2015), pp. 136–145

[WL91] M. Wolf, M. Lam, A loop transformation theory and an algorithm to maximize parallelism. IEEE Trans. Parallel Distrib. Syst. **2**(4), 452–471 (1991)

[Wol96] M. Wolfe, *High performance compilers for parallel computing* (Addison-Wesley, Boston, 1996)

[WLP+14] L. Wu, A. Lottarini, T. Paine, M. Kim, K. Ross, Q100: the architecture and design of a database processing unit, in *Proceedings of the 19th International Conference on Architectural Support for Programming Languages and Operating Systems (ASPLOS)* (2014), pp. 255–268

[Xil11] Xilinx Inc., *Data2MEM User Guide (UG658, v13.3)* (2011), http://www.xilinx.com/support/documentation/sw_manuals/xilinx14_2/data2mem.pdf. Accessed 19 Sept 2012 [Online]

[Xil15a] Xilinx Inc., *MicroBlaze Processor Reference Guide (UG984, v2014.1)* (2015)

[Xil15b] Xilinx Inc., *MicroBlaze Soft Processor Core* (2015), http://www.xilinx.com/tools/microblaze.htm. Accessed 22 July 2015 [Online]

[Xil15c] Xilinx Inc., *OS and Libraries Document Collection (UG643, v2015.2)* (2015)

[Xil15d] Xilinx Inc., *PetaLinux Tools* (2015), http://www.xilinx.com/tools/petalinux-sdk.htm. Accessed 26 July 2015 [Online]

[Xil15e] Xilinx, Inc., *Vivado High-Level Synthesis* (2015), http://www.xilinx.com/products/design-tools/vivado/integration/esl-design.html. Accessed 12 July 2015[Online]

[Xil15f] Xilinx Inc., *Zynq-7000 All Programmable SoC Overview (DS190, v1.8)* (2015)

[Xil15g] Xilinx Inc., *Zynq UltraScale+ MPSoC Product Tables and Product Selection Guide* (2015)

[YFAE14] H. Yang, K. Fleming, M. Adler, J. Emer, LEAP shared memories: automating the construction of FPGA coherent memories, in *Proceedings of the Symposium on Field-Programmable Custom Computing Machines (FCCM)* (IEEE, New York, 2014), pp. 117–124

[YSR09] P. Yiannacouras, J. Steffan, J. Rose, Fine-grain performance scaling of soft vector processors, in *Proceedings of the International Conference on Compilers, Architecture, and Synthesis for Embedded Systems (CASES)* (ACM, New York, 2009), pp. 97–106

[YKL15] M. Yue, D. Koch, G. Lemieux, Rapid overlay builder for Xilinx FPGAs, in *Proceedings of the IEEE 23rd Annual International Symposium on Field-Programmable Custom Computing Machines (FCCM)*, May 2015, pp. 17–20

[ZFJ+08] Z. Zhang, Y. Fan, W. Jiang, G. Han, C. Yang, J. Cong, *AutoPilot: A Platform-Based ESL Synthesis System*, Chap. 6, pp. 99–112; in Coussy, Morawiec [CM08]

[ZL13] Z. Zhang, B. Liu, SDC-based modulo scheduling for pipeline synthesis, in *Proceedings of the International Conference on Computer-Aided Design (ICCAD)* (2013), pp. 211–218

[ZL77] J. Ziv, A. Lempel, A universal algorithm for sequential data compression. IEEE Trans. Inf. Theory **23**(3), 337–343 (1977)

[ZLC+13] W. Zuo, P. Li, D. Chen, L.-N. Pouchet, S. Zhong, J. Cong, Improving polyhedral code generation for high-level synthesis, in *Proceedings of the International Conference on Hardware/Software Codesign and System Synthesis (CODES+ISSS)* (IEEE Press, Piscataway, 2013), pp. 1–10

Index

D. Koch et al. (eds.), *FPGAs for Software Programmers*,
DOI 10.1007/978-3-319-26408-0

CPSIA information can be obtained
at www.ICGtesting.com
Printed in the USA
LVOW02*1022080117
520185LV00001B/36/P

9 783319 264066